Ulrich Reimers

DVB

Ulrich Reimers

DVB

The Family of International Standards for Digital Video Broadcasting

Second Edition

With 261 Figures

 Springer

Professor Dr.-Ing. Ulrich Reimers

Professor Dr.-Ing. Frank Fechter
Dipl.-Ing. Heiko Foellscher
Dr.-Ing. Dirk Jaeger
Dipl.-Ing. Christian Johansen
Dipl.-Ing. Frank Klinkenberg
Dr.-Ing. Uwe Ladebusch
Dipl.-Ing. Volker Leisse

Dipl.-Ing. Claudia Liss
Dr.-Ing. Martin Piastowski
Professor Dr.-Ing. Christof Ricken
Dr.-Ing. Alexander Roy
Dipl.-Ing. Ulrich Schiek
Dr.-Ing. Markus Trauberg
Dipl.-Ing. Andreas Verse

Technische Universität Braunschweig
Institut für Nachrichtentechnik
(Braunschweig Technical University
Institute for Communications Technology)
Schleinitzstr. 22
D-38092 Braunschweig
u.reimers@tu-bs.de

Originally published as a monograph

Library of Congress Control Number: 2004102304

ISBN 3-540-43545-X Springer Berlin Heidelberg New York

Springer is a part of Springer Science+Business Media

springeronline.com

© Springer-Verlag Berlin Heidelberg 2005
Printed in Germany

Typesetting: Camera ready copy from author
Cover-Design: Design & Production, Heidelberg
Printed on acid-free paper   62/3020 Rw 5 4 3 2 1 0

Preface of the 2nd Edition

The first English edition of the book "Digital Video Broadcasting (DVB)" was published at the end of the year 2000. It was based upon the second German edition and reflected the developments in the world of DVB as of the end of 1998. Only a year after the sales of the first English edition began, Springer publishers asked for a new, completely updated version of the book. This request was a result of the very favourable acceptance of the book by its readers. The revision provided me with the opportunity to incorporate those reports of the many new developments in the world of DVB which could not be reflected in the first English edition.

A new team of authors gathered to add new sections to the book. Again they all are or once were researchers working with me at the Institute for Communications Technology (Institut für Nachrichtentechnik – IfN) at Braunschweig Technical University (Technische Universität Braunschweig) in Germany. They contributed to the work of the relevant ad-hoc groups of the Technical Module of the DVB Project, of which I am the chairman, and therefore were able to provide highly informed insider reports about the most recent results of the developments in the world of DVB.

In preparing the second English edition we decided to update chapter 1, the introduction to the DVB world (Reimers) and to considerably enlarge section 5.4 dealing with Service Information (Foellscher). In Chapter 11, which describes DVB-T the standard for terrestrial transmission, we added the latest results of chip and receiver implementations and describe in some detail the capabilities of DVB-T when used by mobile receivers (Liss, Dr. Roy). A new chapter 12 was added in order to describe the use of DVB for data broadcasting (Foellscher). Among the most recent additions to the set of DVB standards are those enabling the offering of interactive services. These are now described in chapter 13 (Leisse, Dr.Piastowski). Chapter 14 deals with the Multimedia Home Platform (MHP) (Klinkenberg, Schiek). The MHP is a software platform to be used in all kinds of terminal devices. It is specified by the most complex of all DVB standards so far and promises to become a real global success. The previous section on Measurement Techniques has now become chapter 15.

I wish to thank Springer publishers, for their support in producing the book and in particular Dr. Dieter Merkle and Ms. Heather King.

I very much hope that this completely revised and amended edition will be received well by its readers.

Braunschweig, Summer 2004   Prof. Dr.-Ing. Ulrich Reimers

Preface

Digital Television ("Digital TeleVision Broadcasting" [DTVB] or "Digital Video Broadcasting" [DVB]) has become one of the most exciting developments in the area of consumer electronics at the end of the twentieth century. The term digital television is not typically used either to describe the digitisation of the production studio or to indicate the advent of digital signal processing in the integrated circuits used in television receivers. Rather, digital television refers to the source coding of audio, data and video signals, the channel coding and the methods for the transport of DVB signals via all kinds of transmission media. The term normally also embraces the technologies used for the return path from the consumer back to the programme provider, techniques for scrambling and conditional access, the concepts of data broadcasting, the software platforms used in the terminal devices, as well as the user interfaces providing easy access to DVB services.

The aim of this book is to describe the technologies of digital television. The description refers to a point in time at which much of the technical development work had taken place. No doubt, in the future there will inevitably be a considerable number of modifications and additions to the list of technical documents describing the technologies used for DVB. The performance data of the DVB systems, specifically of the one used for terrestrial transmission, are still in the process of being evaluated in many countries throughout the world. In some respects this book must therefore be regarded as a report on the present intermediate stage of the development of digital television and of the practical experience gained so far. I nevertheless consider it timely to present such a report, since at the date of its publication DVB will have become a market reality in many countries.

The focus is on the developments within, and the achievements of, the "Moving Pictures Experts Group (MPEG)" in the area of source coding of audio and video signals, followed by an extended description of the work and the results of the "DVB Project", the international body dealing with the design of all the other technical solutions required for the successful operation of digital television. The combination of the specifications designed by MPEG and those designed by the DVB Project has led to the overall system that we can now call digital television. The system for the terrestrial transmission of MPEG signals presented by the "Advanced Television Systems Committee (ATSC)" will not be described in detail in this book. This system will be used in the United States of America, in Canada, Mexico and some other parts of the world. It uses vestigial side-band modulation (VSB) of a single carrier per channel, and sound coding called Dolby AC-3. VSB will be explained in the chapter dealing with modulation techniques.

The book addresses readers who wish to gain an in-depth understanding of the techniques used for digital television. They should already have a good knowledge of the existing analogue television systems and should know the properties of analogue audio and video signals. They should also have some understanding of the techniques used in digital signal origination and digital signal processing.

Chapter 1 is an introduction to the DVB world. It presents an overview of the whole DVB scenario and answers the very fundamental questions concerning the goals of the development of digital television. It describes the state of technical developments and tries to evaluate several scenarios for the introduction of services using the different transmission media.

Chapter 2 recapitulates the fundamentals of the digitisation of audio and video signals. It explains the parameters chosen for the digitisation, such as the quantisation scales and the sampling frequencies, and derives the resulting data rates. On the basis of the figures presented in chapter 2 the fundamentals of source coding of audio and video signals are described in chapters 3 and 4, respectively. The primary goal of source coding is to reduce the data rate for the representation of audio and video signals in such a way that no deterioration of the perceptual quality results or, at most, only a well-defined one. In this way the resources required for the transmission and/or storage of the signals can be limited.

Before the data-reduced signals can actually be transmitted they have to be amalgamated into a system multiplex. Tools for the synchronisation of audio and video, auxiliary data needed for the description of the multiplex and of the programme content conveyed by that multiplex, teletext data, and a great deal of other information need to be added. Chapter 5 explains the multiplexing and the structure of the auxiliary data.

One of the special features of digital signals is that they can be protected against the effects of unavoidable errors occurring during transmission by adding forward error correction before the signals are sent. Chapter 6 describes methods of forward error correction (FEC) in general. It then concentrates on the two methods used in the DVB world, namely, Reed-Solomon coding and convolutional coding.

Digital modulation is dealt with in chapter 7. Here again we highlight those methods which are used in DVB (QPSK, QAM, OFDM, VSB).

The techniques used for the scrambling of digital signals are presented in chapter 8. Owing to the very nature of this topic it is impossible to describe in detail the specific methods which are used for DVB.

The three standards for the transmission of DVB signals via satellite, on cable and via terrestrial networks are described in chapters 9, 10 and 11, respectively. In addition to merely explaining the specific details of the standards, we give the performance data and further information regarding the hardware implementation in the receivers.

Finally, the methods of measuring and evaluating DVB signals as well as audio and video quality are explained in chapter 12.

This book is the result of the joint effort of several researchers working at the Institute for Communications Technology (Institut für Nachrichtentechnik – IfN) at Braunschweig Technical University (Technische Universität Braunschweig) in Germany. The contents are based upon a series of seminars held for interested European industrial professionals since the beginning of 1994. To date, over 250 participants have attended such seminars, in which we not only present the theoretical background of the DVB systems but demonstrate the possibilities of DVB and the existing DVB systems and services. In nearly all cases the authors report about areas in which they themselves have carried out research work. Amongst other things, this work has already generated four doctoral theses. Several of the researchers are – or have been – members of either ad-hoc groups of the DVB Project or of the Moving Pictures Experts Group (MPEG).

I wish to thank the authors for their competency and co-operation in meeting deadlines for the various manuscripts. Ms. Boguslawa Brandt and Ms. Simone Sengpiel prepared a large number of the drawings used in this book and Dipl.-Ing. Christian Johansen was responsible for the final co-ordinating and editing of the whole text. I extend my thanks to all three of them.

The first German edition of this book was published in 1995. It was followed by a revised second edition in 1997. The English version is based upon that second German edition. Of course, several amendments portraying developments between 1997 and 1998 have been included. The translation from German into English was done by Ms. Vivienne Bruns and Ms. Anne Kampendonk. I am most grateful to both ladies for having taken on this very difficult job.

Dr. George Waters, the former Director of Engineering of the European Broadcasting Union (EBU), undertook the final proofreading of the English version. I am especially indebted to him for his invaluable contribution to the quality of the book.

Springer publishers, and in particular Dipl.-Ing. Thomas Lehnert, were competent and active partners in producing the book. My thanks go to them for their guidance and assistance during the publication process of the English version.

Digital television, based on the specifications developed by the DVB Project, would most certainly have remained a dream without the co-operation of countless researchers and engineers in Europe, North America and Asia who were highly enthusiastic about the creation of the digital television paradigm. I have the pleasure of being chairman of the Technical Module of the DVB Project and I wish to join the other authors of this book in thanking all the DVB colleagues and friends for such a great achievement.

Braunschweig, December 1998   Prof. Dr.-Ing. Ulrich Reimers

Table of Contents

1	**Digital Television – a First Summary** (REIMERS)	1
1.1	Definitions and Range of Application	1
1.2	The Genesis of Recommendations for Digital Television	3
1.2.1	Work in the United States of America	5
1.2.2	Work in Europe	6
1.2.3	Work in Japan	8
1.3	Objectives in the Development of Digital Television	9
1.4	Data Reduction as the Key to Success	11
1.5	Possible Means of Transmission for Digital Television	13
1.6	Standards and Norms in the World of Digital Television	17
1.7	The New DVB-Project	18
2	**Digitisation and Representation of Audio and Video Signals**	
	(JOHANSEN)	21
2.1	Sampling and Quantising	21
2.2	Digitising Video Signals	22
2.2.1	ADCs and DACs for Video Signals	24
2.2.2	Representation of Video Signals	26
2.3	Digitising Audio Signals	29
2.3.1	Representation of Audio Signals	30
2.3.2	ADCs and DACs for Audio Signals	30
3	**MPEG Source Coding of Audio Signals** (FECHTER)	37
3.1	Basics of Bit-rate Reduction	37
3.2	Psychoacoustic Basics	39
3.2.1	Threshold of Audibility and Auditory Sensation Area	39
3.2.2	Masking	40
3.3	Source Coding of Audio Signals Utilising the Masking Qualities	
	of the Human Ear	44
3.3.1	Basic Structure of the MPEG Coding Technique	45
3.3.2	Coding in Accordance with Layer 1	49
3.3.3	Coding in Accordance with Layer 2	51
3.3.4	Coding in Accordance with Layer 3	52
3.3.5	Decoding	54

3.3.6   The Parameters of MPEG Audio.   54
3.3.7   MPEG-2 Audio Coding   55
3.4   Summary.   56

4   **JPEG and MPEG Source Coding of Video Signals** (RICKEN) . . .   59
4.1   Coding in Accordance with JPEG   60
4.1.1   Block Diagram of Encoder and Decoder   60
4.1.2   Discrete Cosine Transform   61
4.1.3   Quantisation   63
4.1.4   Redundancy Reduction.   65
4.1.5   Specific Modes   67
4.1.6   Interchange Format   70
4.2   Coding in Accordance with the MPEG Standards   71
4.2.1   Block Diagrams of Encoder and Decoder   73
4.2.2   Motion Estimation   75
4.2.3   Reordering of Pictures   78
4.2.4   Data-rate Control.   79
4.2.5   Special Features of MPEG-1.   80
4.2.6   Special Features of MPEG-2.   83
4.3   Summary.   89

5   **MPEG-2 Systems and Multiplexing** (FOELLSCHER, RICKEN) . . .   91
5.1   Differences between Programme Multiplex and Transport
         Multiplex.   91
5.2   Positioning of Systems in the ISO/OSI Layer Model   92
5.3   End-to-end Synchronisation   94
5.4   Service Information   99
5.4.1   PSI/SI Tables and their Insertion into the MPEG-2 Transport
         Stream   99
5.4.2   Section and Table Structure.   103
5.4.3   Examples of Table Usage: NIT and SDT   105
5.4.3.1 Network Information Table (NIT)   106
5.4.3.2 Service Description Table (SDT)   108

6   **Forward Error Correction (FEC) in Digital Television**
         Transmission (ROY)   111
6.1   Basic Observations   111
6.2   Reed-Solomon Codes.   114
6.2.1   Introduction to the Arithmetic of the Galois Field.   115
6.2.2   Definition of the RS Code and the Encoding/Decoding
         in the Frequency Domain.   121

6.2.3 Error Correction Using the RS Code 122
6.2.4 Examples of Encoding/Decoding in the Frequency Domain . . . 125
6.2.5 Encoding and Decoding in the Time Domain 127
6.2.6 Efficiency of the RS Code 128
6.3 Convolutional Codes 129
6.3.1 Basics of the Convolutional Codes. 129
6.3.2 Examples of Convolutional Encoding and Decoding 131
6.3.2.1 Construction of a Model Encoder 131
6.3.2.2 State Diagram and Trellis Diagram of the Model Encoder 132
6.3.2.3 Example of Encoding with Subsequent (Viterbi) Decoding. . . . 133
6.3.3 Hard Decision and Soft Decision 136
6.3.4 Puncturing of Convolutional Codes 138
6.3.5 Performance of Convolutional Codes 138
6.4 Code Concatenation. 140
6.4.1 Block-Code Concatenation 140
6.4.2 Interleaving 141
6.4.3 Error Correction in DVB 143
6.5 Further Reading 145

7   **Digital Modulation Techniques** (JAEGER) 147
7.1 NRZ Baseband Signal 147
7.2 Principles of the Digital Modulation of a Sinusoidal Carrier Signal 153
7.2.1 Amplitude Shift Keying (2-ASK). 155
7.2.2 Frequency Shift Keying (2-FSK) 157
7.2.3 Phase Shift Keying (2-PSK) 158
7.3 Quadrature Phase Shift Keying (QPSK) 160
7.4 Higher-level Amplitude Shift Keying (ASK) and Vestigial-
       Sideband Modulation (VSB) 163
7.5 Digital Quadrature Amplitude Modulation (QAM) 167
7.6 Orthogonal Frequency Division Multiplex (OFDM) 173

8   **Conditional Access for Digital Television** (VERSE) 181

9   **The Satellite Standard and Its Decoding Technique** (VERSE) . . 187
9.1 The Basics of Satellite Transmission 187
9.1.1 Transmission Distance 187
9.1.2 Processing on Board a Satellite 188
9.1.3 Polarisation Decoupling. 190
9.1.4 Energy Dispersal 191
9.1.5 Signal Reception 191
9.1.6 Reference Data of a Television Satellite with Astra 1D
       as an Example. 192
9.2 Requirements of the Satellite Standard 192

9.3   Signal Processing at the Encoder   194
9.3.1   System Overview   194
9.3.2   Energy Dispersal   195
9.3.3   Error-protection Coding   195
9.3.4   Filtering   197
9.3.5   Modulation   197
9.4   Decoding Technique   198
9.4.1   Demodulator   199
9.4.2   Filtering and Clock Recovery   199
9.4.3   Viterbi Decoder   200
9.4.4   Sync-byte Detector   201
9.4.5   De-interleaver and RS Decoder   202
9.4.6   Energy-dispersal Remover   202
9.4.7   Baseband Interface   203
9.5   Performance Characteristics of the Standard   203
9.5.1   Useful Bit Rates   203
9.5.2   Required Carrier-to-noise Ratio in the Transmission Channel .   204
9.5.3   Antenna Diameter   205
9.6   Local Terrestrial Transmission   205

10   **The Cable Standard and Its Decoding Technique** (JAEGER)   . .   207
10.1   Cable Transmission Based on the Example of a
        German CATV Network   207
10.1.1 Intermodulation   209
10.1.2 Thermal Noise   210
10.1.3 Reflections   212
10.2   User Requirements of the Cable Standard   212
10.3   Signal Processing at the Encoder   214
10.3.1 Conversion of Bytes to Symbol Words   215
10.3.2 Differential Coding of MSBs   215
10.3.3 Modulation   217
10.4   Decoding Technique   220
10.4.1 Cable Tuner   220
10.4.2 IF Interface   221
10.4.3 Recovery of the Carrier Signal   222
10.4.4 Generating the Clock Signal   224
10.4.5 Demodulation of the QAM Signal   225
10.4.6 Differential Decoding of MSBs   228
10.4.7 Conversion of Symbol Words to Bytes   228
10.4.8 Detection of MPEG Sync Bytes   229
10.5   Performance Details of the Standard   230
10.5.1 Determination of Useful Data Rates   230
10.5.2 Carrier-to-noise Ratio Required in the Transmission Channel .   232

10.6   DVB Utilisation in Master Antenna Television Networks 233
10.7   Local Terrestrial Transmission (MMDS). 235

11   **The Standard for Terrestrial Transmission and
        Its Decoding Technique** (LISS, REIMERS, ROY) 237
11.1   Basics of Terrestrial Television Transmission 238
11.2   User Requirements for a System for Terrestrial Transmission
        of DVB Signals 243
11.3   Encoder Signal Processing. 245
11.3.1 Inner Interleaver and Symbol Mapping 245
11.3.2 Choosing the OFDM Parameters 248
11.3.3 Arrangement of the Transmission Frame 251
11.4   Decoding Technique 255
11.4.1 Receiver Classes. 257
11.4.2 Straight Forward Technology – the Classical Approach 258
11.4.2.1 Antenna. 259
11.4.2.2 Tuner 260
11.4.2.3 IF Processing 261
11.4.2.4 DVB-T Decoder Chip 263
11.4.3 Enhanced Technologies for DVB-T Reception 264
11.4.3.1 Antenna Pre-Amplifiers for DVB-T 264
11.4.3.2 One-Chip Silicon Tuner 266
11.4.3.3 Network Interface Module (NIM) Technology 267
11.4.3.4 Antenna Diversity 269
11.5   Hierarchical Modulation 270
11.6   Features of the Standard. 274
11.6.1 Determination of Useful Data Rates 275
11.6.2 Required Carrier-to-noise Ratio in the Transmission Channel . . 278
11.6.3 Features Relevant for Mobile Reception 280

12   **DVB Data Broadcasting** (FOELLSCHER). 285
12.1   Basics of Data Broadcasting 285
12.2   Data Piping 286
12.3   Data Streaming 287
12.3.1 Asynchronous Data Streaming 287
12.3.2 Synchronous Data Streaming 287
12.3.3 Synchronised Data Streaming 288
12.4   Data/Object Carousel 289
12.4.1 Data Carousel 289
12.4.2 Object Carousel 292
12.5   Multiprotocol Encapsulation 295
12.5.1 IP over DVB. 297
12.6   System Software Update. 299

13	**DVB Solutions for Interactive Services** (LEISSE, PIASTOWSKI) . .	303
13.1	Interactive Services.	303
13.2	Network-Independent Protocols for DVB Interactive Services. .	305
13.2.1	Protocol Stack	305
13.2.2	System Model	306
13.2.3	Higher Layer Protocols	307
13.3	Network-Dependent Solutions for PSTN, ISDN, DECT, GSM . .	308
13.3.1	Interaction Channel through PSTN/ISDN	309
13.3.2	Interaction Channel through DECT.	310
13.3.3	Interaction Channel through GSM	311
13.4	Network-Dependent Solutions for DVB-C, DVB-S and DVB-T .	313
13.4.1	Interaction Channel for Cable TV Distribution Systems.	313
13.4.2	Interaction Channel for LMDS	320
13.4.3	Interaction Channel for Satellite Distribution Systems	321
13.4.4	Interaction Channel for Satellite Master Antenna TV	323
13.4.5	Interaction Channel for Digital Terrestrial Television	325
14	**The Multimedia Home Platform (MHP)** (KLINGENBERG, SCHIEK)	331
14.1	The Role of Software Platforms in the Receiver	331
14.2	Some Non-MHP Solutions	332
14.2.1	ATVEF	333
14.2.2	Betanova	334
14.2.3	Liberate	334
14.2.4	Mediahighway	334
14.2.5	MHEG	335
14.2.6	OpenTV	336
14.2.7	Migration Concepts.	337
14.3	MHP 1.0	337
14.3.1	Basic Architecture	338
14.3.2	Transport Protocols	340
14.3.2.1	Broadcast Channel Protocols	340
14.3.2.2	Interaction Channel Protocols	341
14.3.3	Application Model and Signalling.	342
14.3.4	Content Formats	344
14.3.5	Graphics Model.	345
14.3.6	User Interface	347
14.3.7	Security Architecture	347
14.3.8	Minimum Platform Capabilities	350
14.4	MHP 1.1.	350
14.4.1	Enhancements in MHP 1.1	351
14.5	The Future of MHP	356
14.5.1	Ongoing Developments.	356
14.5.2	Aspects of a mobile MHP	356
14.6	The MHP Test Suite	357
14.7	The Globally Executable MHP	358

14.8   Other Java-Based Software Platforms 359
14.8.1 OSGi 359
14.8.2 DAB Java Specification 359
14.8.3 Mobile Information Device Profile – MIDP 360

15   **Measurement Techniques for Digital Television** (JOHANSEN,
        LADEBUSCH, TRAUBERG)   363
15.1   Measurement Techniques for Source-Signal Processing
        in the Baseband 364
15.1.1 Quality Evaluation of Video Source Coding 364
15.1.2 Checking Compressed Audio and Video Signals 366
15.1.3 Checking the MPEG-2 Transport Stream 366
15.1.4 Checking the Functionality of the Decoder 372
15.2   Measurements for Digital Transmission Technology. 373
15.2.1 Representation of the Eye Diagram 374
15.2.2 Measurements Carried out at Modulators and Demodulators . . 375
15.2.3 Measurement of Error Rate 379

16   **Bibliography** 383

17   **Acronyms and Abbreviations** 395

18   **Index** 403

1 Digital Television – a First Summary

1.1 Definitions and Range of Application

The expression "digital" is one of the elements of language which has frequently been used and very often misused in the past few years. "Digital" alone means that there are elements which can be counted on the fingers. In the everyday language of information technology the expression "digital" is often used as a synonym for "sampled, quantized and presented in binary characters". In the field of electronic media technology "digital" is a mark of quality which was first used when the compact disc (CD) was introduced on the market and offered to the general public.

The expression "digital studio" describes the introduction of digital signal forms into radio and television production. The audio and video signals are sampled by means of predetermined sampling frequencies and quantized by predetermined numbers of quantization steps. As a rule, further source coding is not used during processing and distribution, however it may take place in individual pieces of equipment.

The signals which are received by "digital" television receivers in most cases are still analogue input signals (NTSC, PAL, SECAM), which are then internally digitised and processed as digital signals. The presentation of the image, however, remains analogue.

From what has been said, it is evident that the terminology must be carefully defined in order to limit confusion to a minimum. "Digital Television", sometimes termed "Digital TeleVision (DTV)", or "Digital Video Broadcasting (DVB)", usually means the transmission of digitised audio, video and auxiliary information as data signals. However, since these data signals must in most cases be modulated onto continuous-time carrier waves in order for them to fit in the transmitting channel, the actual transmission in digital television uses analogue signals.

In the context of this book the expression "digital television" should in fact be understood as the system for the transmission of audio, video and auxiliary data described above. However, it is not possible to concentrate only on the channel coding and modulation, furthermore necessary system components must be included, in particular the technology for the source coding of audio and video signals and multiplexing. The latter is required for the co-ordination of various elementary streams to form one signal. The expression "multiplexing" is not really quite apt because this simplifies the

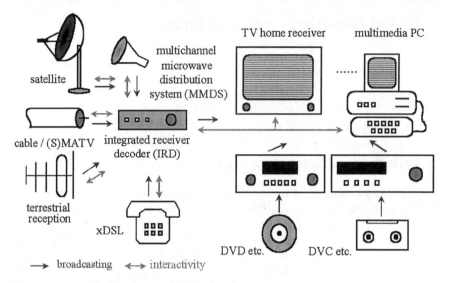

Fig. 1.1. A scenario for the utilisation of DVB in private homes

actual task. In the concept of the Moving Pictures Experts Group (MPEG) the term "systems" is used to indicate that the actual multiplexing is overlaid with a large number of other tasks.

During the development of the system further elements to those named, as belonging to the field of "digital television" have been more or less successfully subsumed under this expression. Fig. 1.1 shows a system scenario which describes the equipment technology of digital television from the viewpoint of the user. Mainly, DVB includes all equipment required for distribution, reception, relevant signal representation and processing. However it excludes studio technology and the actual video display.

The most suitable distribution systems for the transmission of DVB are satellite, cable and satellite master antenna TV (SMATV). Distribution via conventional terrestrial transmission networks has begun in a number of countries and is planned for several others. One of the most innovative possibilities for digital television is the transmission of DVB signals via telephone lines. The utilisation of glass fibre, where this is available, appears to be relatively unproblematic, even for signals with a higher data rate, due to its inherent transmitting capacities. The reduction of the data rate of a television programme to just a few Mbit/s will enable the transmission of a single television programme over a transmission distance of several kilometres via a copper telephone wire between the last switching centre and the subscriber. Finally, in many countries microwave transmission of DVB signals is exploited (Microwave Multipoint Distribution System [MMDS]) to transmit a variety of programmes, similar to that offered via a typical cable network, to private homes. The exceptional advantage of MMDS is, for example, thinly populated regions can be connected where a cable connection would not be financially viable.

For all the mentioned broadcast media, the DVB Project has developed technologies which allow reverse channel compatibility. This means that return information, i.e. from the viewer to the network operator or content provider, can be sent. The utilisation of a reverse channel for the ordering and payment of pay-TV services, or in connection with teleshopping services etc., has become a reality and requires only relatively low data rates.

The DVB decoder (integrated receiver decoder [IRD]) replaces a data modem, in which the digital signals can be demodulated and decoded in order to be displayed. The decoder also edits and evaluates the return-channel data and additional information embedded in the DVB data stream which can, for instance, then be passed onto the display for the creation of a graphic user interface.

Apart from the classic television receiver, the PC or "desktop video workstation" can be utilised as an actual terminal. This in most cases will not replace the television set, but if digital television is to be used for business applications such as stock exchange and banking it will be necessary to have a keyboard and a mouse for interactive direct communication.

Cassettes and video discs like the DVD as well as the hard-disc built into receivers of the personal-video-recorder-(PVR-) type or the hard-disc of a multimedia PC could serve as storage media for digital television.

In recent years all kinds of "multimedia" offers accompanying DVB broadcasts have been made available to TV viewers. The range of such offerings is vast. Electronic program guides are among the simplest. Interactive games and the broadcasting of internet pages are more complex. The scenario in figure 1.1 still gives an adequate portrayal of the user environment in this era of "multimedia services". What it can not depict is the software environment in the receiving devices (such as the multimedia home platform [MHP]) as well as the myriad of communication protocols used.

For the reception of terrestrial DVB signals, various kinds of portable and mobile receivers are being offered. A typical portable receiver is a handheld device having the size of a small laptop computer. Prototypes of receivers have been shown at trade fairs which fit into a shirt pocket. DVB receivers in cars have been demonstrated by numerous companies. From these examples it becomes clear that DVB is able to provide service to users which could not be reached by analogue TV transmissions.

1.2 The Genesis of Recommendations for Digital Television

The evolution of the first concepts of digital television can only be understood in relation to the background of the world-wide development in the field of television engineering in the second half of the 1980s. While in Japan the multiple subsampling encoding (MUSE) system and in Europe, the HD-

MAC system developed within the Eureka 95 project, were ready for the transmission of HDTV via satellite and through cable networks, the broadcasting community in North America felt threatened by these developments, since at that particular time they had nothing competitive to offer. Most particularly the distribution of HDTV via satellite envisaged in Europe and Japan would have endangered the classical commercial business of broadcasters in the USA, if also satellite broadcasting had become the primary means of HDTV transmission. The stability of this commercial business is a result of the large networks acquiring advertising spots with a nation-wide impact, whereas the actual transmission of the programmes within the network is left to the regional partners, so-called affiliates, who in turn secure their income through the sale of commercials of local and regional importance. It is clear that in such a system, satellite distribution which cannot be regionalized would prove to be a destabilising factor. The growth of the cable business, which particularly in the USA resulted in new programmes being offered exclusively over cable, had already drastically changed the broadcasting scene. Therefore it was obvious that North America would start an initiative to develop terrestrial, and thus regionally transmittable, HDTV as an alternative to MUSE and HD-MAC.

The second initiative for the development of a terrestrial transmittable HDTV standard began in Scandinavia [APPELQ], but this time right from the beginning with the goal of breaking through to digital television. Looking back, it is difficult to identify the actual motivation for initiating the HD-DIVINE project. It was probably due to the conviction that HD-MAC was not technically viable, a result of the wish of the national public broadcasting corporations to prevent the Scandinavian governments from granting terrestrial frequencies to commercial broadcasters by proclaiming a great Scandinavian awakening, and pioneering spirit played an important role, too. Back in 1992 HD-DIVINE was able to publicly demonstrate their first, still incomplete system.

At the end of 1991, the first discussions took place in a close circle in Germany with the aim of evaluating the worldwide technical situation of television and to resolve the question as to which real development alternatives were available in Europe. Resulting from these first discussions, the European Launching Group was born in the spring of 1992. This was a group with participants from all sections of the trade who first met unofficially, and not until September 1993 evolved into the International DVB Project [REIMERS 1].

In Japan there was no official research on digital television for a number of years. The success of HDTV with MUSE as a transmission standard could not be endangered. It was only in the summer of 1994 that the Ministry of Posts and Telecommunication (MPT) founded a Digital Broadcasting Development Office, which would be responsible for the co-ordination of all achievements in development. A project structure, now called Advanced Radio Industries and Businesses (ARIB), supports this endeavour.

1.2.1 Work in the United States of America

The replacement of traditional analogue television by digital television was first thought of in 1990. It was recognised as a solution to a national initiative of the Federal Communication Commission (FCC) during 1987, with the objective to develop an HDTV standard in the USA enabling terrestrial and therefore locally confined transmissions of HDTV. At the initial stage the "call for proposals" produced a veritable "gold rush" climate which led to 21 possible systems being submitted, some of which only tried to attain a compatible improvement of NTSC and in this respect showed more similarity to PALplus than to HDTV. Many of the proposals promised HDTV quality but were, nevertheless analogue in the above-mentioned sense and this was particularly true of Narrow-MUSE, a development for the USA by the national Japanese broadcasting authority, NHK. [FLAHERTY].

By 1990 the list of remaining system concepts which could be taken seriously had been reduced to nine. Two of them, both analogue systems, were conceived for parallel transmission of HDTV and NTSC (Simulcast). On the 1st of July 1990, General Instrument was the first company to recommend a fully digital concept. In 1991 there was only a total of five systems left in the race, all of which were intended for the transmission of HDTV. Four of them were digital and one (Narrow-MUSE) was analogue. The results of intensive testing on hardware prototypes between 1991 and 1992 reduced the list even further. Specifically, Narrow-MUSE was eliminated [FCC].

In May 1993 the remaining system developers (the companies AT&T/Zenith, General Instrument, DSRC/Thomson/Philips, and the MIT) stated their readiness to work together on the development of a single proposal, which since then has been known as the Grand Alliance HDTV System. Multiple endeavours to initiate the development of transmission processes for digital television with the current image quality (standard-definition television [SDTV]) were resisted for a long time by the FCC. It was just before finalisation of the specification phase in the summer of 1995 that this quality level was integrated into the system.

The Grand Alliance HDTV System, which was approved by the membership of the Advanced Television Systems Committee (ATSC) as the ATSC Digital Television Standard (A/53) on 16, September 1995 can be described in a somewhat simplified manner as follows [HOPKINS]: the source coding of the image signal is in accordance with the MPEG-2 standard as developed by the Moving Pictures Experts Group (Main Profile @ High Level – see section 4.3). This provision is still maintained when SDTV images are to be encoded instead of HDTV images. The audio coding is in accordance with Dolby recommendation AC-3 and is multichannel-compatible. The forward error protection is multistage. The modulation process utilises vestigial-sideband (VSB) modulation of a single high-frequency carrier (see section 7.4). An 8-

stage modulation (8-VSB) was specified for the terrestrial transmission and a 16-stage modulation (16-VSB) for distribution over cable.

On 24, December 1996 the FCC adopted the major elements of the ATSC Digital Television (DTV) Standard, mandating its use for digital terrestrial television broadcasts in the USA. It did not mandate the specific HDTV and SDTV video formats contained in the standard. On November 1st, 1998 the first terrestrial digital TV service was launched in the USA. As of 15, May 2001 there were 195 DTV stations on the air in the USA. The governments of Canada and South Korea adopted the ATSC standard in 1997 for possible use in their countries.

Not all market partners in the USA were content with the system components chosen by the Grand Alliance. During the work of the Cable Television Laboratories (CableLabs) a cable distribution standard using quadrature amplitude modulation (QAM) was developed which is very similar to the DVB system (DVB-C – see chapter 10). This standard now exists in addition to the VSB-based proposal by the Grand Alliance.

In the summer of 1994 the USA saw the first transmissions from a multi-channel television satellite. Named DirecTV/USSB/DSS, it uses technology which is very similar to that developed by the DVB Project (DVB-S – see chapter 9) by partner companies of the DVB Project and is practically identical to the European standard [BEYERS].

1.2.2 Work in Europe

The DVB Project (in the beginning a European industry consortium) is the focal point of the development of digital television for most of the world. This project came into being from the experience that developments in the complex field of electronic media can only be successful when all the important organisations working in this field participate in such a development and when the commercial interests are allowed to carry the same weight as technical considerations in the definition of technological objectives.

Figure 1.2 shows the organisational structure of the project. The number of member companies is 260, representing 32 countries from all over the world. The members are either content providers, hardware manufacturers, network operators or regulatory bodies from the various countries. The Commission of the European Communities (CEC), the European Broadcast ing Union (EBU), associations and standardisation institutes also have a special status for participation.

The commercial working group (Commercial Module) of the DVB Project is responsible for the definition of commercial requirements for the new systems from the viewpoint of the users. These requirements form the basis of work within the Technical Module. After completion of the development, the Commercial Module verifies the specifications for the new systems and passes them on to the Steering Board for final decision.

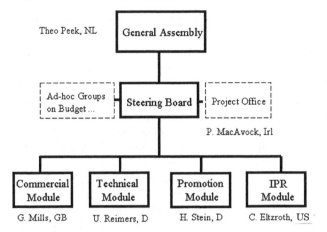

Fig. 1.2. Organisational chart of the DVB Project (as of 2002)

By means of a cooperation contract with the standards institutes ETSI and CENELEC, an integration of specifications from the DVB Project into the regular standardisation procedures of both institutes is ensured.

At the time of the first work on DVB in Europe, the Moving Pictures Experts Group (MPEG), a working group reporting to the standardisation institutions ISO and IEC which are active worldwide, was already working on a solution for the source coding of video and audio signals and had already started the design of the respective systems level (MPEG-2). The proposed system known as MPEG Audio for the source coding of audio signals, in mono and stereo, was already in the final phase of standardisation. The DVB Project decided that in order for the technological solution used by the DVB Project to find a wide international basis, digital television should utilise the MPEG standard. This decision led to an intensive cooperation between many European organisations and MPEG with the result that the research conducted in some places in Europe – with a different approach to image coding – was put aside in favour of a worldwide standard. The DVB Project's strategic aim to support a single worldwide solution was partly responsible for the decision taken by the Grand Alliance in the USA to also select MPEG-2 for the source coding of image signals. Japan also adopted MPEG-2 for source coding.

The first important result of the DVB Project emerged in the second half of 1992 under the leadership of the author. This was the report to the European Launching Group on the "Prospects for Digital Terrestrial Television" [WGDTB], which was presented in November 1992. This report showed how, and with which aims, a DVB system for Europe could be developed. The report was relatively heavy weighted in favour of terrestrial transmission and towards HDTV as the probable quality objective. In this respect it was a product of its time and took into account the fact, that at the end of 1992 the official European development policy was still centred on the satellite transmission of HD-MAC.

Even during the production of the above-mentioned report the DVB Project was able to profit enormously from the existence of research consortia which were sponsored by funds from the EC (RACE dTTB, SPECTRE, Sterne) or from the German Federal Ministry for Research and Technology (HDTV-T), or which were organised by a private enterprise (HD-DIVINE). During the further development of the work, the said consortia continued playing an important role, particularly in the definition of the standards for the terrestrial transmission of DVB (DVB-T – see chapter 11).

The first complete system specification was the recommendation for satellite transmission (DVB-S) adopted by the technical module of the DVB project in November 1993. [REIMERS 3, REIMERS 4]. In December the Steering Board agreed with this recommendation and in November 1994 by unanimous decision of all member states of the ETSI, this became the European Telecommunications Standard, ETS 300 421. In January 1994 the specification for DVB distribution via cable (DVB-C – ETS 300 429) followed and since then numerous other specifications have been developed and adopted.

In the following chapters many of the results of the work of the DVB Project are described in detail. The complexity of the DVB scenario, however, does not permit the inclusion of all the DVB results. At the end of 2002, CENELEC and ETSI listed some 44 DVB-based standards and some 19 guidelines for the implementation of these standards. Some examples from this long list such as the specifications of technologies to be used to interface individual pieces of equipment, the technologies to be used in network headends or the solutions for in-home networks will not be described in this book.

1.2.3 Work in Japan

As already mentioned, in Japan work on the development of digital television officially commenced in the summer of 1994. However, unofficially this topic had already been examined in many places since a lot of European subsidiaries of Japanese equipment manufacturers were already members of the DVB Project. Therefore early reports from Japan (personal communication) showed a great correspondence between Japanese and European conceptions. As in Europe, MPEG-2 was chosen for the source coding and for the system level. Quadrature phase shift keying (QPSK) was also adopted as the modulation procedure for the satellite transmission [NHK]. Quadrature Amplitude Modulation (QAM) was used for the distribution in cable networks. In practice this means that all "communication satellite" services in the 12.2 GHz to 12.5 GHz bands use either the DVB standard for satellite broadcasting (DVB-S – see chapter 9) or the DirecTV system introduced in the USA. In those cable networks where digital TV was introduced beginning in 1996, the decision was in favour of the DVB standard (DVB-C – see chapter 10).

The development of an additional, national, Japanese system for digital broadcasting began in 1994 and was jointly carried out by the Association of Radio Industries and Businesses (ARIB) and the Digital Broadcasting Experts Group (DiBEG). The system was designated as Integrated Services Digital Broadcasting (ISDB). It consists of a satellite component (ISDB-S), a cable component (ISDB-C) and a terrestrial component (ISDB-T). Services using ISDB-S started in December 2000. ISDB-S is used on "broadcasting satellites" in the 11.7 GHz to 12.2 GHz bands. Instead of QPSK, the modulation scheme used by DVB-S, the ISDB-S system uses 8-PSK (see chapter 7) and Trellis coding. The first installations of ISDB-C in cable networks were completed in December 2000. ISDB-C is very similar to DVB-C and differs only in a very limited number of technical details.

ISDB-T is a terrestrial system for digital broadcasting which is very closely related to the DVB standard for terrestrial transmission (DVB-T – see chapter 11). The modulation method COFDM (see chapter 7) is used in both DVB-T and ISDB-T. The most important new idea incorporated in ISDB-T is the possibility to split the frequency band of a single TV channel (6 MHz in Japan) into 13 segments of 432 kHz width. Using only one segment, it is possible to transmit a digital radio service. Using several segments, SDTV can be offered. Using all 13 segments, HDTV services can be transmitted. The first pilot transmissions of ISDB-T started in April 1999. The official launch of the digital terrestrial services took place in December 2003.

1.3 Objectives in the Development of Digital Television

One of the most important questions asked with regards to the development of DVB is about its commercial reason. Why should a system for entertainment television be commercially successful, in a saturated market, competing with a multitude of existing television standards, some of which have been in use for many years and therefore can address a great number of compatible receivers?

At the beginning, the DVB Project compiled a catalogue of possible goals, which could be described as classical or typical of broadcasting. [REIMERS 2]:

1. Digital television might enable the transmission of very high-quality HDTV images, possibly even via future terrestrial broadcasting networks.
2. DVB might enable the broadcasting of programmes of contemporary technical quality (standard definition television – SDTV) using narrowband channels for transmission, or it might enable an increase of the number of programmes offered within existing transmission channels.

3. DVB might be the method of broadcasting to low-cost pocket TV receivers, equipped with built-in receiving antennas or short rod antennas, which would guarantee stable reception for a number of television programmes.
4. Television receivers in vehicles (trains, busses or cars) might be served by DVB with broadcasts of a superb quality, i.e. DVB might enable stable reception in moving vehicles even over difficult radio channels and at high speeds.
5. Moreover, as a data transmission technique, DVB might retain the typical characteristics of digital technology, such as the stability of the reception within a very clearly defined coverage area, the possibility of simple distribution over telecommunication lines as one service among many, and the possible integration into the world of personal computers.

As work on DVB has progressed, the objectives have changed considerably. HDTV has not been forgotten at all, but it has lost its role as a primary objective. The servicing of portable receivers has remained an objective during the development of DVB-T, the standard for terrestrial transmission, but for several countries it is not as important as originally envisaged. From the extensive list of optional parameters for the terrestrial standard (see chapter 11) it is possible to choose operational modes suitable for portable reception. Finally, mobile reception was not a part of the original user requirements of the DVB system, although this requirement is currently evolving in some countries. The DVB system specification for terrestrial transmission (DVB-T) has, in the meantime, demonstrated its capability to provide stable mobile reception up to very high speeds.

In the course of time „data container" has become the key concept with regards to the definition of the objectives of DVB [REIMERS 5]. This concept illustrates the idea which underlies the design of the transmission standards for all methods of transmission. A data container is defined by the fact that a maximum amount of data per unit of time can be transmitted in it quasi error free (QEF). It does not matter what kinds of data are transmitted as long as they are packetised and supplemented with additional data, such as synchronisation information, in accordance with the rules of the various DVB standards. With this background, the question about the reason for DVB can now be answered as follows:

1. DVB enables a multiplication of the number of television programmes which can be broadcast on one transmission channel or in one data container.
2. DVB makes the broadcasting of radio programmes possible and enables data transmission for entertainment and business purposes.
3. DVB makes a flexible choice of image and audio quality possible, including the choice of HDTV as long as the resulting data rate does not exceed the capacity of the data container.

4. For use in connection with pay services there are very secure coding methods which ensure that unauthorised access to such services is extremely difficult, if not impossible.
5. DVB standards for return transmission from the viewer to the network operator or content provider enable full interactive services to be introduced.
6. The DVB Multimedia Home Platform (MHP) provides an open and interoperable software platform for enhanced services like enhanced broadcasting or even full internet access from a TV receiver.
7. DVB-T offers the possibility to address receivers in all kinds of environments from the classical TV sets in the living room via portable devices in ones shirt pocket to TV receivers built into vehicles.
8. Furthermore, as a data transmission technique, DVB incorporates typical characteristics for the utilisation of digital technology, such as the stability of the reception within a clearly defined coverage area, the possibility of simple distribution over telecommunication lines, as one service among many others, and the possible integration into the world of personal computers.

1.4 Data Reduction as the Key to Success

The data containers already mentioned in section 1.3 are available in various capacities, in accordance with the transmission medium. Specifically, for example, in satellite transmission and distribution on cable networks, the capacity is in the range of 40 Mbit/s (see chapters 9 and 10), while for terrestrial transmission the capacity could be in the range of 13 Mbit/s to 20 Mbit/s (see chapter 11).

Regarding the digitisation of colour television signals in compliance with ITU-R Recommendation BT.601 [ITU 601], it is obvious that without an efficient reduction in the data rate not even a single television programme, let alone many, can be transmitted within a data container. Recommendation 601 calls for a data rate of 216 Mbit/s for a complete television image including blanking intervals, or 166 Mbit/s when blanking intervals are buffered and can be ignored (see chapter 2). Hence data-rate reduction is indispensable (see chapter 4).

Similar considerations apply to audio signals. The original data rate for stereo signals in CD quality is 1.4 Mbit/s (see chapter 2). Therefore they would require an exorbitantly large part of the capacity of a data container, unless here, too, efficient data-rate reduction is achieved (see chapter 3).

In figure 1.3 there is an illustration of data-rate reduction for video signals. There are three groups of input data rates on the ordinates, values which are produced in television studios before data-rate reduction. In each case these are net values, data rates describing the video signals without blanking intervals. Current digital studios for standard-definition (SDTV)

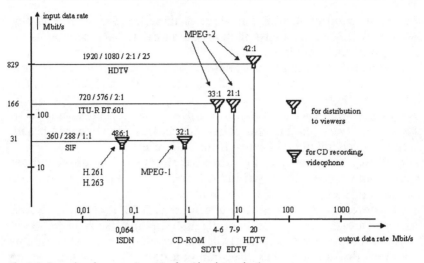

Fig. 1.3. Examples of compression ratios for video data reduction.

and enhanced-definition television (EDTV) use 166 Mbit/s (see above). One important proposed HDTV studio format uses 829 Mbit/s. If, right from the start, video services are to be produced with a greatly reduced video quality (which would be roughly comparable to a medium-quality VHS recording), it would seem that the utilisation of original video material with reduced video quality would suffice (MPEG-1 source input format – SIF, see chapter 4). The parameters of the luminance of SIF images are 288 lines, with 360 pixels per line, and scanning is progressive, which means that an image is represented by 25 complete frames per second. In contrast to what is required in today's studio standards, the chrominance components are added to the luminance signals line-sequentially at a ratio of 4:2:0. The resulting data rate of quantized source signals of 8 bits per pixel is therefore 31.1 Mbit/s.

The funnels in figure 1.3 symbolise the data-rate compression by means of the source coding techniques introduced in chapter 4. Significant examples of output data rates after completion of compression are indicated. "ISDN" stands for the compression by means of the ITU-T Recommendations H.261 or H.263 [ITU H.261, ITU H.263], which also aim towards videophone applications with the ISDN data rate of 64 kbit/s. The expression "CD-ROM" symbolises the use of data reduction to produce video signals for the storage on CD formats such as CD-ROM. The MPEG-1 standard is used as the compression method. "SDTV" (standard-definition television) refers to a video quality which does not warrant an error-free reproduction, but the sum of whose artefacts in most of its picture content is comparable to the present PAL television. It seems that 4 – 6 Mbit/s are sufficient for SDTV when, as with all higher data rates, one of the standards known under the generic term of MPEG-2 is employed. "EDTV" (enhanced-definition television) stands for virtually transparent video transmission, at a quality capable of being produced in a studio in accordance with the ITU-R Recommendation BT.601.

One needs to be very careful in interpreting the values indicated. At a given data rate the achievable video quality varies greatly with the complexity of the scene that was encoded. The quality of the encoder used is very important and it is now very typical to employ statistical multiplexing of the various video signals, which are to be transmitted within a single data container. Therefore it may be possible to deliver SDTV quality at an average data rate below 3 Mbit/s, if enough video signals with different scene content are jointly encoded in such a way that the data rate momentarily allocated to each individual signal is chosen in accordance to the scene complexity.

1.5 Possible Means of Transmission for Digital Television

Most people, when thinking of the technical means of transmission for television programmes, probably think first of terrestrial transmission, which is wireless broadcasting of programmes from antenna masts standing on the ground.

However, already during the phase of planning the introduction of Digital Audio Broadcasting (DAB), the allocation of broadcasting frequencies for new terrestrial services – at least in Europe – proved to be exceptionally problematic. The same is also true for television, because it is neither possible, at short notice, to clear channels, now in use for analogue transmissions for the exclusive use of digital television, nor are there unused frequency resources which could be allocated for television broadcasting. In this respect the terrestrial transmission of DVB was dependent on finding new technical concepts for the broadcasting standard DVB-T and enabling the utilisation of the few frequencies available to reach a greater number of members of the public. In addition, a long-term strategy had to be found for the reallocation of frequency resources. Another option was to first introduce digital television in areas where spectrum capacity could be found, which resulted in the creation of "geographic islands".

For the broadcasting of analogue TV programmes in the United States of America there are approximately 1,700 television transmitters in use [JANSKY]. This represents such a minimal utilisation of the spectrum capacity allocated to television broadcasting, that the work of the Grand Alliance targeted the introduction of DTV into existing gaps of the frequency spectrum, the so-called "taboo-channels" [JANSKY]. These taboo-channels could, in principle, be used for the transmission of analogue NTSC signals, but in reality they were not used because of possible interference with either a co-channel transmitter somewhere in the region or a transmitter using an adjacent channel, or because of possible interference with the local oscillator fre-

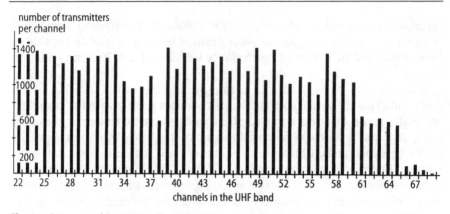

number of transmitters
per channel

channels in the UHF band

Fig. 1.4. Occupancy of the UHF band for television transmitters in Europe

quency used in the receiver. The ATSC DTV standard is in a position to allow broadcasts in some of these channels which are at present unused. This
means that – although the service areas of the NTSC transmitter and that of
the ATSC transmitter remain comparable in size, the interference effect of
the ATSC transmitter on an existing NTSC receiver must be reduced, for example by reducing the radiated power in comparison with an NTSC transmitter, however at the same time the robustness of the digital system must
be sufficient enough to enable interference-free ATSC reception despite the
presence of the NTSC transmitter.

In comparison with the USA the situation in many parts of Europe is dis-
tinctly more complex. It is typical for the national broadcasting services in
European countries to take on the task of total coverage for the general public. This means that there are a great number of transmitters with varying
power levels. Figure 1.4 shows the European utilisation of the UHF band
(channels 22 to 69) for television transmitters of all power levels in 1991
[WGDTB]. It is easy to see that in channel 22 alone there are almost as many
transmitters as in the whole of the United States of America. At first it looks
as if the only reserves are within the range of channels 34 to 38 and above
channel 60. Actually, different allocations in both these ranges exist in some
European countries. Most particularly the channels above 60 are still being
used in many countries for military services.

In Europe a recommendation for a new distribution of the available frequency spectrum was developed by a specialist group from the European
Radio communications Office [ERO] on the basis of an analysis of the present allocation of frequency bands to the various types of users. This recommendation, which was presented in autumn of 1995 during the European
Conference of Postal and Telecommunications Administrations [CEPT], includes the proposal that, by the year 2008, in all regions where channels
above channel 61 are not yet utilised for radio broadcasting, these channels
ought to be made available for DVB. By the year 2020 large parts of the UHF

spectrum should then be utilised by DVB, while at the same time the number of transmitters for analogue television should be reduced.

In Chester, United Kingdom, in the summer of 1997, more than 30 European countries signed a multilateral co-ordination agreement which includes all necessary rules and technical parameters that enabled the start of frequency planning for terrestrial digital television in all of Europe.

The first country to introduce terrestrial digital television based on the DVB standard DVB-T was the United Kingdom. Immediately before that introduction, the company BSkyB commenced digital satellite broadcasting based on DVB-S on 1st October 1998. Digital terrestrial was launched on 15th November 1998. Six multiplexes were operated by four so-called multiplex operators. Later digital cable based on DVB-C was being rolled out in a number of franchise areas. In November 2001, there were around 8.3 million digital homes in the UK. The official subscriber figures were: 5.5 million homes using DVB-S, 1.6 million homes using DVB-C and 1.2 million homes using DVB-T.

A planning conference which will revise the results of the "European VHF/UHF Broadcasting Conference" (Stockholm 1961) is planned for 2004 and 2005. It will at least cover the whole "European Broadcasting Area" but may include all of Africa as well as some neighbouring countries. This conference will be held in two sessions. The first session will take place in May of 2004. Among other things, it will be used to create the planning criteria and determine the planning method. The second session is planned for 2005. During that session all agreements will have to be negotiated.

The introduction of DVB by satellite and cable is much less affected by the necessity of long-term planning than the introduction of terrestrial transmission. Therefore, the mainstay of the introduction strategy in Europe was satellite distribution. The initial stages in this strategy go back to the spring of 1996. The ASTRA 1E satellite, in its orbit position of 19,2° east, had been ready for the DVB transmission since October 1995. Figure 1.5 shows a possible scenario for the distribution of DVB with the satellite taking the central position.

Following the multiplexing of individual compressed audio, video and auxiliary data signals into programmes 1 to n, these are combined to form a single serial data stream (the data container), which is then transferred to the satellite uplink after suitable processing. The satellite then broadcasts the data container to the direct-to-home (DTH) receiver, to cable head ends and to the locations of terrestrial transmitters. The signals are received in the cable head ends, and demodulated, although they may not be completely decoded. Next, if required, programmes are re-multiplexed within a container and then distributed further within the cable network by means of a specific modulation technique (DVB-C). Of course, cable programmes can also be distributed to the cable head ends by other means, for example, via telecommunication links. The procedure for cable networks is also applicable to

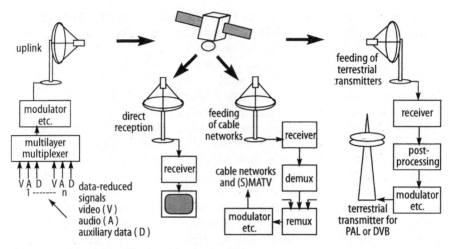

Fig. 1.5. Possibilities of a satellite-based distribution of DVB

community antenna systems. At sites with terrestrial transmitters for analogue or digital television, satellite signals can also be received, decoded and
demodulated. They can then be converted to any other transmission standard such as PAL or SECAM, before transmission to the public [ROY].

Figure 1.6. shows the status of the introduction of digital television in the
various parts of the world.

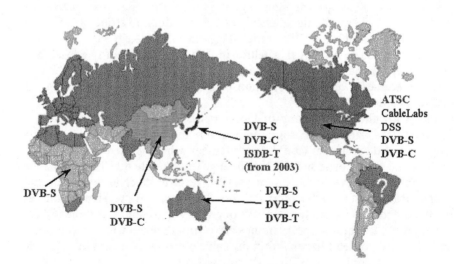

Fig. 1.6. Status of the introduction of digital television in the various parts of the world

Satellite services using DVB technology (DVB-S) are available nearly any-where in the world. DVB over cable (DVB-C) exists in many countries. Terrestrial DVB (DVB-T) was introduced officially in Australia, Finland, Germany, the Netherlands, Singapore, Spain, Sweden, Taiwan and the United Kingdom. Several other countries made the decision to introduce DVB-T and are currently installing the networks. The status of the introduction of digital television in Canada, the USA and South Korea is described in section 1.2.1. The status of the introduction of ISDB-T in Japan is mentioned in section 1.2.3.

1.6 Standards and Norms in the World of Digital Television

The development of digital television has led to technical specifications for a myriad of system components, ranging from the description of coding algorithms to that of modulation procedures, transmission parameters, hardware components, device interfaces, protocols and techniques for interactive channels and software platforms. Many organisations, such as MPEG, the DVB Project, the Digital Audio Visual Council (DAVIC) and the TV Anytime forum, participate in the development of specifications. The transformation of the specifications into standards, however, does not lie within the competence of these groups, but is the responsibility of European or even world standardisation institutes. The International Standardisation Organisation (ISO), the International Electrotechnical Commission (IEC), the European Telecommunications Standards Institute (ETSI) and the Comité Européen de Normalisation Électrotechnique (CENELEC) all play an important role in the DVB context. The task of creating unity in the system concepts and standards for digital broadcasting which are being developed in various parts of the world (ATSC, DVB, ISDB) is undertaken by the International Telecommunications Union (ITU).

The Moving Pictures Experts Group (MPEG) is a body reporting to both the ISO and IEC organisations. The ISO/IEC Standards (IS 11 172, IS 13 818 [Parts 1,2,3]) are a result of the work of MPEG.

A cooperating contract has been agreed between the DVB Project and the already-mentioned organisations ETSI and CENELEC, to the effect that the specifications originating from the Project are passed on to the Joint Technical Committee (JTC) Broadcast which is jointly run by the European Broadcasting Union (EBU), ETSI and CENELEC. This JTC then decides to which standardisation institute each specification will be delivered. Figure 1.7 lists the documents of the DVB Project which have been or are currently being handled by the institutes. Instead of listing the detailed names of standards or guidelines, the figure indicates the "territories" in which these documents reside. Details can be found on the two websites www.dvb.org and www.etsi.org.

	Specifications	Guidelines
MPEG-Coding	0	2
Service Information	1	2
Teletext, Subtitling, Data Broadcast	5	1
Transmission to the Home	9	2
Interactive Services	10	4
In-Home Networks, Interfaces etc.	12	3
Conditional Access	2	2
Multimedia Home Platform	3	0
Measurement Technologies	0	2
Digital Satellite Newsgathering	2	1
Total	44	19

Fig.1.7. List of DVB documents handled by ETSI and CENELEC (end of 2002)

1.7 The New DVB Project

DVB was formed to drive the future of TV broadcasting and to respond to the commercial opportunities created by a new delivery infrastructure. This was driven by market demand and the realisation of commercial value particularly through pay TV, advertising supported and public service broadcasting. The answer at the time was encapsulated by the word "digital" which brought promises of high quality and economic multi-channel programming. The word "digital" is as crucial to the future of broadcasting, content distribution and home entertainment as ever. However, "digital" is much more than a means offering more choices. It brings about great flexibility, but it also adds the complexities of managing the underlying stability of the consumer television experience. There is an accelerating requirement to maintain and improve the consumer offering in the face of diverse and competitive market developments. The document "A Fresh Vision for a new DVB" was approved by the DVB General Assembly in December 2000. It defines the foundations on which a new building is being erected. The commercial and the technical framework of this new building is defined, based on the following vision:

"DVB`s vision is to build a content environment that combines the stability and interoperability of the world of broadcast with the vigour, innovation and multiplicity of services of the world of the Internet."

The DVB Promotions and Communications Module (PCM) uses the following terminology when positioning DVB now and for the future:

"DVB is the point to multipoint data delivery mechanism with a guaranteed quality of service. DVB will enable the implementation of open and interoperable pathways to the consumer and thus offer a multiplicity of services and content for exploitation in the home, on the move, and in the office."

Figure 1.8 shows the new reference model of the workdone in the DVB Project. The cursive text indicates areas which were dealt with up till now or which will be dealt with in the future. Traditionally, the DVB Project developed specifications addressing the technology of broadcast networks in order to transport digitised audio and video to TV receivers, set top boxes and high-fidelity audio receivers. Later specifications for interaction channels were added. The completion of the specification for data broadcasting opened up DVB networks for the transmission of graphics, photos and data – for example by encapsulating internet-protocol-(IP-)based services in DVB data streams. At that time the users of personal computers and laptops started to take advantage of DVB services. The advent of the MHP indicated another milestone since, with the MHP, software applications can be run on all sorts of terminal devices.

The ongoing development of the Multimedia Home Platform is one of the key activities of the new DVB. Among other things the embedding of the MHP into digital broadcasting environments which are not completely DVB compliant, like for example many of the cable networks in the USA, is an important ongoing activity. One other significant new area for DVB is the development of specifications for the transport of "DVB content" over telecommunications networks (both fixed and mobile). A quite enlightening way of describing the technical activities within the Technical Module of the DVB Project is to list just a few of its new ad-hoc groups: "Audio/Video

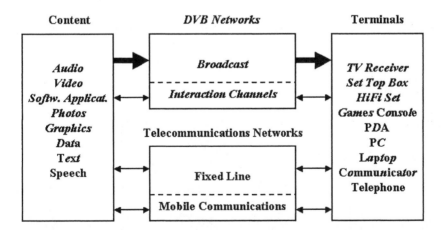

Fig. 1.8. New reference model for the work of the DVB Project

Content Formats (for the transport of DVB content over IP-based networks)", "CBMS–Convergence of Broadcast and Mobile Sevices (defining the technologies necessary to construct hybrid networks with a broadcast and a co-operating mobile telecommunications branch)", "Internet Protocol Infrastructures (for the efficient transport of DVB data streams over IP-based networks)", "Copy Protection Technology (including copy management and rights management)", "DVB-H (andheld) (for the transmission of high-rate data streams to battery-powered personal communications devices)".

After some ten years of work, the DVB Project is therefore still extremely active in developing new solutions for the ever-changing world of the electronic media.

2 Digitisation and Representation of Audio and Video Signals

The sources of the audio and video signals, such as microphone and camera, produce analogue signals, i.e. signals of an unlimited range of values, progressing continuously in the time domain. For digital processing the signals must be discretely sampled at regular intervals, quantised and coded. These processes are subsumed under the term 'digitisation' and are performed by analogue-to-digital converters (ADCs). Reconversion into analogue signals is carried out by digital-to-analogue converters (DACs).

Between these two converters the signals are transmitted in bit-parallel or bit-serial form and further processed. Certain signals and/or periods of time within periodically recurring signal patterns (frames) can be used specifically for synchronisation and for the transmission of auxiliary information. The signal shapes and data rates resulting from this procedure are here classed under the term 'signal representation'.

2.1 Sampling and Quantising

The frequency of the sampling clock obeys Nyquist's criterion which postulates that the sampling rate of the analogue signal must be at least twice the highest frequency to be sampled, so as to avoid aliasing in the digitised signal [SCHÖNFD 1]. The analogue signal must be prefiltered accordingly. Usually a sampling frequency higher than the Nyquist limit (oversampling) is chosen to increase the distance between the baseband spectrum and the repeat spectra and so reduce the implementation requirements for the filter.

The number of steps required for the quantisation and coding of the analogue signal is determined by the amplitude resolution of the digitisation process. This entails 2^b quantisation steps of the same size for linear, binary coding with b bits. The attainable level range is limited at the lower end by the quantisation error of one step and at the upper end by the limiting effect of the ADC which sets in as soon as the quantisation characteristic of the latter is overdriven. This level range is also referred to as system dynamics.

Under very general conditions relative to a stochastic analogue signal to be quantised, the power of the inherent quantisation error can assume the value of $Q^2/12$ if Q represents the size of one of the quantisation steps. If the chosen step Q is small enough, this quantisation error can be conceived as a uniformly distributed random sequence with zero mean value and with a white spectrum which is not correlated with the original signal [SCHÜSSLER]. This may also be called quantisation noise and can be used to define a signal-to-noise ratio. With deterministic signals, such as those often found in video applications, these requirements are not satisfied in every case, owing to which the quantisation error can no longer be described in such a general way. Even so, the above assumption offers a sufficient estimation of the disturbing effect of the quantisation error.

The following is valid for video signals when only the range of the picture signal S_V is digitised:

$$S_V / N_Q = b\,(6\ \mathrm{dB}) + 10{,}8\ \mathrm{dB}\ , \tag{2.1}$$

for audio signals, when referred to the r.m.s. value of S_A:

$$S_A / N_Q = b\,(6\ \mathrm{dB}) + 1{,}8\ \mathrm{dB}\ . \tag{2.2}$$

The different constants in both formulae can be derived from S_V (peak-to-peak value of the signal amplitude) and S_A (r.m.s. value of the signal amplitude) [SCHÖNFD 1]. The equivalent r.m.s. value of the quantisation error in both cases is $N_Q = Q / \sqrt{12}$ in the sense already explained.

The quantisation error can also depend on the frequency of the analogue signal. The layout of the sampler in the ADC (transitional behaviour, aperture jitter) is important here.

2.2 Digitising Video Signals

The standardised sampling rate of 13.5 MHz for the R, G, B or Y baseband signals [ITU 601], together with the quantisation of 8 bits per sample specified for broadcast transmission, leads to a data rate of $H_0 = 108$ Mbit/s for each of these signals. For each of the chrominance signals C_B and C_R the sampling rate is reduced to 6.75 MHz and the data rate to 54 Mbit/s. The Y sampling rate of 13.5 MHz is defined as an integral multiple of the line frequency for the 625-line standard as well as for the 525-line standard. This results in a duration of the 'active' line of 720 samples in both standards. Furthermore, this choice facilitates the design of the pre-filter required in front of the ADC, as already mentioned. The same applies to the chrominance components.

For an RGB transmission (format $4:4:4$) this results in a total data rate of $H_{0\,Total} = 324$ Mbit/s or, for a $Y/C_B C_R$ transmission, due to the reduced sam-

Table 2.1. Digitisation characteristics and data rates of video signals in accordance with [ITU 601]

Signals	Clock [MHz]	b [bit]	H_o [Mbit/s]	H_{oTotal} [Mbit/s]	Format
R	13.5	8	108		4 : 4 : 4
G	13.5	8	108		ITU 601
B	13.5	8	108	324	
Y	13.5	8	108		4 : 2 : 2
C_B	6.75	8	54		ITU 601
C_R	6.75	8	54	216	

Table 2.2. Pre-filter specifications in accordance with [ITU 601]

	Y/RGB Signals	$C_B C_R$ Signals
Pass band: Insertion loss	ascending from ±0.01 dB at 1 kHz to ±0.025 dB at 1 MHz ±0.025 dB/1 ... 5.5 MHz; ±0.05 dB/5.5 ... 5.75 MHz	ascending from ±0.01 dB at 1 kHz to ±0.05 dB at 1 MHz ±0.05 dB/1 ... 2.75 MHz
Group delay time	ascending from ±1 ns at 1 kHz to ±3 ns at 5.75 MHz	ascending from ±2 ns at 1 kHz to ±6 ns at 2.75 MHz ±12 ns/2.75 ... 3.1 MHz
Stop band: Insertion loss	≥12 dB/6.75 ... 8 MHz ≥40 dB/8 ... 13.5 MHz	≥6 dB/3.375 ... 4 MHz; ≥40 dB/4 ... 6.75 MHz

pling rate of 6.75 MHz for the chrominance components (format 4 : 2 : 2), a rate of H_{oTotal} = 216 Mbit/s. The results for both formats are compiled in table 2.1.

A quantisation of 10 bits per sample is also possible as an option for signal transmission and processing in a studio environment and within equipment, in which case the data rates will increase accordingly. Of late, a sampling rate of 18 MHz, instead of 13.5 MHz, has been suggested for an aspect ratio of 16 : 9, in order to obtain a horizontal resolution which is adapted to this picture format [ITU 601]. In this way the data rates stated increase by the factor of 4/3.

The specifications for filters before of the ADCs for the sampling rates of 13.5 or 6.75 MHz are shown in table 2.2.

Filters after the DACs generally have lower requirements to fulfil as most receivers in the frequency range under consideration already have a low-pass characteristic. For equipment with subsequent analogue signal processing the filter characteristics have to be designed accordingly.

Due to the sample-and-hold function of the DAC the output signal has a spectrum which decreases with higher frequencies in accordance with an si-

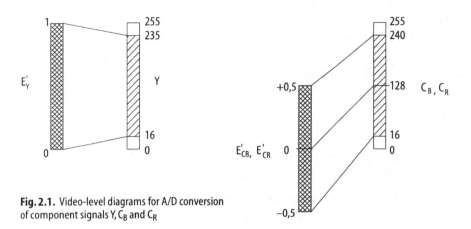

Fig. 2.1. Video-level diagrams for A/D conversion of component signals Y, C_B and C_R

function [SCHÖNFD 1]. The compensation of this frequency response can take place passively in the post-filter or actively by means of an equaliser.

The level ranges for the conversion of analogue component signals into the digital Y/C_BC_R component signals are smaller than the system dynamics available for the digitisation as shown in figure 2.1. The values 255 and 0 are reserved solely for the coding of the synchronising signals. In addition, at the upper and lower ends of the level range some quantising steps are reserved as a protection against overdriving. Generally, the amplitudes are converted to the binary code, with the luminance signal Y not being offset and the bipolar chrominance signals C_B and C_R being offset by their addition to the centre value of the system dynamics.

The correct control of the level ranges requires fixed signal values for 'black' or 'uncoloured'. This is effected by means of appropriate circuits directly in front of the ADC. Depending on the particular requirements, these can be simple clamping circuits or more complex control circuits to retain the black or uncoloured value with changing signal averages such as have been used for many years in television engineering. The feedback of digitised reference values into the aforementioned control circuits can also be used to compensate for tolerances in analogue signal processing, including that of the ADC [IRWIN].

2.2.1 ADCs and DACs for Video Signals

For video ADCs, almost the only technique still used nowadays is the parallel method or flash conversion. The block diagram of such a type of converter is shown in figure 2.2 [SCHÖNFD 1, TIETZE]. The analogue input signal $u_1(t)$ is fed to a chain of 2^b-1 comparators which receive their references from the taps of a voltage divider that divides the reference voltage U_{REF} into 2^b-1 parts. The comparators are activated by the sampling clock and in this way

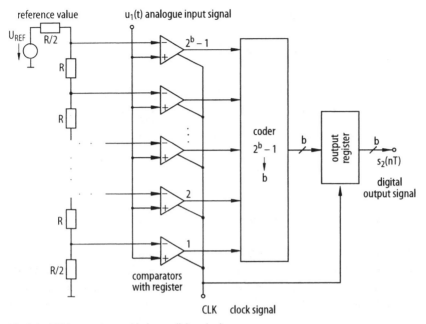

Fig. 2.2. ADC in accordance with the parallel method

function as samplers and quantisers at the same time. Their 2^b-1 output signals represent the digital value in the form of a thermometer scale, the digital value is converted by a subsequent coder into one of the usual binary codes with b bits in the digital output signal $s_2(nT)$.

The hold function for the sampling value is performed by registers at the outputs of the comparators (not shown in figure 2.2) as well as by the output register of the ADC.

Each of the b bits of the input signal in video DACs controls a current source which, according to the weight of the respective bit, contributes to the sum current at the output. Thus the magnitude of this sum current varies in accordance with the value of the digital input signal.

In the circuit diagram of figure 2.3 the individual currents which are allocated to the bits of the digital input signal $s_1(nT)$ are graded by powers of two with the help of an R-2R resistance network. The switches $b_7 \ldots b_0$ connect the currents either to the analogue output $i_2(t)$ or to the ground, depending on their bit pattern [SCHÖNFD 1]. The correct function of the device depends mainly on the simultaneous current switching of the circuit; a temporal misalignment of individual currents can lead to correspondingly large transients ('glitches') in the sum current $i_2(t)$ and therefore to take-over errors in the analogue output signal $u_2(t)$.

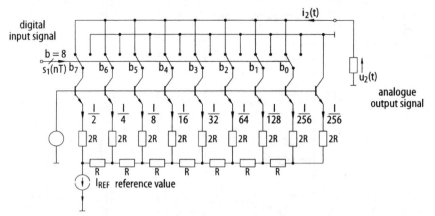

Fig. 2.3. DAC with an R-2R resistance network

Other models of video DACs use equal individual currents which, when added, are weighted by powers of two by an R-2R resistance network [TIETZE]. Hybrid forms between this variant and the one depicted in figure 2.3 are also known.

2.2.2 Representation of Video Signals

Figure 2.4 [ITU 656] shows the temporal relationships between analogue and digital video signals and the multiplex schedule of the video signals. The blanking of the digital signals begins prior to the rising-edge centre of the analogue synchronising signal, the duration of this blanking (144 samples for the 625-line standard, 138 samples for the 525-line standard) being shorter than that of the analogue signal. The fact that the sampling frequency and the duration of the 'active' lines for both standards are the same warrants that the temporal relationships between analogue and digital signals are almost independent of the standard.

At the beginning and at the end of the 'active' line the digital signal contains synchronising words (video timing reference signals), referred to as 'start of active video' (SAV) and 'end of active video' (EAV). Each of these words requires four time slots of the $Y/C_B C_R$ multiplex. Included is information about the actual field, the vertical blanking interval, and the beginning (EAV) or end (SAV) of the blanking. In addition, four further bits are coded such that they can correct one bit error and detect two bit errors in this information at the receiving end.

The multiplex schedule, as depicted in the last line of figure 2.4, applies to the parallel transmission of the eight or ten data streams, depending on the bit values. Apart from this, an additional ninth or eleventh line for the multiplex clock (27 MHz) is included so that no clock recovery at the receiving end

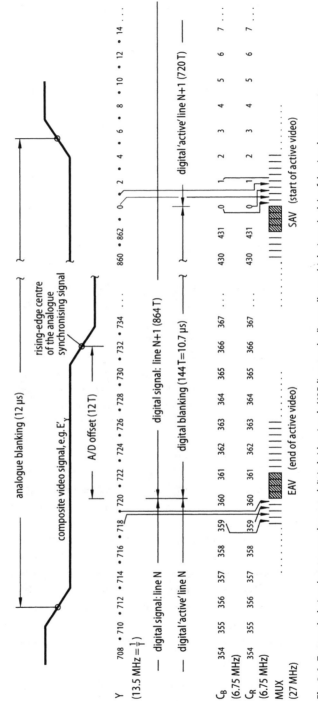

Fig. 2.4. Temporal relations between analogue and digital video signals (625-line standard) as well as multiplexing schedule of the signal components

Table 2.3. Digitisation characteristics and data rates of video signals with irrelevance reduction

Signals	Clock [MHz]	Values/ [line]	Lines	H_o [Mbit/s]	H_{oTotal} [Mbit/s]	Format
R	13.5	864	625	108		4 : 4 : 4
G	13.5	864	625	108		ITU 601
B	13.5	864	625	108	324	
Y	13.5	864	625	108		4 : 2 : 2
C_B	6.75	432	625	54		ITU 601
C_R	6.75	432	625	54	216	
Y	13.5	720	576	83		4 : 2 : 2
C_B	6.75	360	576	41.5		only active image
C_R	6.75	360	576	41.5	166	
Y	13.5	720	576	83		4 : 2 : 0
$C_B C_R$	6.75	360	576	41.5	124.5	only active image
Y	6.75	360	288	20.7		4 : 2 : 0, SIF
$C_B C_R$	3.375	180	288	10.4	31.1	only active image

is necessary. Directly after the end of the blanking (SAV), data words with the signal components C_B, Y, C_R, Y are transmitted in the succession here indicated. This sequence repeats itself until the beginning of the next blanking (EAV) and in this way ensures the allocation of two luminance signal values to the corresponding two chrominance signal values.

For the serial transmission (digital serial components, DSC) an additional coding, after the parallel-to-serial conversion of the data stream, is required with which, apart from the spectral shaping of the signal, support of the clock recovery and the word synchronisation at the receiving end is achieved. The bit rate of the serial signal is 270 Mbit/s; it is mainly used for signal distribution in the studio [ITU 656].

Possible data rates of video signals are listed in table 2.3, taking into account the effect of simple methods of data reduction. Irrelevance reduction is what this is about, i.e. omission of information which is not required for the relevant application. The frame frequency of the signal formats in table 2.3 is always 25 Hz and the word width 8 bits. By eliminating blanking intervals during transmission it is possible to reduce the total data rate H_{oTotal} from 216 to 166 Mbit/s, and by line-sequential C_B/C_R transmission, to 124.5 Mbit/s (format 4 : 2 : 0). Finally, a subsampling by a factor of 2 in the horizontal and vertical directions results in a further reduction of the data rate H_{oTotal} by a factor of 4, to 31.1 Mbit/s (source input format, SIF). This signal format refers to a progressive picture scanning without line interlacing, independent of the later type of picture display.

Table 2.4. Digitisation characteristics and data rates for the transmission of audio signals

Standard	Clock [kHz]	H_{oMono} [kbit/s]	$H_{oStereo}$ [kbit/s]	Uses
DSR	32	512	1024	Digital satellite radio
CD	44.1	706	1412	Audio CD
AES/EBU	48	768	1536	Professional audio studio

2.3 Digitising Audio Signals

The sampling parameters and the data rates of the more usual digital audio systems are shown in table 2.4. Throughout, a quantisation of 16 bits per sample is assumed for the transmission.

The clock frequencies of 32 and 48 kHz are derived from multiples of the sampling frequency of telephone signals (8 kHz). The numerical ratio of the clock frequency of 44.1 kHz to the line frequencies of the TV systems for both 625-line and 525-line scanning is relatively simple. The reason for this lies in the fact that in order to produce moulds during the mastering of CDs the digital audio signals have to be stored temporarily for the playing time of the carrier (for approx. 70 minutes). It is often video recorders that are used for this purpose, therefore it is convenient to choose an audio clock frequency which is quasi-compatible with the line frequencies.

How the chosen quantisation affects the levels is illustrated in figure 2.5. The system dynamics, for a 16-bit and a 20-bit system, is given in (2.2). A

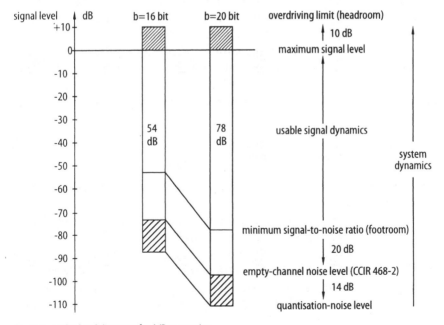

Fig. 2.5. Audio-level diagrams for A/D conversion

headroom of 10 dB is allowed for overdriving tolerance since the driving control for audio signals is not easy to handle and an abrupt limiting must definitely be avoided [ZANDER]. The aurally compensated evaluation of the quantisation noise (with some tolerance) leads to an empty-channel noise level of 14 dB above the quantisation noise level [JAKUBOW, HESSNM]. The remaining differences of 74 dB or 98 dB represent the actual usable level ranges. If an S/N ratio of 20 dB is allowed at the lowest passages, then a usable signal dynamics of 54 dB and 78 dB is obtained.

The level diagrams illustrate why a quantisation of 20 bits or more is appropriate in the professional studio. On the other hand, during transmission the effective quantisation can be reduced to between 14 and 12 bits by non-linear quantisation (companding) or by a dynamic adjustment of the level range to the quasi-instantaneous value of the signal without the transmission quality being, as a rule, audibly impaired [HESSENM].

2.3.1 Representation of Audio Signals

Apart from the bit-parallel processing of digital audio signals (which chiefly takes place inside equipment), bit-serial transmission has been adopted in the professional studio and for consumer use. One of the signal formats used (AES/EBU) is described in [IEC 958]. It is structured in subframes of 32-bit lengths, each of which, apart from one audio-signal sample (of a length of up to 24 bits), contains the synchronising characters, the parity bit and one information bit. Two of these subframes together form a frame and transport the dual-channel or stereo sound. A block is formed by 192 frames in which the channel status is transmitted by the sum of all information bits. Information about source and destination, about parameters, the allocation of the signals and, further, a time code and error-protection bits are included. The data rates of these bit-serial audio signals prior to modulation (biphase mark) are shown in table 2.5 and are dependent on the sampling frequencies of the source signals.

2.3.2 ADCs and DACs for Audio Signals

There are no guiding values for pre- and post-filters applied to audio signals, as recommended for video signals in accordance with table 2.2. The stop-band attenuation of these filters ought in principle to achieve the value of the signal-to-quantisation-noise ratio. In the case of a clock frequency of 48 kHz, a pre-filter cut-off frequency of 15 kHz and $b = 16$ bits, a Cauer low-pass filter of grade 9 is required to meet the above requirement. This very conveniently chosen example shows that this type of pre-filter – if necessary with phase equalisation – is costly to implement. This cost can, however, be con-

Table 2.5. Data rates for bit-serial audio signals

Clock [kHz]	b_{Max} [bit]	H_{oSer} [Mbit/s]
32	24	2.048
44.1	24	2.8224
48	24	3.072

siderably reduced by oversampling and digital signal processing, which will be discussed later. The same applies to the post-filter.

The weighting process is widely used in ADCs for audio signals (figure 2.6). By means of a register all b bits of the digital output signal are tentatively arranged in succession, commencing with the most significant bit (MSB). The result, after the D/A conversion, produces a comparison signal, which is fed to a comparator, the other input of which contains the sampled and retained analogue input signal. The greater-smaller decisions of the comparator determine the final value of the output signal. For this the clock frequency of the successive-approximation register must be at least *b* times higher than the sampling frequency of the input signal [TIETZE].

Progress in semiconductor technology allows the use of a considerably higher sampling frequency for the converter than would be required in accordance with the Nyquist theorem. An advantage of this oversampling is the separation of the frequency locations of the baseband spectrum and the sampling spectrum. In this way the requirements placed on the steepness of the stop-band slopes of the pre- and post-filter are reduced accordingly, as already mentioned in 2.1.

A further advantage is the now possible partial relocation of the required signal filtering to the area of digital signal processing, where even structures of a higher order can be realised with a favourable effect – for instance, without phase distortion – and at low cost.

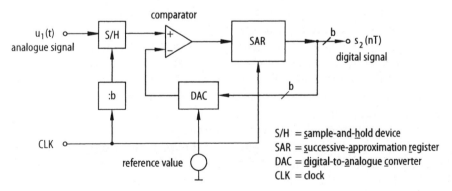

Fig. 2.6. ADC in accordance with the weighting process

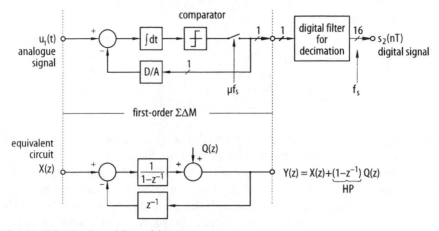

Fig. 2.7. ADC with a sigma-delta modulator

Finally, the oversampling can be utilised for the exchange of the speed and the resolution of a converter. As a topical application of the above-mentioned advantages of advanced semiconductor technology the sigma-delta modulator ($\Sigma\Delta$M), which has received wide acclaim in audio technology as an ADC, will be presented below.

The $\Sigma\Delta$M, the block diagram of which is shown in the upper part of figure 2.7, can be derived from the well-known delta modulator (ΔM) by changing the position of a module – the integrator – within the circuit [PARK, CANDY]. A bipolar signal, which has been derived from the binary output signal of the ADC, is subtracted from the actual analogue signal value $u_1(t)$. The resulting difference signal is fed via the integrator to a comparator, which acts as an ADC with a one-bit resolution. The actual sampling process can be accomplished through the activation of the comparator by the clock signal, which is represented here by a switch.

The whole circuit can be conceived as a sampled control loop keeping the continuously integrated difference between input and output signal as small as possible. As the output signal can only assume two states, it undergoes a pulse-density modulation (PDM) and thus a change of its mean value, which tries to follow the amplitude of the input signal. The error occurring on account of the quantisation with 1 bit can only be reduced by oversampling, which leads to a diminishing of the changes in the input signal between two successive clock periods.

As already mentioned, $\Sigma\Delta$M and ΔM are similar in their structure and also in their function. However, the transfer of the integrator from the feedback loop (ΔM) to the forward loop ($\Sigma\Delta$M) leads to a different response in the frequency domain. In the equivalent circuit (figure 2.7, lower part) the $\Sigma\Delta$M circuit is shown after undergoing a z-transformation with the integrator repre-

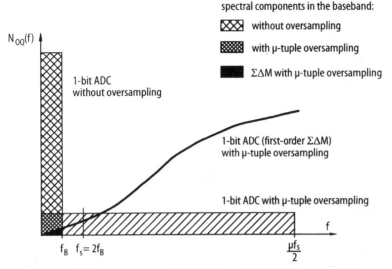

Fig. 2.8. Spectra of quantisation errors of 1-bit ADCs for audio signals with oversampling

sented by the operation $1/(1-z^{-1})$ [PARK, AGRAW]. At this point the integrator causes the quantisation error $Q(z)$ in the output signal $Y(z)$ to be weighted with a high-pass characteristic $(1-z^{-1})$, whereas the input signal $X(z)$ remains unweighted. As opposed to this, both components in the ΔM are unweighted in their frequency responses after demodulation.

The spectra of the quantisation errors in figure 2.8 show these effects. A 1-bit ADC in the baseband f_B with a Nyquist sampling frequency of $f_s = 2f_B$ distributes the power density N_{oQ} of its quantisation error evenly in this area with a correspondingly high value. Oversampling, for the same converter, results in a distribution, equal in area, over a µ-tuple-extended spectrum but with a correspondingly low value. After the filtering-out of the baseband the residual noise-power density is reduced by a µ-tuple, i.e. the oversampling factor. In the same way the ΔM functions as an ADC.

With the $\Sigma\Delta M$ as a 1-bit ADC it is possible to further reduce this noise-power density, as shown schematically in figure 2.8. This is due to the spectral noise shaping resulting from the high-pass characteristic [PARK], so that after filtering out the baseband, only a triangular-shaped area remains as noise-power density. However, this spectral noise shaping of the $\Sigma\Delta M$ of the first order cannot suffice for practical applications with an effective resolution of, for example, 16 bits. By cascading several such circuits it is possible to obtain higher $\Sigma\Delta M$ orders, which make the required noise reduction in the baseband possible [CANDY].

As shown in figure 2.7, a digital filter follows the $\Sigma\Delta M$, which by decimation makes the required number of bits available in the output signal while simul-

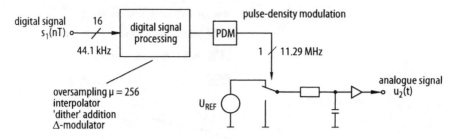

Fig. 2.9. 1-bit DAC for audio signals with oversampling

taneously reducing the external clock frequency to f_s. Usually this operation is carried out by a number of series-connected subfilters.

The DACs for audio signals often work on the same principle as those for video signals as depicted in figure 2.3. But the high-resolution requirement causes problems for the control of the resistor tolerances and the very small individual currents in the range of 10^{-8} A. Therefore different variants of this type of converter are in use which, due to internal splitting into several signal branches and by temporal averaging out of the errors which occur as a result ('dynamic element matching'), achieve a technically better solution [SKRI-TEK].

The use of an analogue signal as reference current I_{REF} is an interesting application of this DAC. In this case the converter functions as an attenuator, the attenuation of which is determined by the digital input signal. This arrangement is well-suited to a remote-controlled level- and volume adjustment. The 'fineness' of the step-by-step characteristic of such adjusting devices is determined by the number of bits b in the DAC.

As already mentioned, oversampling is also advantageous for DACs. The corresponding inverse application of the principle governing the ADC leads to a 1-bit DAC in accordance with figure 2.9 [BIAESCH]. To start with, a large number of new values between the present values of the input signal $s_1(nT)$ are generated by an interpolation process. By this method the differences between these interpolated values become smaller, so that a quantisation with a resolution of only 1 bit is sufficient in the case of a correspondingly large oversampling (here $\mu = 256$). As with the ΔM, a PDM signal is generated from the 1-bit sequence, which controls a switch as a 1-bit DAC. An additional 'dither' bit results in a perceivable reduction of the quantisation noise at very low signal levels. The subsequent analogue integrator smooths the form of the output signal $u_2(t)$. An advantage of this concept – apart from the easy implementation as an IC – is the relatively high linearity at small amplitudes.

Symbols in Chapter 2

B	colour component signal: blue
b	number of bits with which a signal is digitised
b_i	bit value i
C_B	digital colour difference signal: blue
C_R	digital colour difference signal: red
E'_{CB}	analogue colour difference signal: blue; γ pre-corrected
E'_{CR}	analogue colour difference signal: red; γ pre-corrected
E'_Y	analogue luminance signal, γ pre-corrected
f	frequency in general
f_B	limiting frequency of the baseband signal
f_s	sampling frequency
G	colour component signal: green
H_o	data rate of a single signal
H_{oTotal}	data rate of a total signal
H_{oSer}	data rate of a serial signal
I_{REF}	reference current
$i(t)$	analogue signal in the time domain (current)
n	running variable, integer
N_{oQ}	power density of quantisation-error signal
N_Q	r.m.s. value of quantisation-error signal
Q	size of one quantisation step
$Q(z)$	quantisation-error signal, transferred to the z plane
R	colour component signal: red
S_A	audio-signal amplitude (r.m.s. value)
S_V	video-signal amplitude (peak-to-peak value)
$s(nT)$	digital signal at sampling instants nT
T	clock period, inverse of f_s
t	time in general
U_{REF}	reference voltage
$u(t)$	analogue signal in the time domain (voltage)
μ	oversampling factor
$X(z)$	input signal, transferred to the z plane
Y	digital luminance signal
$Y(z)$	output signal, transferred to the z plane
z	variable of z-transformation

3 MPEG Source Coding of Audio Signals

The digital television of the future will also have digital associated sound. In this way an audio quality can be achieved which is far better than that obtained with the FM transmission system used in analogue television. However, in order to keep a digital television transmission within the realms of a realistic bandwidth, a suitable bit-rate reduction of the audio source signal is required. Moreover, the transition to a digital associated sound should offer the content providers the opportunity to choose between various data rates and thus between various qualities of audio signals. Furthermore, several alternative methods (such as dual-channel audio, stereo, surround-sound etc.) should be available to the content provider. It goes without saying that certain compatibility requirements must be satisfied so that each receiver is in a position to decode the audio signal regardless of which system or bit rate has been chosen by the content provider.

The techniques for the source coding of audio data which meet these requirements have been worked out by the MPEG audio group, a subsidiary group of the ISO/IEC JTC1/SC29/WG11. The standards developed by this group are referred to by experts as the MPEG standards and designated as IS 11172-3 (MPEG Audio) and IS 13818-3 (MPEG-2 Audio). According to the decisions taken by the DVB Project the coding of the associated audio in digital television must comply with the above standards to be discussed in this chapter. After a survey of the bit-rate reduction the psychoacoustic basics of the coding techniques will be explained. Section 3.3 will deal mainly with the coding and decoding in accordance with MPEG. Following that, there will be an overview of the coding of surround-sound in accordance with MPEG-2.

3.1 Basics of Bit-rate Reduction

Figure 3.1 shows the block diagram of the transmission of information with bit-rate reduction. The information source supplies an analogue audio signal which is transformed into a digital signal by sampling and quantisation. The usual studio sampling rate of 48 kHz and a 16-bit quantisation result in a bit rate of 768 kbit/s for a monophonic channel. In order to transmit this signal with the least possible energy input, the bit rate is further reduced in a source

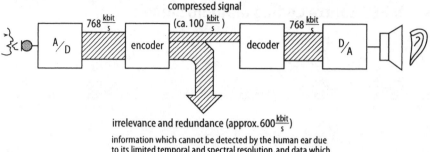

Fig. 3.1. Principle of bit-rate reduction for audio signals

coder. The decoder at the receiving end converts the bit-rate-reduced signal back into a PCM signal, which is then changed into an analogue signal, amplified, and used to drive a loudspeaker.

The art of source coding requires that the bit-rate reduction be carried out in such a way that the information sink (in the case of audio coding: the human ear) cannot detect any deterioration in the sound quality or that the negative effects caused by the bit-rate reduction are at least kept to a minimum. In the literature of information technology, two basic approaches are identified: redundancy reduction and irrelevance reduction.

The redundancy of a signal is a measure of the predictability of portions of that signal. To reduce the redundancy, one needs to know the statistical properties of the information source. Redundant signal portions can be reduced at the encoder and restored at the receiver so that the original audio signal can be completely reconstituted. However, generic audio signals – unlike specific ones such as speech signals – hardly ever contain the quantity of statistical correlation necessary for obtaining high compression rates by means of a simple redundancy reduction.

An irrelevance reduction, on the other hand, takes advantage of the limited capacity of the information sink in order to compress the data. In the case of audio signals those signal portions which the human ear – owing to its limited capacity for discrimination – cannot perceive in amplitude, time and spectrum, are eliminated in the encoder. Contrary to the redundancy reduction, the irrelevance reduction is an irreversible process. However, since this reduction cannot be perceived by the human ear, the ear is not aware of the degeneration of the signal. As an irrelevance reduction takes advantage of the limited receptivity of the sink, this type of compression of audio data requires a thorough understanding of the perception limits of the human ear. For this reason some findings with regard to the capacity of the human ear will be discussed in the following.

3.2 Psychoacoustic Basics

First of all, the perception of stationary sounds will be described in which no influence of the time structure is ascertainable. This is the case for sound with a duration of not less than 200 ms [ZWICKER]. In this section the threshold of audibility and the auditory sensation area, which describe the perception of single sinusoidal tones, will be explained. If a sound event is comprised of more than one sinusoidal tone, individual sounds may be masked by others. This masking effect is explained in section 3.2.2. Finally there is a description of time-dependent events in the perception of sounds.

3.2.1 Threshold of Audibility and Auditory Sensation Area

The threshold of audibility is that sound-pressure level of a sinusoidal tone that can just be perceived. It is dependent on the frequency. The sound pressure level L can be defined as follows:

$$L = 20 \log \left(\frac{p}{p_0} \right) , \tag{3.1}$$

where p represents the sound pressure and p_0 can be defined as:

$$p_0 = 2 \cdot 10^{-5} \frac{N}{m^2} = 20 \, \mu Pa . \tag{3.2}$$

In order to determine the threshold, test subjects were required – in a subjective study – to set the sound level of a sinusoidal source to the point at which it just ceased to be perceptible. This sound-pressure level was then recorded, determined for various frequencies and averaged over a large number of subjects. These levels form the threshold of audibility shown in figure 3.2[1]. The threshold of pain was reached at a sound-pressure level of 130 dB. The range between the threshold of audibility and the threshold of pain is called the auditory sensation area. Further, curves of the same loudness perception can be seen in figure 3.2. These curves show that the loudness perception and the threshold of audibility depend to a great extent on the frequency, which is an indication that for audio coding it is recommendable to treat different parts of the spectrum differently.

However, most sounds which occur in nature are not comprised of isolated sinusoidal tones. The perception by the human ear of more complex sounds must therefore be examined.

---

[1] To be precise, the median of all measured values is taken. The reason for this is explained in [ZWICKER].

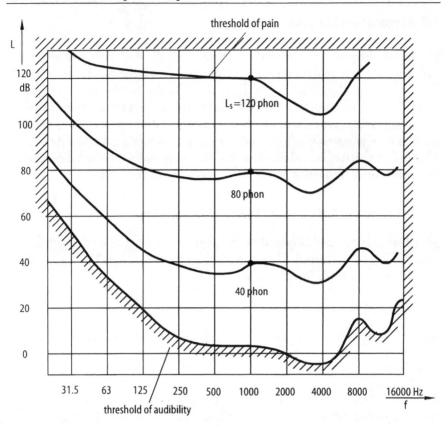

Fig. 3.2. Auditory sensation area

3.2.2 Masking

Masking is an effect known from everyday life. For instance, an audio event (such as music) which is easily heard in a quiet environment can be imperceptible in the presence of noise (as that caused by a pneumatic drill). In order for the useful sound to remain perceptible in spite of the noise it is necessary for the level to be considerably higher than in a quiet environment. The concept of "masking thresholds" is used for the quantitative description of the masking effects. A masking threshold is defined as the sound-pressure level to which a test sound (as a rule, a sinusoidal test tone) must be reduced to be only just audible beside the masking signal [ZWICKER]. The masking thresholds and the method of their determination are described below.

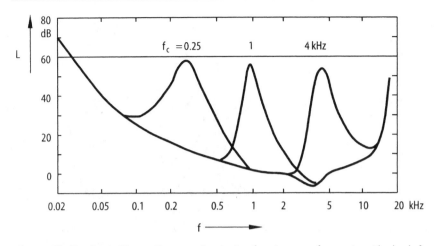

Fig. 3.3. Masking threshold caused by narrow-band noise of varying centre frequencies with a level of $L_G = 60$ dB according to [ZWICKER]

3.2.2.1 Masking by Stationary Sounds

First of all, we will discuss the masking by sound whose time structure can be disregarded. The time structure of a sound can be disregarded if the duration of the sound is longer than 200 ms. In order to define the masking threshold caused by narrow-band noise, the subjects are exposed to a sinusoidal tone in addition to the narrow-band noise. The sound-pressure level of the sinusoidal tone is progressively reduced until it becomes just inaudible. This procedure is repeated with several hundred test subjects for a range of frequencies between 20 Hz and 20,000 Hz. The determined sound-pressure levels – averaged over a number of subjects – constitute the masking threshold.

Figure 3.3 depicts the masking threshold by means of a masking narrow-band noise level of 60 dB, at centre frequencies of 250 Hz, 1 kHz and 4 kHz with bandwidths of 100 Hz, 160 Hz and 700 Hz resp. All tones at a level below these thresholds are masked by the narrow-band noise and are therefore imperceptible to the human ear. As can be seen, the course of these masking thresholds is highly dependent on frequency. This entails that the course of the masking thresholds depends on the centre frequency of the narrow-band noise and that the distance of the maxima from the 60-dB line increases with increasing frequency. However, the frequency of the masker is not the only influence on the form of the masking threshold. Another parameter is the level of the masker. The masking threshold, caused by narrow-band noise of varying sound-pressure levels L_G at a centre frequency of 1 kHz and a bandwidth of 160 Hz, is shown in figure 3.4. Apart from the frequency and the level of the

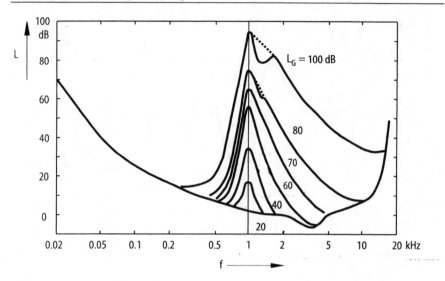

Fig. 3.4. Masking threshold caused by narrow-band noise of varying levels according to [ZWICKER]

masker, a third parameter has to be taken into account, namely the tonality of the masker. If the masking tone is sinusoidal, the resulting masking threshold will differ from that depicted in figure 3.4.

Until now, only the masking by stationary sound has been assessed, i.e. by sinusoidal tones or narrow-band noises with a long duration. In the following section the effects observed in sound events of a very short duration will be discussed.

3.2.2.2 Time-dependent Masking Effects

In this section we will examine the question to which degree masking effects occur in maskers of a very short signal duration. In order to determine this, subjects were each exposed to a masking impulse of white Gaussian noise with an amplitude of L_{WN} for a duration of 0.5 s. This masking impulse was followed, at an interval of t_v, by a pressure impulse of a duration of 20 µs (see figure 3.5a). The subjects had the task of adjusting the amplitude of this Gaussian impulse so that it was just imperceptible. The results of this subjective study are shown in figure 3.5b. The masking threshold still has approximately the same level, up to about 10 ms after switching off the masking noise, as when the masking noise and the Gaussian impulse are presented simultaneously. It is only after these approx. 10 ms that the masking threshold falls, reaching the threshold of audibility after approx. 200 ms.

However, masking not only occurs when the Gaussian impulse is offered during or after the masker but also prior to it. This effect – in the literature de-

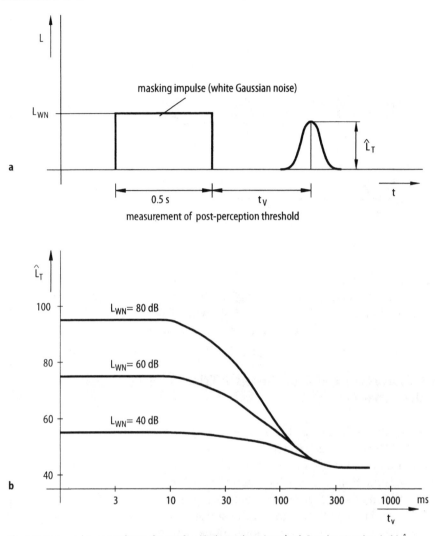

Fig. 3.5. Post-masking according to [ZWICKER]. **a** Masker and test impulse; **b** Post-hearing threshold $\hat{L}_T$ at different amplitudes of the masker and different distances to the masker

scribed as pre-masking – is only detectable when the length of the interval between the Gaussian impulse and the switching on of the masker is no more than 20 ms.

The masking effects are summarised in figure 3.6. The masking of sound events which are presented to the human ear at the same time as a masker are also referred to as "simultaneous masking". However, the masking effect remains after the masker has been switched off and disappears after approx. 200 ms have elapsed. This is referred to as "post-masking". Even sounds

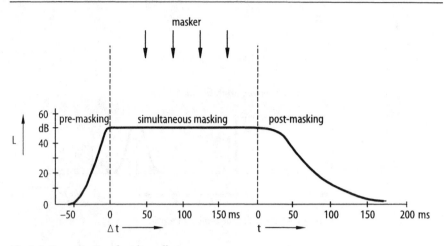

Fig. 3.6. Summarisation of masking effects

presented before the masking noise commences can be masked. But this effect, which is called "pre-masking", is effective for just about 20 ms.

In the following sections we shall show that these masking effects can be exploited for a very effective irrelevance reduction in audio signals.

3.3 Source Coding of Audio Signals Utilising the Masking Qualities of the Human Ear

This section will discuss the source coding of audio signals in accordance with MPEG standards [ISO 11172] and [ISO 13818]. The explanation of the basic structure of an MPEG audio coder will be followed by the description of the bit-rate reduction in accordance with MPEG layer 1. Layer 1 was developed with the aim of keeping the implementation requirement for the encoder to a minimum. There will then follow a description of layer 2 which compared to layer 1 is more complex but enables higher compression of an audio signal while retaining the same quality. To complete the picture, the third layer in the MPEG standard will be briefly dealt with. The compression rate in layer 3 is higher again than that of layer 2, involving a yet higher implementation requirement. However, the "Guidelines on the implementation and usage of Service Information (SI)" of the DVB Project [ETR 211] make no provision for the use of layer 3 in digital television.

Those developing the MPEG standards took care to greatly reduce the expenditure involved in the decoding of audio signals, as compared to the cost incurred in the encoding process. In distribution services, such as TV broadcasting, low expenditure is of great importance so as to make possible an economical manufacture of the receiving equipment. The decoding of the bit

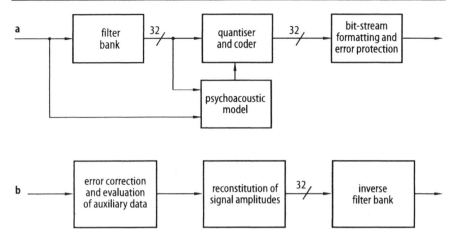

Fig. 3.7. Principle of bit-rate reduction in accordance with the MPEG standard. **a** Encoder; **b** Decoder

stream is described in section 3.3.5. To conclude, the audio coding in accordance with MPEG-2 is outlined. This is a coding for surround-sound, which is both forward and backward compatible with the standard for MPEG audio.

In order to ensure compatibility between receivers from different manufacturers two items only need to be laid down in the standard, viz, the decoding procedure and the shape of the resulting signals. In contrast to this, the description of the MPEG coder in the standard is of a purely informative nature and the design may vary from one manufacturer to another.

3.3.1 Basic Structure of the MPEG Coding Technique

The basic structure of an MPEG audio coder is shown in figure 3.7a. The audio signal to be coded is first passed through a filter bank, which breaks it down into 32 frequency bands of equal bandwidth. The ISO/MPEG standard provides for the use of a polyphase filter bank, the coefficients of which are listed in the standard. Simultaneously in each of these 32 channels the signal is critically subsampled, which means that the sampling rate is reduced to the thirty-second part of the sampling rate used in the digitisation. These signals are then fed into a quantiser which quantises the signals in each individual band in such a way that, whilst the number of steps is kept as low as possible, the quantisation noise still remains below the masking threshold so that noise is not perceptible to the human ear. It goes without saying that the permissible degree of quantisation cannot be determined without a proper knowledge of the psychoacoustic laws described in section 3.2. A particular element of the coder – called "psychoacoustic model" in figure 3.7a – evaluates the input signal and, taking into account the masking effects, computes the permissible quantisation for each subband. Subsequently, the quantised sam-

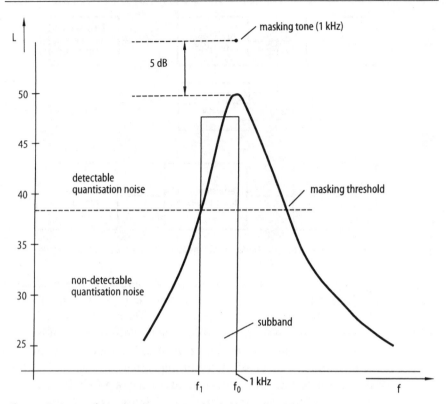

Fig. 3.8. Derivation of the optimal quantisation threshold according to [THEILE]

ples and a number of auxiliary data which are required for the reconstitution of the audio signal by the decoder are formatted into a bit stream and furnished with an error protection.[2]

The operations at the decoder (figure 3.7b) start with a correction of transmission errors, followed by an interpretation of the auxiliary data and a reconstitution of the samples of the audio signals. The reconstituted 32 subbands are then conveyed to an inverse filter bank and there combined into one frequency band.

The achievable compression ratios depend to a great extent upon a suitable determination of the quantisation thresholds in each subband. Therefore the determination of these thresholds shall here be described in detail. First, a 1-kHz tone is assumed as masker (figure 3.8). The most critical case exists when this tone is to be found in the upper band limit of the corresponding subband, since the masking thresholds show a steeper curve when coming from low frequencies. In this case the masking effects in the subband in ques-

---

[2] The error protection is optional.

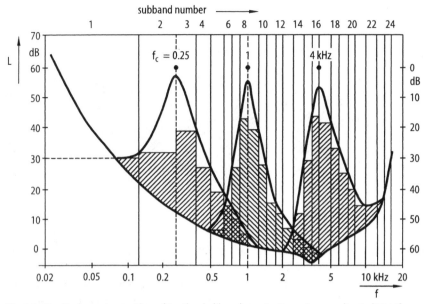

Fig. 3.9. Amplitude spectrum with masking thresholds and quantisation noise according to [STOLL 1]

tion are lowest. The intersection of the masking threshold with the lowest band limit now defines a level underneath which no audio signals are perceptible, so that these need not be reproduced at the decoder. Therefore the quantisation of this subband can be chosen roughly enough up to the point at which the quantisation noise reaches this value (broken line in figure 3.8).

However, in general we are not concerned with one single masker, but with a music or speech signal which is far more complex. Figure 3.9[3] shows an example of three sinusoidal tones which function as maskers. These do not only mask signal parts in the subbands in which the masking noise itself is detected, but also in neighbouring bands. In figure 3.9 the auditory sensation area which is masked by these three tones is hatched in. The quantisation noise may cover this area without it being perceptible to the listener. In individual subbands, under particular circumstances, even all signal parts may lie below the masking threshold, so that these do not need to be transmitted. It is apparent from figure 3.9 that the aim of coding is to shape the spectrum of the quantisation noise so that it is masked by the signal. With the aid of this spectrum-shaping – in contrast to white Gaussian noise – far more quantisation noise is permissible without its becoming perceptible to the human ear.

---

[3] A representation of 32 bands of the same width, as specified in the MPEG standard, leads to an unclear picture due to the logarithmic abscissa. Therefore the diagram does not show 32 bands of equal width.

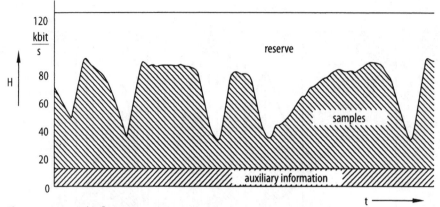

Fig. 3.10. Dynamic bit-flow reserve

Therefore, with no change in the subjective quality of the signal, this type of encoder requires a considerably lower bit rate than a PCM encoder.

From figure 3.9 it can also be seen that the extent to which the quantisation noise is masked by the audio signal, and therefore imperceptible, depends heavily on the amplitude and spectral distribution of the masking noise itself. However, the permissible quantisation noise also indirectly determines the required bit rate of the signal so that the bit rate is bound to vary as a function of the extent of the masking. Figure 3.10 shows the required time-dependent bit rate H in the case of a natural microphone signal. On the one hand, the bit stream contains the auxiliary information necessary for the reconstitution of the subbands and for controlling the decoder. The bit rate required for this is almost constant. On the other hand, the bit stream contains the quantised samples, the bit rates of which are very much dependent on the quantisation noise allowed. As most information channels are defined by a constant maximum permissible bit rate, one would give away part of the available capacity when using a signal with a time-dependent data rate. For this reason the quantisation steps are always increased to a number which ensures that the capacity of the channel is completely utilised. Therefore the maximum values for the quantisation noise, as shown in Figure 3.9, can always be underrun. The advantages of this approach are explained below.

If a quantisation-noise signal is allowed to reach the masking threshold, as in figure 3.9, then no further processing of this signal can take place without quantisation noise becoming perceptible. This is even true for a simple change in the frequency response by means of an equaliser. If the frequencies of the spectrum shown in figure 3.9 were to be pre-emphasised upwards of 5 kHz, the quantisation noise would exceed the masking threshold. The same considerations apply when a signal which has been coded in this form is decoded and then recoded ("cascading"). This recoding is accompanied by a

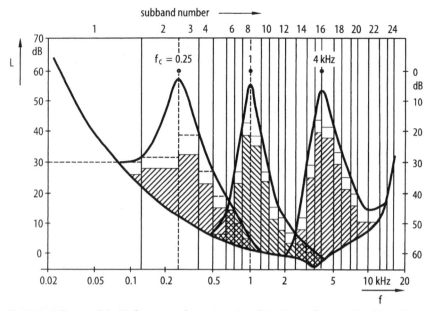

Fig. 3.11. Utilisation of the bit-flow reserve for maximisation of the distance from masking threshold to quantisation noise in accordance with [STOLL 1]

new quantisation and therefore by an increase in the quantisation noise above the masking threshold.

In order to allow a further processing of the signal – at least to a certain extent – the quantisation is fixed in a way that there is a margin between the masking threshold and the quantisation noise (figure 3.11[4]). For this, the margin is increased in an iterative process until the maximum available data rate of the channel has been exhausted. Therefore, the larger the bit rate the greater the post-processing capacity of the signal.

Following the discussion of the basic principles of psychoacoustic coding according to the MPEG standard, the MPEG encoder and decoder will now be described in detail.

3.3.2 Coding in Accordance with Layer 1

Figure 3.12 shows the block diagram of an audio coder in accordance with MPEG layer 1. The filter bank which divides the signal into 32 subbands of equal bandwidth has already been described in section 3.3.1. In these subbands a critical subsampling of the signal takes place. Subsequently, the maximum value – the scaling factor – of a block of 12 samples is determined, the

---

[4] See footnote 3.

digital audio signal (PCM)
(768 kbit/s)

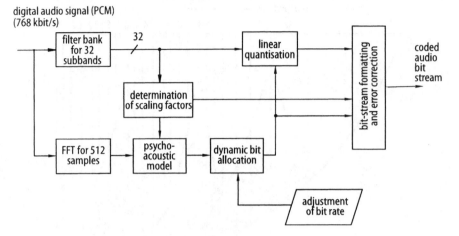

Fig. 3.12. Block diagram of a coder in accordance with layer 1 of the MPEG standard (mono)

samples corresponding to 8 ms signal duration at a 48-kHz sampling rate. This scaling factor is quantised and transmitted with an amplitude resolution corresponding to 6 bits. The 12 samples are then divided by the scaling factor, and the resulting signal is subjected to quantisation. For computing the required number of quantisation steps one needs to apply the masking threshold provided by the psychoacoustic model. All 12 samples are then subjected to the same quantisation. This is permissible because, due to the pre- and the post-masking, the effect of a masker extends over the entire period in question, which lasts between 8 and 12 ms, depending on the sampling rate.

In parallel with the subband analysis the signal is subjected to a Fourier transform with 512 samples, so that an even higher resolution is obtained for the spectrum of the signal. The local maxima are then determined in the spectrum, and the neighbourhood of these maxima is evaluated in order to determine whether these are tonal or non-tonal components of the signal. This evaluation is necessary since the tonality of a signal has a considerable impact on the shape of the masking thresholds (see section 3.2.2.1). In this way it is possible to determine, from the masking thresholds, the maximum sound level in the subbands.

On the basis of the masking thresholds and the given bit rate the number of the quantisation steps can be so determined as to maximise the margin between the masking threshold and the quantisation noise. The quantised samples are combined with the scaling factors and the bit allocation to form a bit stream ("bit-stream formatting"), the bit allocation indicating the number of bits for each sample.

These data are preceded by a header which contains information for the controlling of the decoder. Figure 3.13 shows the construction of the bit

data frame with 384 PCM samples
(equal to 8 ms at a 48-kHz sampling frequency)

header	error protection	bit allocation	scaling factors	samples	auxiliary data
12 bits synchronising signal and 20 bits system information	16 bits (optional)	4 bits each	6 bits each	2 ... 15 bits each	

Fig. 3.13. Audio bit stream in accordance with layer 1 of the MPEG standard (mono)

stream at the output of an MPEG-layer-1 encoder. The data frame shown is defined as part of the stream which contains all the relevant information required for the decoding. At the front of the frame is the 32-bit header, which starts with a synchronising word and contains the subsequent system information describing the audio signal in detail. A following error protection is optional and serves to protect the most important data, the bit allocation, and part of the header against transmission errors. A convolutional code [ISO 11172] has been provided for this error protection. Next, there follows the bit allocation, with the number of bits required for each of the samples, and the scaling factors, which are quantised with 6 bits each. The samples as such are represented by between 2 to 15 bits according to the permissible quantisation noise. There can be further data added to the frame which are not necessarily interpreted by an MPEG decoder. Among other things, a compatible extension leading to a surround-sound system is made possible by these auxiliary data. This possibility will be dealt with in section 3.3.7.

3.3.3 Coding in Accordance with Layer 2

Contrary to the layer-1 encoder, the layer-2 encoder (figure 3.14) combines 36 PCM samples into a block, which, at a sampling frequency of 48 kHz, corresponds to a section of 24 ms of the audio signal. In this case one scaling factor per subband might prove to be no longer sufficient, as the effect of the premasking only lasts for a maximum of 20 ms. This must be taken into account in the case of major temporal changes in the signal ("drum beats"), which would require two or three scaling factors per subband and per block. But if the signal is not subject to major temporal changes, one scaling factor per subband will suffice. There is a selection unit in the encoder which determines the required number of scaling factors. The next step consists in determining the maxima of the samples of a block. These values form the scaling factors. Apart from the scaling factors themselves, information about their number has to be transmitted to the decoder.

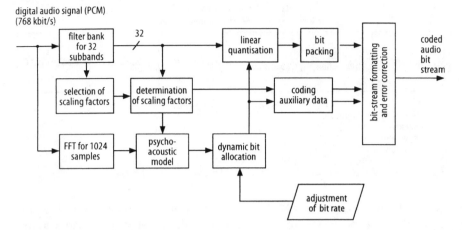

digital audio signal (PCM)
(768 kbit/s)

Fig. 3.14. Block diagram of a coder in accordance with layer 2 of the MPEG standard (mono)

To determine the tonality of the maskers, a Fourier transform on 1024 samples is carried out in the case of layer 2 so that a higher spectral resolution is obtained than with layer 1. The determination of the masking threshold and the formatting of the bit stream are both performed in the same way as in layer 1.

The fixing of the number of quantisation steps also differs from that in layer 1. In layer 2 it is only in the lower subbands that all values between 2 and 15 bits are admissible for the representation of the samples. In the upper subbands the possibilities of representation are limited. Thus, at a bit rate above 50 kbit/s and at a sampling rate of 48 kHz, only 0, 3, 5 or 65,535 quantisation steps are admissible in subbands 23 to 26. Therefore, as only four different quantisations are possible, only 2 bits are required for the transmission of the bit allocation.

As the signal energy in the upper bands is generally very low, a greater number of quantisation steps is seldom required. The required data rate is therefore kept to a minimum in that only 0, 3, 5 or 65,535 quantisation steps are allowed and only 2 bits are thus required for the bit allocation. At a sampling rate of 48 kHz the subbands 27 to 31 are not transmitted since the frequencies involved are above 20 kHz and are therefore no longer audible.

Figure 3.15 shows the frame of an MPEG-layer-2 bit stream. This differs from that of layer 1 only in that it comprises 1152 bits and that in addition it contains an identification of the number of scaling factors used in the bit stream.

3.3.4 Coding in Accordance with Layer 3

To complete the picture, the coding in accordance with MPEG layer 3 should also be presented. This is an algorithm developed with the aim of increasing

frame with 1152 PCM samples
(equal to 24 ms at a 48 kHz sampling frequency)

header	error protection	bit allocation	label for number of scaling factors	scaling factors	samples	auxiliary data
12 bits synchronising signal and 20 bits system information	16 bits (optional)	4 bits each for lower 3 bits each for middle 2 bits each for upper subband	2 bits each	6 bits each	2 ... 15 bits each	

Fig. 3.15. Audio bit stream in accordance with layer 2 of the MPEG standard (mono)

the compression as compared to layer 2, which, however, entails a higher implementation requirement. Figure 3.16 shows the block diagram of the coder.

Layer 3 requires an additional division of the signal spectrum into 576 subbands, which is achieved by means of a modified discrete cosine transformation (MDCT). Since the pulse duration of a filter increases the narrower the chosen bandwidth, this additional division of the frequency band has the disadvantage that the resolution of the coder in the time domain is reduced to an unacceptable level. Therefore a signal-adaptive changeover from high time-resolution to low frequency-resolution and vice versa is provided. This ensures a more precise definition of the masking thresholds and therefore a better adaptation of the quantisation noise to the limits of perception.

Further, layer 3 provides for a non-linear quantisation, which results in a favourable adaptation to the human ear, since the latter also shows a non-

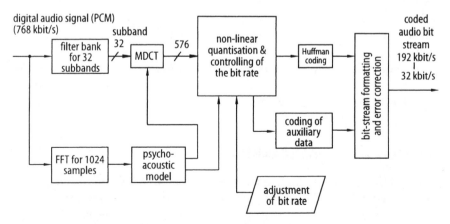

Fig. 3.16. Block diagram of a coder in accordance with layer 3 of the MPEG standard (mono)

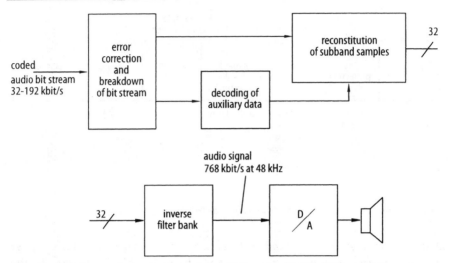

Fig. 3.17. Block diagram of a decoder in accordance with layers 1 and 2 of the MPEG standard (mono)

linear behaviour with regard to the perception of changes in sound level [ZWICKER].

By using a Huffman encoding procedure the redundancy still contained in the signal can be reduced so that an even lower bit rate can be achieved while maintaining an audio signal of good quality.

3.3.5 Decoding

At the decoder, first the error correction is performed and then the bit stream is broken down into the samples and the auxiliary information (figure 3.17). Through the evaluation of the auxiliary information (bit allocation, scaling factors) the individual samples can be reconstituted for the 32 subbands. By means of an inverse filter bank the subbands can then be combined again.

Since there is no psychoacoustic model to be evaluated at the decoder, the technique is much simpler than at the encoder. As noted above, it is of great importance, most particularly for radio distribution services, to have a simple decoding technique.

3.3.6 The Parameters of MPEG Audio

In this section the most important parameters for MPEG audio coding will be summarised. In accordance with the implementation guidelines of the DVB Project [ETR 211] a receiver must have an MPEG-compatible audio decoder which supports all operational modes stipulated in the standard. An MPEG decoder must be able to decode mono signals and stereo signals as well as two completely independent channels ("dual-channel sound"). A further operational mode is joint stereo. This mode is based on the knowledge that in intensity stereophony for frequencies above 2 kHz only the envelope and not the fine structure of the signal contributes to the stereo effect. For this reason one can concentrate on the coding of the aggregate signal for frequencies above 2 kHz and need only transmit the scaling factors for both channels.

The sampling rates recommended for the audio signal to be transmitted are 32 kHz, 44.1 kHz and 48 kHz. The total bit rate of a coded audio signal lies between 32 kbit/s and 384 kbit/s for layer 1, and between 32 kbit/s and 448 kbit/s for layer 2. In the standard there are 14 bit rates each defined within these limits which must be supported by every decoder.

Further, a decoder conceived for layer 2 must also be able to decode a layer-1 bit stream. Accordingly, a decoder for layer 3 must be compatible with layers 1 and 2.

The operational modes and data rates described must be supported by an MPEG-compatible decoder and therefore also by a decoder for a digital television set ("integrated receiver decoder"). The content provider thus has considerable freedom to determine the audio coding to be used and therefore also the type and quality of the audio signal.

Apart from the above-mentioned operational modes a content provider can also offer surround-sound. In November 1994, during the second phase of the MPEG activities, a separate international standard was adopted for the coding of a surround-sound signal. This will be outlined in the following section.

3.3.7 MPEG-2 Audio Coding

A loudspeaker configuration with one central loudspeaker and two surround loudspeakers – in addition to the usual left and right channels for stereo – is specified by MPEG for the surround-sound system. As three frontal and two surround loudspeakers are used, this arrangement is referred to as 3/2 stereo. Figure 3.18 shows the loudspeaker configuration recommended by ITU-R in [CCIR 10].

During the development of the coding system great emphasis was placed on backward compatibility, which means that a surround signal must be decodable by an MPEG decoder – of course as a stereo signal. This backward compatibility makes it possible for a content provider to transmit surround-sound, and for viewers with simple MPEG decoders and no surround-sound decoders to have the programme decoded (though only in stereo quality). Moreover, the MPEG-2 audio coding technique is forward compatible, which means that an MPEG-2 decoder can also decode an MPEG data stream.

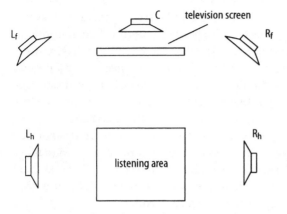

Fig. 3.18. Loudspeaker configuration for a 3/2 stereo system

To ensure backward compatibility both stereo signals L_0 and R_0 are computed from the 5 channels L_f, R_f, L_h, R_h, and C by a linear equation as follows:

$$L_0 = L_f + xC + yL_h$$
$$R_0 = R_f + xC + yR_h \; .$$

The coefficients x and y typically have the value 0.71 [STOLL 2].

Compatibility can now be achieved by formatting the bit stream of a surround signal in accordance with that of layer 2 (figure 3.15), where the channels L_0 and R_0, which constitute the compatible stereo signal, are placed at the front part of the frame, whereas the other three channels are transmitted in that part of the frame which is used for auxiliary data. The MPEG decoder does not evaluate these auxiliary data and only decodes the stereo signal. The surround decoder, however, recognises from the header that the signal possesses further surround channels and evaluates these.

By means of a suitable data-rate reduction which also reduces redundancies and irrelevancies between the channels it is possible to code a surround signal without the data rate increasing to five times that required for a mono channel.

3.4 Summary

The sound for digital television is subjected to a bit-rate reduction in accordance with the MPEG standard. This coding technique reduces the amount of data by exploiting the limited spectral and temporal discrimination of the human ear. The technique is based on a breakdown of the spectrum into 32 subbands and on a block formation with 12 or 36 samples. For each of these blocks quantisation is carried out on the basis of a psychoacoustic model by exploiting the masking quality of the human ear. This quantisation can be performed such that the generated quantisation noise is shaped, in terms of spectrum and time, in such a way as to be masked by the signal and therefore to remain inaudible. At the output of the coder the bit stream is formatted so that the decoder can reconstitute the samples.

With MPEG-layer-2 audio coding it is possible to achieve a bit-rate compression to approx. 100 kbit/s without coding artefacts becoming perceptible (transparent quality). Furthermore, the various operational modes and data rates to which an MPEG audio decoder is suited ensure a significant adaptation of the encoding parameters to the requirements of the user.

MPEG layer 3 was developed with the aim of obtaining higher compression than with layer 2, which, however, increases the implementation requirement. Joint stereo coding permits an exploitation of redundancies between the two stereo channels. Audio coding in accordance with MPEG 2 constitutes a forward and backward compatible extension of the MPEG audio standard for coding surround-sound.

Symbols in Chapter 3

C	centre channel
f	frequency in general
f_c	centre frequency
f_l	lower cut-off frequency
f_u	upper cut-off frequency
H	data rate
L	noise-pressure level
L_f	front surround signal, left
L_G'	noise-pressure level for narrow-band noise
L_h	back surround signal, left
L_o	stereo signal after matrixing, left
L_s	loudness
$\hat{L}_T$	noise-pressure level for a post-hearing threshold
L_{WN}	noise-pressure level for white Gaussian noise
p	noise-pressure level
p_0	reference noise pressure
R_o	stereo signal after matrixing, right
R_f	front surround signal, right
R_h	back surround signal, right
t	time in general
t_v	time interval
x	matrix coefficient
y	matrix coefficient
Δt	time interval

4 JPEG and MPEG Source Coding of Video Signals

In state-of-the-art television a digitised video signal typically uses a net bit rate of 166 Mbit/s (cf. section 2.2). If this data rate were transmitted without being compressed, considerably more bandwidth would be required than for the present analogue procedure. That is why the use of data compression techniques is indispensable for the development of a digital transmission standard for television signals. As with audio signals, a mere redundancy reduction (cf. section 3.1) would only result in a small average compression factor. With the aid of an irrelevance reduction which takes into account the characteristics of the human visual system, rejecting all imperceptible image contents, considerably higher reduction factors can be attained (with no change in the subjective quality of the image). This decreases the bandwidth requirement far below that for analogue transmission of a similar quality, thus making this procedure commercially very interesting.

A series of standards has been established for the coding of images, as shown in table 4.1 in historical order. First, the so-called JPEG (Joint Photographic Experts Group) standard was developed as ISO/IEC IS 10918 [ISO 10918], which was conceived for the efficient storage of still pictures [WALLACE, PENNEBK]. Owing to the early availability of reasonably priced ICs, a further utilisation offered itself for moving images. However, this "Motion JPEG" (M-JPEG) was not standardised and does not warrant compatibility of equipment purchased from different manufacturers.

Table 4.1. International standards for image coding

Standard	Range of application	Data rate
ISO/IEC IS 10918 "JPEG"	storage of stills; Motion JPEG: studio applications	not defined
ITU-T H.261 "p64"	ISDN, video conferencing	p * 64 kbits/s
ISO/IEC IS 11172 "MPEG-1"	CD-ROM, multimedia	up to 1.5 Mbit/s
ISO/IEC IS 13818 "MPEG-2"	television transmissions, studio applications	MP@ML: up to 15 Mbit/s 4:2:2@ML: up to 50 Mbit/s

By comparison, the standard ITU-T H.261 [ITU H.261], also known in some parts of America under the name of "p64", was conceived from the outset for the coding of moving images. It was optimised for visual telephony and similar applications, which can be offered via narrow-band ISDN networks at data rates of 64 kbit/s or integral multiples thereof.

With ISO/IEC IS 11172 [ISO 11172] MPEG-1 a technique was developed which was conceived for application in the field of multimedia [HUNG] where only a limited storage capacity or data rate is available (particularly for CD-ROMs) and therefore quality requirements have to be modest. Compromises that had to be reached in this matter include the limitation to the SIF format (half the spatial resolution compared to ITU-R BT.601 resolution), progressive sampling, and the 4 : 2 : 0 chrominance format (see section 2.2).

The extension of MPEG-1 in the direction of higher quality and, connected with this, higher data rates, is the MPEG-2 standard (ISO/IEC IS 13818 [ISO 13818]), which enables the transmission and storage of television signals based on interlaced scanning [KNOLL, TEICHNER].

The image-coding standards relevant to digital television will be discussed in the following sections.

4.1 Coding in Accordance with JPEG

4.1.1 Block Diagram of Encoder and Decoder

Figure 4.1 shows the block diagram of a JPEG encoder which is here discussed in detail. First of all, the input image is divided into blocks of 8×8 pixels. A two-dimensional discrete cosine transformation (DCT) which transforms the sampling values of the image into the so-called spatial frequency domain is applied to these blocks. Here the individual frequency components of the image appear in the form of coefficients, so that the subsequent quantisation can be carried out in varying degrees of coarseness, graded according to spatial frequencies, in order to enable an adaptation to the human visual system. The quantisation is the only lossy step in the algorithm (irrelevance reduction); all other steps (apart from rounding errors during computation of the

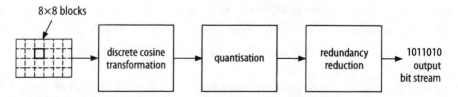

Fig. 4.1. Block diagram of the JPEG encoder

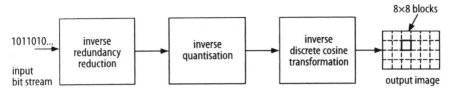

Fig. 4.2. Block diagram of the JPEG decoder

DCT) are completely lossless and can be revoked by the respective inverse operation. In order to further reduce the data rate, the quantisation is followed by a redundancy reduction which is essentially a combination of run-length coding and Huffman coding (see section 4.1.4).

The corresponding inverse processing steps, as shown in figure 4.2, are performed in the decoder. With the correctly chosen quantisation table (see section 4.1.3), the output image, which is constructed by putting the 8×8 pixel blocks back together, barely differs visually from the original image.

4.1.2 Discrete Cosine Transform

The discrete cosine transform can be conceived as a transformation of the original block of 8×8 pixels in the spatial domain into an equally sized block with 64 coefficients in the spatial frequency domain. Figure 4.3 shows such a transformation using an example taken from [HUNG]. The circular area, which can be conceived as a "pizza", with pixels differing in value from those surrounding it, can no longer be recognised as such in the spatial frequency domain (result of the DCT, in this case rounded to integers). It can be seen that the energy of the transformed block is concentrated at low frequencies. The higher the spatial frequencies f_x and f_y the smaller, in general, the coefficients. Apart from this, many coefficients are equal to zero, which proves very helpful for the data reduction.

Using the formula for a two-dimensional DCT with 8×8 pixels,

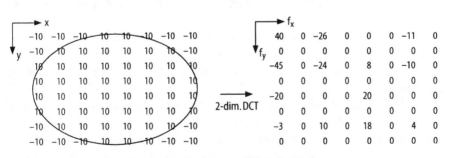

Fig. 4.3. Discrete cosine transformation (DCT) of an example 8×8 pixel block

$$G(f_x, f_y) = \tfrac{1}{4} C(f_x) C(f_y) \sum_{x=0}^{7} \sum_{y=0}^{7} g(x, y) \cos\left((2x+1) f_x \tfrac{\pi}{16}\right)$$

(4.1)

$$\times \cos\left((2y+1) f_y \tfrac{\pi}{16}\right)$$

where $C(f) = \begin{cases} \tfrac{1}{\sqrt{2}}, & \text{if } f = 0 \\ 1, & \text{if } f > 0 \end{cases}$

with: f_x, f_y = spatial frequencies
      $G(f_x, f_y)$ = DCT coefficients
      $C(f)$ = constant
      x, y = spatial co-ordinates
      $g(x, y)$ = video signal

we see that if the input signal $g(x,y)$ is quantised with 8 bits, 11 bits would be required for the integral representation of the DC coefficient $G(0,0)$ (if it is rounded to an integer, which is equivalent to the finest possible quantisation with a quantisation step size = 1 in the subsequent step; see section 4.1.3). This is due to the adding up of the 64 sampling values ($\cos(0) = 1$) and the scaling with the constant $\tfrac{1}{8} = \tfrac{1}{4} C(0) C(0)$. All other coefficients can equally be represented by 11-bit integers, since the values of the cosine functions in connection with the scaling factor or in front of the summation sign warrant the adherence to the 11-bit range of values. The transformation of 64 integers with an 8-bit precision to 64 integers with an 11-bit precision does not, of course, represent a decrease in the bit rate; on the contrary, it constitutes an increase. However, taking into consideration that the majority of the coefficients are very small and that the coarser quantisation in general leads to a narrowing of the range of numbers, a considerable decrease in the amount of data will be achieved in the course of the subsequent redundancy reduction.

Figure 4.4 illustrates the basis functions of the DCT. It shows blocks which, subsequent to the DCT, produce precisely one coefficient not equal to zero. The blocks are arranged according to the spatial frequency of this coefficient, which increases to the right and downwards. In this way the left uppermost block consists of only one uniform grey-scale value, which is generated by $G(0,0) \neq 0$. $G(f_x, f_y) = 0$ holds for all other f_x, f_y. Correspondingly, $G(7,7) \neq 0$ holds for the block in the bottom right corner. Each random block of a real image can be compiled by overlaying the 64 DCT basis functions with varying amplitudes. The coefficients $G(f_x, f_y)$ indicate just these amplitudes (including preceding sign).

The DCT is not the only transform which can be used for data-rate reduction techniques. Slightly higher reduction factors can be achieved with the Karhunen-Loeve transform. However, the requirements in terms of hardware and software are incomparably higher. The choice of DCT as an almost ideal

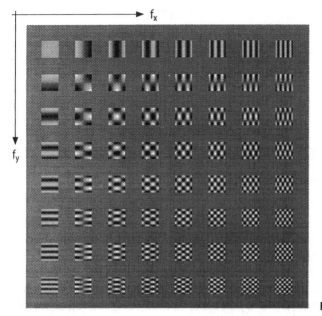

Fig. 4.4. DCT basis functions

transform for the JPEG standard and also for the MPEG standards was a trade-off between cost and benefit. The same applies to the fixing of the block size at 8×8 pixels. Larger blocks would lead to a minimally higher effectiveness of the data reduction, but the computation requirement increases geometrically in relation to the side length of the block. An early implementation of the algorithm in reasonably priced ICs could only be ensured with the 8×8 DCT.

4.1.3 Quantisation

Following the transformation into the frequency domain, the quantisation of the coefficients $G(f_x, f_y)$ can be carried out in accordance with the equation

$$G_Q(f_x, f_y) = \text{round}\left(\frac{G(f_x, f_y)}{Q(f_x, f_y)} \right) \tag{4.2}$$

with: f_x, f_y = spatial frequencies
       $G_Q(f_x, f_y)$ = quantised DCT coefficients
       $G(f_x, f_y)$ = unquantised DCT coefficients
       $Q(f_x, f_y)$ = quantisation step size

in which the quantised coefficients $G_Q(f_x, f_y)$ are computed by being divided by the required quantisation step size $Q(f_x, f_y)$ and then rounded to integers.

f_x							
16	11	10	16	24	40	51	61
12	12	14	19	26	58	60	55
14	13	16	24	40	57	69	56
14	17	22	29	51	87	80	62
18	22	37	56	68	109	103	77
24	35	55	64	81	104	113	92
49	64	78	87	103	121	120	101
72	92	95	98	112	100	103	99

f_y (vertical axis)

Q (f_x, f_y), luminance

f_x							
17	18	24	47	99	99	99	99
18	21	26	66	99	99	99	99
24	26	56	99	99	99	99	99
47	66	99	99	99	99	99	99
99	99	99	99	99	99	99	99
99	99	99	99	99	99	99	99
99	99	99	99	99	99	99	99
99	99	99	99	99	99	99	99

f_y (vertical axis)

Q (f_x, f_y), chrominance

Fig. 4.5. Examples of possible quantisation tables

This results in a linear quantisation characteristic. The denominator is generally different for each spatial frequency (f_x, f_y) and therefore an adaptation to the dynamic range of the human visual system can be achieved.

Figure 4.5 shows one table each of quantisation step size $Q(f_x, f_y)$ for luminance and for chrominance, which were determined by means of a psychophysiological experiment [LOHSCH]. In order to determine the values for the luminance the basis functions of the DCT are mixed with a background with an average grey-scale value. By operating a manual adjustment the subjects reduce the amplitude of the basis functions just so far as to be no longer perceptible. The detected 'just noticeable difference', the smallest identifiable difference, is used as the corresponding quantisation step size.

As can be seen, the quantisation tends towards being coarser the higher the spatial frequency. This is due to the fact that the human eye can perceive fine details with minimal dynamic content. Thus, in the case of luminance, the 11-bit amplitude resolution of the direct-current component (DC coefficient) $G(0,0)$ resulting from the DCT is limited to 7 bits – and hence to 128 amplitude steps for $G_Q(0,0)$, – as a consequence of its being divided by 16, whereas G_Q (7,7) can only assume $2^{11}/99 = 21$ discrete values.

The quantisation tables shown in figure 4.5 are mentioned in the informative annexe to the JPEG standard, but are not stipulated for the compression. After all, they only represent the purely empirically determined optimum for the quantisation of luminance and chrominance of typical natural images of the sampling format of ITU-R BT.601 [ITU 601] with 720×576 pixels, when viewed at a distance of four times the image height. The tables actually used must be transmitted in the compressed data stream in order to enable the decoder to invert the quantisation.

To illustrate the interworking of DCT and quantisation, figure 4.6 shows an example taken from [WALLACE]. Numerical block a shows the sampling values of the original image and b shows the resulting DCT coefficients. Following quantisation with the aid of block c, the quantised coefficients in d are

139	144	149	153	155	155	155	155
144	151	153	156	159	156	156	156
150	155	160	163	158	156	156	156
159	161	162	160	160	159	159	159
159	160	161	162	162	155	155	155
161	161	161	161	160	157	157	157
162	162	161	163	162	157	157	157
162	162	161	161	163	158	158	158

a samples of the original

235,6	-1,0	-12,1	-5,2	2,1	-1,7	-2,7	-1,3
-22,6	-17,5	-6,2	-3,2	-2,9	-0,1	0,4	-1,2
-10,9	-9,3	-1,6	1,5	0,2	-0,9	-0,6	-0,1
-7,1	-1,9	0,2	1,5	0,9	-0,1	0,0	0,3
-0,6	-0,8	1,5	1,6	-0,1	-0,7	0,6	1,3
-1,8	-0,2	1,6	-0,3	-0,8	1,5	1,0	-1,0
-1,3	-0,4	-0,3	-1,5	-0,5	1,7	1,1	-0,8
-2,6	1,6	-3,8	-1,8	1,9	1,2	-0,6	-0,4

b DCT coefficients

16	11	10	16	24	40	51	61
12	12	14	19	26	58	60	55
14	13	16	24	40	57	69	56
14	17	22	29	51	87	80	62
18	22	37	56	68	109	103	77
24	35	55	64	81	104	113	92
49	64	78	87	103	121	120	101
72	92	95	98	112	100	103	99

c quantisation table

15	0	-1	0	0	0	0	0
-2	-1	0	0	0	0	0	0
-1	-1	0	0	0	0	0	0
0	0	0	0	0	0	0	0
0	0	0	0	0	0	0	0
0	0	0	0	0	0	0	0
0	0	0	0	0	0	0	0
0	0	0	0	0	0	0	0

d quantised coefficients

240	0	-10	0	0	0	0	0
-24	-12	0	0	0	0	0	0
-14	-13	0	0	0	0	0	0
0	0	0	0	0	0	0	0
0	0	0	0	0	0	0	0
0	0	0	0	0	0	0	0
0	0	0	0	0	0	0	0
0	0	0	0	0	0	0	0

e coefficients after inverse quantisation

144	146	149	152	154	156	156	156
148	150	152	154	156	156	156	156
155	156	157	158	158	157	156	155
160	161	161	162	161	160	157	155
163	163	164	163	162	160	158	156
163	164	164	164	162	160	158	157
160	161	162	162	162	161	159	158
158	159	161	161	162	161	159	158

f reconstructed samples

Fig. 4.6. An example of DCT and quantisation

subjected to the redundancy reduction as described in the following section. This completely reversible process has been omitted in the example for reasons of clarity. Following the inverse quantisation e and the inverse DCT, the reconstructed image f is obtained, which differs from the original to such a small extent that the difference cannot be detected by the viewer. Without accepting these invisible differences, i.e. by performing a so-called lossless coding, it would only be possible, on average, to achieve a data-rate reduction by a factor of two or less.

4.1.4 Redundancy Reduction

Following the quantisation of the DCT coefficients there is a redundancy reduction which further reduces the amount of data for the representation of the corresponding block. A prerequisite for this is the rearranging of the coefficients in zigzag order, as shown in figure 4.7. The two-dimensional matrix is mapped into a one-dimensional field such that the coefficients are ordered from the lowest frequency to the highest. This typically causes the larger coefficients to appear at the beginning of the field, while the small coefficients, as well as many consecutive zeros, appear at the end. This circumstance, which is illustrated in the "pizza" example in figure 4.7, is very convenient for the subsequent combination of run-length coding and Huffman coding.

As shown in figure 4.8, the alternating-current (AC) coefficients are coded such that the number of consecutive zero coefficients and the next coefficient not equal to zero are combined to form a pair of numbers. From the sequence 0,0,–2 results the pairing (2,–2), from 0,0,0,0,–1 the pairing (4,–1). Finally,

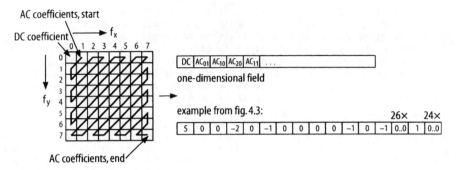

Fig. 4.7. Reordering of the coefficients in zigzag order

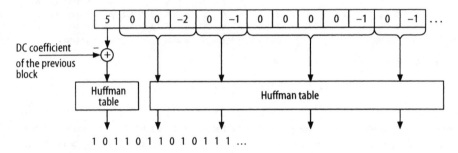

Fig. 4.8. Redundancy reduction

simply by referring to a coding table, one will find the most frequent pairs of numbers represented by few bits, while more bits will represent less frequent pairings. This method of coding, after Huffman, makes for a minimisation of the amount of data.

As the DC coefficient typically carries the greatest share of energy of the block (and is therefore only seldom equal to zero), it is subjected to a special treatment. By subtraction from the DC coefficient of the previous block its value will, on average, be considerably reduced. The resulting difference value is then also subjected to a correspondingly adapted Huffman coding.

The bit stream generated in this way, together with some signalling information such as quantisation tables and details concerning the resolution, etc., is then stored and transmitted as the JPEG bit stream. The algorithm described above typically entails a bit-rate reduction factor of between four and eight, but for most of the image content, despite this reduction factor, the difference between the image reconstructed in the decoder and the original is subjectively almost indiscernible.

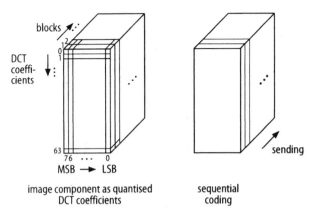

image component as quantised DCT coefficients

sequential coding

Fig. 4.9. Sequential coding in accordance with JPEG

4.1.5 Specific Modes

The algorithm described is also referred to as "baseline JPEG". For most applications it constitutes a suitable method for data reduction. For some specific applications, however, specific modes are defined in the JPEG standard, which are described in the following.

Figure 4.9 shows the so-called sequential coding of quantised coefficients, which is identical to the already described baseline algorithm. The left side of the figure depicts the required amount of data as a rectangular prism on which the edge lengths are determined by the number of blocks per picture, the number of coefficients per block, and the number of bits required per coefficient. For convenience, it has been assumed that all coefficients are represented by 8 bits; in reality, of course, this number varies from coefficient to coefficient depending on the quantisation.

In sequential coding, the individual blocks are transmitted in sequence or are sequentially read out of a storage medium. The disadvantage of this procedure is that when data are transferred slowly (for example, during research in image data bases which are accessed via networks over considerable distances) the image is built up from top left to bottom right. Therefore it can take quite a long time for the viewer to receive a rough outline of the image and to decide whether or not it is the image sought. This costs the user of a data bank considerable time and unnecessarily wastes transmission capacity.

In this case help can be found in the so-called "progressive coding", which is shown in figure 4.10. The DC coefficients of all blocks are transmitted in a first run, and this is followed by the spectral selection in which the AC coefficients are transmitted in the order of increasing frequency. Thus, after the transmission of only a few bits, the viewer obtains a rough idea of the image as a whole, which at the start consists of 8×8 blocks of constant brightness and colour and which, through the inclusion of the other DCT basis functions, ap-

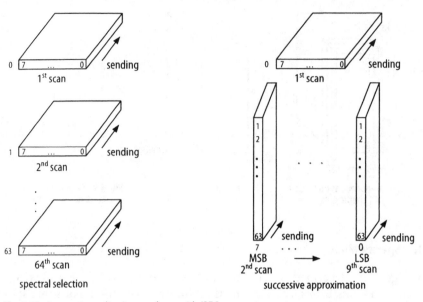

Fig. 4.10. Progressive coding in accordance with JPEG

pears in more and more detail in the course of the data transmission. Should the image turn out not to be the one required, this transmission can be broken off at any time; only the desired image needs to be completely transmitted. This saves time and transmission capacity.

A second example of progressive coding, the "successive approximation", can be seen in the right-hand side of figure 4.10. Following the transmission of the DC component, first the most significant bits (MSBs) of the AC coefficient, and then the least significant bits (LSBs) are transmitted. In this way an increasing image quality is obtained.

For some applications even the small differences between the reconstructed image and the original, as depicted, for example, in figure 4.6, are not acceptable. In medicine, for instance, the smallest distortion in an X-ray could influence a diagnosis. In order to supply the tools for such applications the JPEG standard offers a lossless mode. As shown in figure 4.11, this mode does not use transformation, as even the small rounding errors which occur inevi-

						selection value	prediction
						0	no prediction
	C	B				1	A
	A	X				2	B
						3	C
						4	A+B–C
						5	A+((B–C)/2)
						6	B+((A–C)/2)
X = actual pixel						7	(A+B)/2

Fig. 4.12. Lossless coding in accordance with JPEG, possible predictions

tably in the DCT-based mode of operation would cause a disturbance. Instead, based on the knowledge of the neighbouring pixels, a prediction of the actual pixel is made. The resulting difference from the actual value is coded with the aid of a Huffman table. Seven predictors, as listed in figure 4.12, are available for the prediction. In each case the most suitable prediction is chosen by the encoder and signalled to the decoder.

The data reduction factor obtained with lossless coding has an average value of <2 for natural images, while for the baseline algorithm, with almost the same subjective quality, an average factor of 4 to 8 is achieved. For this reason the lossless mode can only be considered as a solution for special applications. For example, for X-rays which have a high proportion of black areas the reduction factor obtained is much higher than two.

A further possibility of achieving a gradual increase in the resolution of an image is offered by the "hierarchical coding" which can be paired with all other JPEG modes. The relevant block diagram is shown in figure 4.13. The image to be coded is subsampled horizontally and vertically with the same factor, which must be a power of two (2^N). The resulting image is coded with a non-hierarchical JPEG encoder and transmitted. At the same time it is reconstructed at the decoder and oversampled by a factor of two so that it is avail-

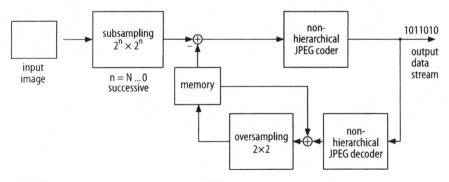

Fig. 4.13. Hierarchical coding in accordance with JPEG

able as a prediction for the next coding step, which uses the image now only subsampled by the factor 2^{N-1}. This continues until the best available resolution of the image (subsampling factor $2^0 = 1$) has been transmitted. In this way a successively increasing resolution is obtained during transmission, which can also be used in connection with lossless coding.

Concerning the use of the JPEG standard for transmitting moving pictures, none of these special coding modes is of any significance; firstly, because the transmission must take place in real time anyway and therefore a successively increasing resolution of the image quality would not be appropriate and, secondly, because with lossless coding a satisfactory data-rate reduction cannot be achieved. As mentioned before, M-JPEG was not standardised, which signifies that it is an adaptation of the baseline algorithm to moving pictures. If the term "JPEG" is used, then what is typically meant is baseline JPEG.

4.1.6 Interchange Format

In order to ensure compatibility and interchangeability between application environments, the JPEG standard defines an interchange format which on the one hand determines the structure of the bit stream, and on the other hand also determines the subsampling matrix. Figure 4.14 shows the relationship between the various components (with different resolutions) of the input image. The samples of the highest-resolution component (usually the luminance Y) are arranged in an orthogonal matrix. In the case of lower-resolution components (for example, chrominance signals C_B and C_R) the subsampling factor must be horizontally and vertically equal and the location of the samples must be midway between those of the higher-resolution component. The same is true for lowest-resolution components, as can occur in specialised applications.

This definition is in accordance with the usual format in computer applications like, for example, Postscript Level 2, or Apple Quicktime, but not with Recommendation ITU-R BT.601 which relates to television. As stated by ITU, the chrominance for the $4:2:2$ format is only subsampled horizontally and the chrominance sampling matrix is otherwise co-located with the luminance sampling matrix. If, in M-JPEG applications, one were to adhere to the ITU

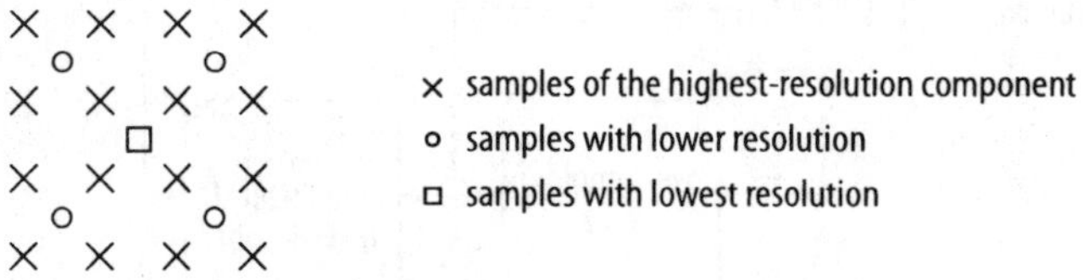

× × × ×
  o   o
× × × ×   × samples of the highest-resolution component
     □   o samples with lower resolution
× × × ×   □ samples with lowest resolution
  o   o
× × × ×

Fig. 4.14. Interchange format in accordance with JPEG

recommendations as well as to the JPEG format, a costly format conversion entailing quality losses would be required before the encoder and after the decoder. Therefore, most applications operating with the JPEG algorithm for the transmission and storage of moving pictures adhere to the ITU recommendation and infringe the JPEG format. Regarding such applications, this procedure should not cause any problems as long as data are not exchanged among codecs from different manufacturers.

4.2 Coding in Accordance with the MPEG Standards

The JPEG standard was originally conceived for still pictures and is therefore not completely suitable for data reduction of moving pictures. Firstly, the lack of a standard for M-JPEG means that there are considerable incompatibilities between the methods used by different manufacturers. Secondly, JPEG does not take advantage of the similarity of successive moving pictures, and therefore the reduction factor is limited to unnecessarily low values. Thirdly, JPEG neither provides for a coding of the associated audio information nor for a multiplexing of video and audio signals.

For this reason the Moving Pictures Experts Group defined an algorithm for coding moving pictures – including associated audio – which does not have the above shortcomings. The first step in this direction was the development of the MPEG-1 standard which was designed for use with computers and multimedia, particularly for the storage of video on CDs with a data rate of 1.15 Mbit/s. In order to obtain an acceptable image quality at this very low data rate it was necessary to include a number of limitations in the standard, e.g. that the resolution of the image to be coded must not be higher than 352×288 pixels, that the frame rate must not exceed 30 Hz, and that only progressive image sampling must be supported. An upper limit of 1.5 Mbit/s was set for the data rate. Adherence to this so-called 'constrained parameter set' ensures that any MPEG-1 decoder can decode the data. Owing to the early availability of suitable components, it was only a short time after the completion of the standard that applications came on the market which went beyond the established parameters and made it possible to also code images in the ITU-R BT.601 format, with 720×576 pixels (often referred to as MPEG-1+, MPEG-1.5 or similar). However, since the treatment of interlaced scanning is not specified in the MPEG-1 standard, applications using this type of scanning must be seen as individual solutions, all of which became obsolete with the completion of the MPEG-2 standard in November 1994.

MPEG-2 includes an extension of the MPEG-1 standard for the transmission of television signals, therefore it takes interlaced scanning into consideration and offers a succession of quality steps and options which are usu-

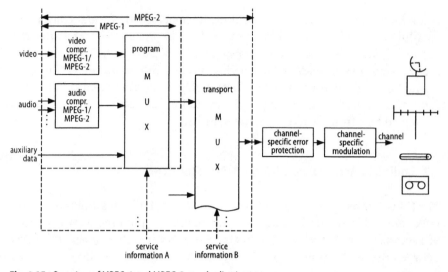

Fig. 4.15. Overview of MPEG-1 and MPEG-2 standardisation ranges

ally represented in the profile/level table, which is described later. The Main Profile at Main Level (MPML), which is intended for the transmission of standard television signals and allows for data rates of up to 15 Mbit/s, has shown in MPEG tests that an image quality comparable to PAL can be achieved even at bit rates of about 6 Mbit/s and that, at a code rate of 9 Mbit/s, the reconstructed image is almost indistinguishable from that of the original (so-called visual transparency). Originally, a third level, MPEG-3, was planned for the MPEG algorithm, to allow the coding of HDTV signals. However, as the coding of high-definition images has been incorporated in the profile/level table of MPEG-2 there is no requirement for MPEG-3.

The work of the MPEG group nevertheless continued in the direction of higher data-compression factors or extremely low data rates, e.g. below 64 kbit/s, and was subsumed under the name of MPEG-4. This "very low bit-rate coding", which might be applied in video conferencing or mobile services, will no longer exclusively rely on the DCT-based technique, as used in MPEG-1 and MPEG-2, but, with the aid of object-oriented coding, tries to achieve much higher reduction factors, albeit at a lower-grade quality. The use of MPEG-4 in broadcast services is currently not being envisaged and therefore MPEG-4 will not be discussed here.

Figure 4.15 gives an overview of the MPEG-1 and MPEG-2 standardisation range. In the MPEG standards not only the video-data reduction is defined, but also the audio-data reduction which was discussed in detail in chapter 3. It includes, moreover, the multiplexing of video, audio and other information, such as videotext, which are to be transmitted in a communal bit stream.

MPEG-1 only offers the multiplexing of one single programme, which is distinguished by the fact that all components (video, audio and auxiliary data) have one common time base. With MPEG-2, on the other hand, several programmes can be combined to form one common data stream. All other transmission parts outside the dashed lines representing the various blocks, i.e. the error protection adapted to the channel characteristics, the modulation used for the transmission, and the service information (see chapter 5), are not included in the MPEG standards. In Europe these elements are the responsibility of the European Digital Video Broadcasting (DVB) Project and have already been defined.

In the next section, the algorithm, i.e. the principle which is used by MPEG-1 and MPEG-2 for video coding, will be explained, followed by a discussion of the differences between the two standards.

4.2.1 Block Diagrams of Encoder and Decoder

The MPEG standards, too, make use of the discrete cosine transformation for spatial decorrelation. In addition, however, they also exploit the similarity of successive pictures and therefore, with no change in the quality, achieve considerably higher compression factors. Figure 4.16 shows the principle of differential coding, on which the data-rate reduction in the temporal direction is based. In the encoder the decoded signal, which is delayed by time factor τ, is subtracted from the actual signal in order to decorrelate the distribution of the amplitude values. The decoder inverts this step. τ can represent the duration of a pixel (in this case the neighbouring pixels are subtracted in the encoder), the duration of a line (corresponding to the subtraction of pixels in neighbouring lines), or the duration of an image (corresponding to the subtraction of equally located pixels in a succession of pictures). The last case has been put into effect in MPEG, as the similarities between the images are exploited by the DCT.

As can be seen, there is a complete decoder in the feedback branch of the encoder so that exactly *that* signal is used for the subtraction which is later generated in the decoder and added to the incoming signal. In this way the coding process can work completely without loss (redundancy reduction) as long as the range of numbers has not been limited by way of rounding errors

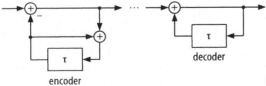

encoder   decoder

Fig. 4.16. Principle of differential coding

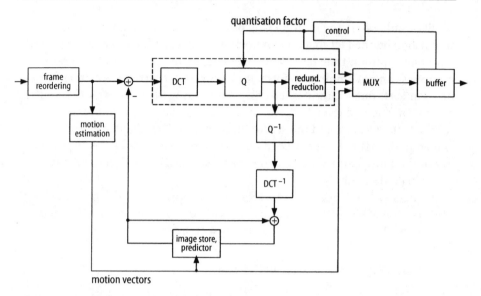

Fig. 4.17. Block diagram of an MPEG video encoder

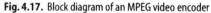

or quantisation. If the input signal is quantised with 8 bits and thus assumes values in the range of 0 to 255, then values between −255 and +255 (9 bits) can occur at the output of the encoder in the most extreme case. Simple methods, which achieve a data reduction solely by performing the differential coding, limit the difference signal in its word length to, for instance, 6 bits (range of values −32 ... + 31), but accept artefacts (slope overloads) on all steep signal transitions (so-called differential pulse code modulation [DPCM]). For MPEG the difference values are fed into the subsequent DCT with an unchanged 9-bit word length. The gain in terms of data reduction results from the decorrelation.

Figure 4.17 shows the block diagram of an MPEG video encoder, the components of which will be discussed in detail in the following paragraphs. The steps DCT, quantisation and redundancy reduction that can be seen in the centre, framed by dashed lines, are also found in JPEG. In addition to this, the encoder features differential coding with an image store for time delays by the value of τ. This prediction is supported by motion estimation which seeks the best possible match in the previous image for the image to be coded in each block. As the decoder requires the motion information in order to invert the prediction, the calculated motion vectors are also transmitted in the bit stream. In order to also exploit the similarity with successive pictures, the coding is preceded by a reordering of the sequence of pictures.

The second important new functionality included in MPEG-2, as compared to JPEG, is the control of the quantiser to regulate the transmission of a

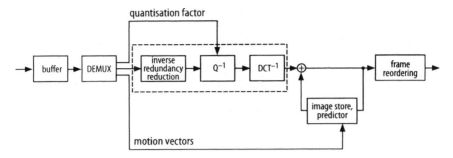

Fig. 4.18. Block diagram of an MPEG video decoder

constant data rate. At the output of the encoder there is a buffer which absorbs the incoming data of varying rates and passes them on at a constant data rate. If there is any danger of the buffer overflowing, the buffer will cause the quantisation factor (see section 4.2.4) to produce a coarser quantisation so that less data are transferred to the buffer. A constant data rate ensues as a consequence.

The corresponding decoder is shown in figure 4.18. The data which arrive at a constant data rate are absorbed by the input buffer and transferred at varying rates to the demultiplexer which separates the coded image data from the required auxiliary information (especially the quantisation factor and the motion vectors). The inverse quantisation follows the inverse redundancy reduction which evaluates the quantisation factor transmitted in the bit stream. The inverse DCT transforms the coefficients back into the spatial domain whereupon the predicted values are added. For this step the motion vectors are required which were calculated by the encoder and transmitted in the data stream. Finally the decoded images are put back in the right order.

It is obvious from the block diagrams that the construction of the decoder is considerably simpler than that of the encoder. In particular no motion vectors have to be defined, as these are generated by the encoder. The motion estimation is, however, the step which requires the most computing power.

4.2.2 Motion Estimation

To illustrate the motion-estimation-supported prediction an example is given in figure 4.19. This shows a ball which has moved minimally to the right and downward from image 1 to image 2. For coding purposes the images were broken down into so-called macroblocks with 16×16 pixels (corresponding to 2×2 DCT coding blocks). To code the thickly drawn macroblock in image 2,

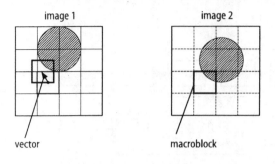

vector   macroblock   **Fig. 4.19.** Example of motion estimation

the difference between the corresponding macroblocks in image 2 and image 1 is computed. There remains in the top right-hand corner of the resulting macroblock a relatively large difference value because the ball has not yet appeared in the corresponding macroblock in image 1. Accordingly, a relatively large number of bits is required for coding. A considerably better prognosis can be made by computing the difference from a macroblock which has been shifted in the right direction, as shown in figure 4.19. However, in this case the calculated motion vector must also be transmitted so as to enable the decoder to invert the process.

The selection of macroblocks as the basic element for motion estimation offers the advantage that the same vector can be used for the luminance components as for the chrominance components. If the chrominance is subsampled horizontally and vertically by the factor 2 (4:2:0 coding, see section 4.1.6 and 4.2.5), one macroblock is made up of four luminance blocks as well as of one block each of the two colour difference signals, as shown in figure 4.20. A corresponding definition holds for formats 4:2:2 (as in ITU-R BT.601) and 4:4:4 (no subsampling), which are provided only by MPEG-2.

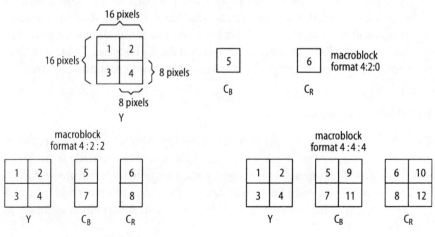

Fig. 4.20. Possible macroblock structures

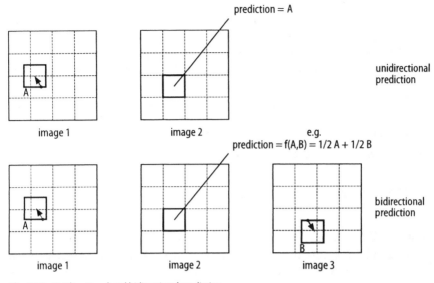

Fig. 4.21. Unidirectional and bidirectional prediction

The MPEG standards do not specify which algorithm is to be used to calculate the motion vectors. Usually a block-matching technique is used [MUSMANN], by which, for example, the macroblock is shifted to all possible positions within a given search area, and the position for coding is chosen where the difference to the actual macroblock is the smallest. In "full-search block matching" all positions are actually checked in steps of half a pixel (realisation by interpolation), whereas hierarchical procedures commence with larger steps, locating a provisional minimum, and then continue with smaller steps in a search area around the provisional minimum. In any case, applying the search algorithm requires a very large computational effort, which means that for hardware realisations the search area is often considerably reduced. However, the size of the search area plays a decisive role in determining the quality of the reconstructed picture. For example, if the search area is restricted to ±10 pixels horizontally and ±5 pixels vertically, then, if wide-range shifting of more than 10 pixels occurs horizontally from image to image (e.g. due to a fast camera swing) the prediction fails. The reduced coding efficiency resulting from this causes a considerable quality loss.

Prediction using future images has also been included in MPEG in order to further increase the coding efficiency. Figure 4.21 shows what is to be understood by "bidirectional prediction". As opposed to unidirectional prediction, the macroblock in image 3 which follows image 2, the image which is to be coded, is additionally searched. The prediction can thus, for example, result from the mean value between macroblock A and macroblock B. If image 1 or image 2 does not yield a usable prediction (which could be the case, for example, in cut images), the encoder can decide that only the other prediction must be used. Both orientations of the prediction could fail in the case of fast movements. In such a case the encoder is free to revoke the prediction and save the bits for coding the vectors. Even though bidirectional prediction means doubling the expense, it was included in the standard because it considerably increases the coding efficiency. A bidirectionally predicted macroblock can on average be coded with only about half the amount of data required for a unidirectionally predicted macroblock.

4.2.3 Reordering of Pictures

Bidirectional prediction not only doubles the computational effort of the motion vector calculation but also necessitates the already-mentioned reordering of the pictures. The upper part of figure 4.22 shows an extract from a frame sequence. If one were to use a prediction for all pictures, it would not be possible to find a starting point for the decoding, as each picture is dependent on the one before and after it. For this reason one picture, at suitable intervals, is coded without the help of prediction (so-called I-pictures, intraframe-coded). If an I-picture is inserted every 12 frames, then, when the television receiver is switched on or the programme changed, it takes a maximum of half a second before the decoding can begin.

Between the I-pictures, so-called P-pictures, which have been unidirectionally predicted from the preceding I-picture or P-picture, are inserted as supporting points. The interleaved bidirectionally predicted pictures (B-

viewing succession

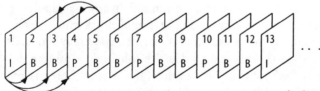

I = intraframe-coded picture
P = predicted picture
B = bidirectionally predicted picture

transmission succession

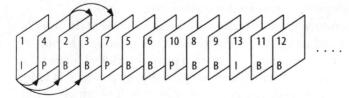

Fig. 4.22. Reordering of pictures

pictures) result from the I-pictures or P-pictures which appear before and after. As can be seen from the interdependencies of the pictures, marked by arrows, the decoder must always recognise the subsequent P-picture or I-picture before the interleaved B-pictures can be decoded. The same applies to the encoder since there is a complete decoder in its feedback branch. That is why the sequence of the pictures is changed before transmission. The framing I-pictures and P-pictures are always transmitted before the interleaved B-pictures. Therefore, the encoder has to have a total of four image stores and to carry out the following steps for two consecutive B-pictures: coding the I-picture and retaining the original for the prediction; merely retaining the originals of the B-pictures for the time being; coding the P-picture and retaining the original for the prediction of B-pictures; then coding the B-pictures. The decoder, by comparison, requires only two image stores, irrespective of the number of consecutive B-pictures to be decoded, as it only needs to retain the framing I-pictures and P-pictures; the B-pictures can be displayed directly after decoding and removed from storage. The decoding of the B-pictures requires extra storage space, the extent of which depends on the implementation.

The determination of the sequence of the I-pictures, P-pictures and B-pictures (so-called group-of-pictures structure) depends solely on the decision of the encoder. The decoder can process all sequences since it never requires more than just a little over two image stores for the decoding. The complexity of the encoder as well as the image quality achieved depend essentially on the structure of the group of pictures. The structure shown in figure 4.22 has proved to be a viable trade-off between cost and performance; however, this structure is by no means mandatory.

4.2.4 Data-rate Control

As already mentioned, a constant data rate is achieved by controlling the quantiser. If there is a danger of the output buffer of the encoder overflowing – for instance, because a picture content with fast movement and thus with a lot of high frequencies has to be coded or because the prediction malfunctions on account of the limitation of the search area – the quantisation step is increased, resulting in a coarser quantisation and possibly in a visible deterioration of the picture quality. When the situation eases, the quantisation step can again be decreased. It is on the quantisation that the amount of data input to the buffer depends. For a minimum quality of the reconstructed picture to be guaranteed, the data-rate control does not affect the quantisation of the DC component of intraframe-coded pictures. The DC coefficient is always divided by eight. On the other hand, the DC coefficients for P-pictures and B-pictures, just as the AC coefficients, are quantised with a step size taken from a similar table to the one used for JPEG. This, in addition, is followed by a further quantisation which performs the data-rate control.

DC coefficient for I-pictures:

$$G_Q(0,0) \quad = \text{round}\left(\frac{G(0,0)}{8}\right) \tag{4.3}$$

Otherwise:

$$A(f_x, f_y) \quad = \text{round}\left(\frac{G(f_x, f_y)}{Q(f_x, f_y)}\right)$$

$$G_Q(f_x, f_y) = \text{round}\left(\frac{A(f_x, f_y)}{Q_F/8}\right) \tag{4.4}$$

with: f_x, f_y = spatial frequencies
      $G_Q(f_x, f_y)$ = quantised DCT coefficients
      $G(f_x, f_y)$ = unquantised DCT coefficients
      $A(f_x, f_y)$ = auxiliary quantity
      $Q(f_x, f_y)$ = entries in the quantisation table
      Q_F = quantisation factor

Here Q_F is the quantisation factor for the data-rate control, which varies from macroblock to macroblock and which can be set by the buffer control to a value between 1 and 31. By this means entries in the quantisation table are variegated by a common factor which can be set in steps of 1/8. If the setting $Q_F = 31$ does not suffice to prevent a buffer overflow, so-called "skipped macroblocks" occur, i.e. macroblocks which are coded with 0 bits and are therefore skipped. In this case the decoder sets all motion vectors and all coefficients to zero, resulting in the macroblock of the previous image simply being repeated. Of course such macroblocks are generally very conspicuous as they do not fit exactly. In contrast, if the image content is so simple that even at $Q_F = 1$ too few data are produced, then an emptying of the buffer is avoided by inserting so-called stuffing bits.

Contrary to the JPEG standard, quantisation tables were agreed for MPEG-1 and MPEG-2, and these can be seen in figure 4.23. In this way the bits for signalling these tables can be saved. However, should the encoder discover that variant quantisation tables would lead to a higher coding efficiency, then it has the option to transmit those tables in addition to the others.

4.2.5 Special Features of MPEG-1

MPEG-1 was designed for use with computers and multimedia and particularly for the recording of videos on conventional CDs. To keep within the given limited data rate, the resolution of the image to be coded had to be re-

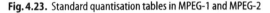

f_x								f_x							
8	16	19	22	26	27	29	34	16	16	16	16	16	16	16	16
16	16	22	24	27	29	34	37	16	16	16	16	16	16	16	16
19	22	26	27	29	34	34	38	16	16	16	16	16	16	16	16
22	22	26	27	29	34	37	40	16	16	16	16	16	16	16	16
22	26	27	29	32	35	40	48	16	16	16	16	16	16	16	16
26	27	29	32	35	40	48	58	16	16	16	16	16	16	16	16
26	27	29	34	38	46	56	69	16	16	16	16	16	16	16	16
27	29	35	38	46	56	69	83	16	16	16	16	16	16	16	16

$Q(f_x,f_y)$ for I-pictures $Q(f_x,f_y)$ for P- and B-pictures

Fig. 4.23. Standard quantisation tables in MPEG-1 and MPEG-2

stricted to the so-called "source input format" (SIF) (see sections 2.2 and 4.1). This is to be understood as essentially half the resolution compared to an image conforming to Recommendation ITU-R BT.601 in conjunction with the 4 : 2 : 0 colour representation. In figure 4.24 the sampling structures of the two formats are compared. Whereas the chrominance in ITU-R BT.601 is only subsampled horizontally by a factor of two and whereas apart from that the sampling grid coincides with that of the luminance, the chrominance in the SIF is subsampled in both directions and its samples lie midway between those of the luminance (as with the JPEG interchange format, see section 4.1.6).

To transmit or store a television signal with MPEG-1 it is necessary to perform a format conversion before the encoder and after the decoder. This is illustrated by the transmission chain in figure 4.25, where "channel" can also be interpreted as a recording on CD. By way of an example the individual steps

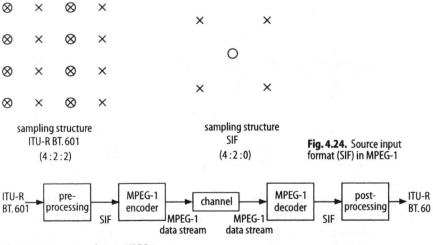

sampling structure   sampling structure
ITU-R BT.601   SIF   **Fig. 4.24.** Source input
(4 : 2 : 2)   (4 : 2 : 0)   format (SIF) in MPEG-1

ITU-R BT.601 → pre-processing → SIF → MPEG-1 encoder → MPEG-1 data stream → channel → MPEG-1 data stream → MPEG-1 decoder → SIF → post-processing → ITU-R BT.60

Fig. 4.25. Processing chain in MPEG-1

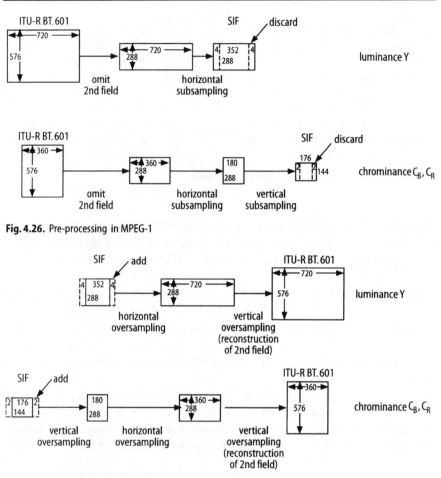

Fig. 4.26. Pre-processing in MPEG-1

Fig. 4.27. Post-processing in MPEG-1

for the pre-processing and post-processing are shown in detail in figures 4.26 and 4.27. In changing the format to SIF, in the easiest case, one field is completely omitted for the time being. The use of a decimation filter at this stage would result in the motion phases of the two fields being blurred. This would cause artefacts during movement in the image. Should a filter be used at this stage to lessen the vertical aliasing which occurs by omitting a field, then this should be done adaptively in those image regions where movement is only minimal.

Afterwards, the horizontal subsampling is performed. This time, though, a suitably designed decimation filter should be used which, with regard to chrominance, ensures that the required position of the sampling points is observed. Some hardware implementations do not include filters at this point or elsewhere. They carry out the subsampling by simply omitting pixels. This, however, causes aliasing, which has a negative effect on the coding quality. For the chrominance components another vertical subsampling follows (here again, if possible, with the correct filter in terms of systems theory) with the object of achieving the 4 : 2 : 0 format. Since because of the macroblock structure the number of luminance pixels per line (just as the number of lines) must be divisible by 16 and the number of chrominance pixels per line must be divisible by 8, some pixels in the right and left margins of the picture will be discarded (delineated by broken lines in figure 4.26). The input signal for MPEG-1 coding, with a progressive sampling structure and a 25-Hz frame frequency, thus has a resolution of 352×288 luminance pixels and of 176×144 chrominance pixels.

Following the decoding process, the corresponding inverse steps must be carried out to reconstruct the original format. Horizontal oversampling takes place subsequent to the addition of black pixels to the left and right margins of the picture and subsequent to vertical oversampling of the chrominance components. The reconstruction of the second field can also be performed by oversampling; however, the second motion phase which was eliminated by the encoding can, of course, not be reconstructed. This would be possible by means of costly motion-adaptive processing, but even then only partially.

4.2.6 Special Features of MPEG-2

The considerable limitations of the MPEG-1 standard do not recur in MPEG-2. MPEG-2 provides for the coding of signals in accordance with ITU-R BT.601 as well as for the use of the 4 : 2 : 2 chrominance format and even of the 4 : 4 : 4 format. The main profile proposed by the European DVB Project for television broadcasting utilises the 4 : 2 : 0 format which, however, differs from the 4 : 2 : 0 definition in MPEG-1 in the position of the chrominance sampling grid in relation to that of the luminance. Apart from this, the interlaced scanning is taken into account. The left part of figure 4.28 shows the 4 : 2 : 0 format with a progressive sampling grid, where the luminance sampling points are marked by crosses and the chrominance sampling points by circles. The grid differs from its counterpart in MPEG-1 and JPEG in that the vertical samples of the chrominance lie midway between the luminance lines but that horizontally they remain unchanged as compared to ITU-R BT.601. For MPEG-1 a horizontal filter has to be employed to enable the subsampling. So this same filter can be used for the horizontal phase shifting. Thus a format is created which conforms with the usual conventions in the field of computers. Right from the beginning, however, MPEG-2 was aimed at applications in television engineering, where a sampling structure like that depicted in figure 4.28 is used and where a horizontal shifting of the sampling grid would mean

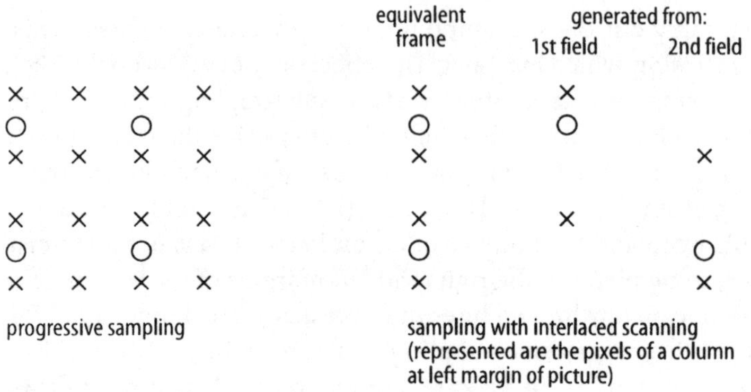

				equivalent frame	generated from:	
					1st field	2nd field
×	×	×	×	×	×	
O		O		O	O	
×	×	×	×	×		×
×	×	×	×	×	×	
O		O		O		O
×	×	×	×	×		×
progressive sampling				sampling with interlaced scanning (represented are the pixels of a column at left margin of picture)		

Fig. 4.28. Sampling structure 4 : 2 : 0 in MPEG-2

extra costs. For sampling using interlaced scanning the position of the samples does not change; the assigning of the chrominance lines to the fields is indicated in the right half of figure 4.28, which shows the first column of an image as part of a frame and as part of the two fields.

The high profile of the MPEG-2 standard also permits the 4 : 2 : 2 format, which is shown in figure 4.29, whereas the 4 : 4 : 4 format, as depicted in figure 4.30, has been provided for but is not contained in the profiles up to now. For future applications the standard can be supplemented by further profiles with the option of a 4 : 4 : 4 format. The 4 : 2 : 2 format conforms to ITU-R BT.601, whereas for the 4:4:4 representation the luminance and chrominance grids are congruent and superimposed on each other.

The interlaced scanning also affects the performance of the DCT. In MPEG-2 the encoder can choose between the field mode and the frame mode, as shown in figure 4.31. The two fields of the picture are represented by black and white picture lines. In the frame mode each macroblock is divided into

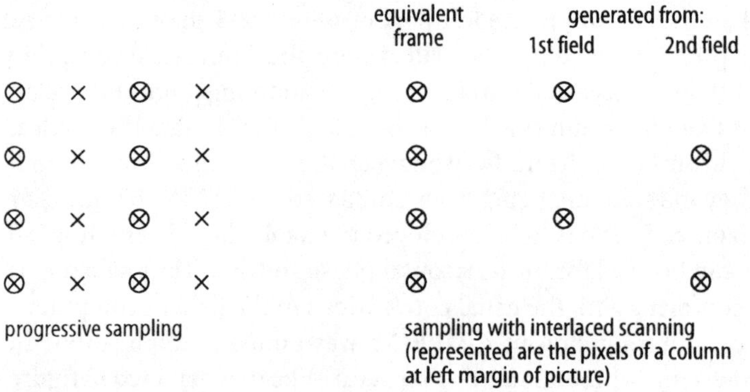

				equivalent frame	generated from:	
					1st field	2nd field
⊗	×	⊗	×	⊗	⊗	
⊗	×	⊗	×	⊗		⊗
⊗	×	⊗	×	⊗	⊗	
⊗	×	⊗	×	⊗		⊗
progressive sampling				sampling with interlaced scanning (represented are the pixels of a column at left margin of picture)		

Fig. 4.29. Sampling structure 4 : 2 : 2 in MPEG-2

| | equivalent frame | generated from: | |
| | | 1st field | 2nd field |

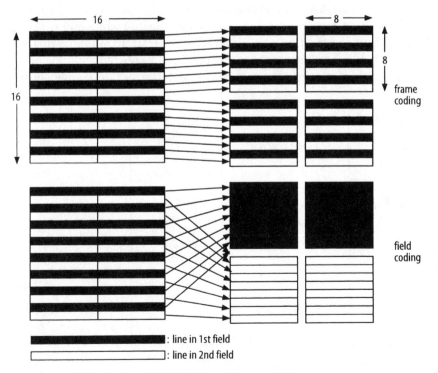

Fig. 4.30. Sampling structure 4 : 4 : 4 in MPEG-2

8×8 DCT blocks in such a way that each 8×8 block contains both motion phases. For fast movements in the picture, however, this is a disadvantage since the breaking up of vertical edges in the image, caused by interlaced scanning, leads to high vertical frequencies and therefore to low coding efficiency. In this case it is more convenient to rearrange the fields as shown in

Fig. 4.31. Possible distributions of 8×8 DCT blocks within a macroblock

the lower part of the representation. The chrominance results in the 4:2:0 format having two coding blocks of the size 8×4 instead of one coding block of the size 8×8. For a static image or an image with little movement the frame coding is generally more efficient as spatially neighbouring lines are usually more closely correlated. The decision as to which method is to be used is made by the encoder.

MPEG-2 differs from JPEG and MPEG-1 in that an optional non-linear quantisation of DCT coefficients is also possible. As a rule, this results in a higher coding efficiency. There is also the possibility of a modified run-length coding, which can be better adapted to the interlaced scanning using a different zigzag order and a second Huffman table.

A further innovation in the MPEG-2 standard are the different types of scalability, the meaning of which is explained below. The degradation of a digital television transmission is generally very abrupt. Even a small increase in the bit-error rate can cause a transition from perfect reception to total disruption. If the power budget of the transmission path does not include enough margin, e.g. if too small a receiving antenna is used for a satellite transmission, changeable weather conditions can cause a situation in which the picture continually vanishes and reappears. In the case of terrestrial transmissions it can happen that one household at the edge of a coverage area can still receive a certain programme in perfect quality while others in the next street have no reception whatsoever (cf. chapter 11). To minimise this unpleasant behaviour an intermediate stage can be introduced with the aid of scalability tools, as shown in figure 4.32. At a low bit-error rate completely un-

low bit-error rate

medium bit-error rate

high bit-error rate

SNR scalability   spatial scalability

Fig. 4.32. Scalability in MPEG-2

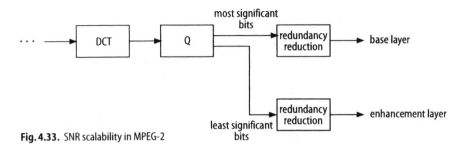

Fig. 4.33. SNR scalability in MPEG-2

disturbed reception is possible. If the bit-error rate increases to a higher value, then, in the case of SNR (signal-to-noise ratio) scalability, the picture received will be noisy but still acceptable. However, this is not noise as we understand it, but noise due to quantisation errors, combined with signal and block structures. By comparison, in the case of spatial scalability, at the increased bit-error rate, a lower-resolution image is received. It is only with even higher bit-error rates that the reception is completely disrupted. If it is envisaged to exploit the possibilities arising from the utilisation of scalability in source coding in a digital television system, "hierarchical modulation" (cf. section 11.5) will be required. The European DVB Project does not include either SNR scalability or spatial scalability in its system specifications.

Figure 4.33 shows the implementation of SNR scalability using a section of an MPEG-2 encoder. After quantisation, the DCT coefficients are divided into most significant and least significant bits and subjected to a separate redundancy reduction. In the channel encoder the so-called base layer with the most significant bits is then furnished with a better error protection for transmission than the least significant bits of the enhancement layers. Hence, in the case of a deterioration of the transmission characteristics of the channel it is the less important bits of the coefficients that can no longer be received, a fact which becomes noticeable as noise-like interference.

An encoder for spatial scalability is quite differently constructed (see figure 4.34). The image to be coded is first subsampled horizontally and vertically by a factor of two and then fed into an MPEG-2 encoder for this reduced resolution. The resulting base layer is then furnished with a high-grade error protection in the channel encoder. In the encoder the base layer is immediately decoded and, after reconversion to the original resolution, used as prediction for the enhancement layer, which is then furnished in the channel encoder with a lower-grade error protection. A particular advantage of spatial scalability is that it enables the compatible transmission of HDTV signals.

By combining both scalabilities it is possible, moreover, to achieve a multistage degradation. Two other types of scalability should be mentioned here:

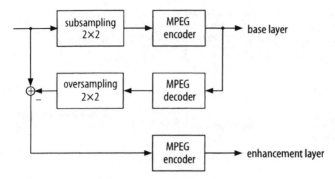

Fig. 4.34. Spatial scalability in MPEG-2

temporal scalability, which reduces the temporal resolution, i.e. the frame frequency, and data partitioning, similar in character to SNR scalability. Neither will be further discussed here.

The profile/level table for MPEG-2 mentioned at the beginning will now be examined more closely. Figure 4.35 shows its construction. The so-called profiles, under which certain syntactical elements and hence certain algorithmic peculiarities are subsumed, are arranged from left to right. The levels, limiting certain parameters like resolution and data rates to their corresponding maximum values, are arranged from bottom to top.

The upper limits for definition and data rate can be seen in the boxes. The numbers in brackets refer to the lower layer or layers of the scalable coding. The parameters for the cancelled boxes have not been defined.

levels / profiles	simple profile	main profile	SNR scalable profile	spatial scalable profile	high profile
high level		1920×1152 pixels 80 Mbit/s			1920×1152 pixels (960×576) 100(80.25) Mbit/s
high-1440 level		1440×1152 pixels 60 Mbit/s		1440×1152 pixels (720×576) 60(40.15) Mbit/s	1440×1152 pixels (720×576) 80(60.20) Mbit/s
main level	720×576 pixels 15 Mbit/s	720×576 pixels 15 Mbit/s	720×576 pixels 15(10) Mbit/s		720×576 pixels (352×288) 20(15.4) Mbit/s
low level		352×288 pixels 4 Mbit/s	352×288 pixels 4(3) Mbit/s		
	(main profile, without B-pictures)	(4:2:0, no scalabilty)	(main profile, + SNR scalability)	(SNR profile, + spat. scalability)	(spatial profile, +4:2:2 coding)

Fig. 4.35. Profiles and levels in MPEG-2

The low level has been conceived for the coding of television pictures with reduced definition, similar to the already-known SIF in MPEG-1 (LDTV: low-definition television). However, due to the specified upper limit of 4 Mbit/s for the data rate, a considerably better quality can be achieved with MPEG-2 than with MPEG-1. The main level is used for the coding of signals in the standard definition of contemporary television (SDTV: standard-definition television, comparable to PAL; EDTV: enhanced-definition television, comparable to ITU-R BT.601). For the coding of HDTV (high-definition television) two levels are available: the high-1440 level and the high level, the latter providing a number of pixels per line appropriate for the 16:9 aspect ratio. However, the transmission of 16:9 signals is possible at all levels (and with all profiles).

The simple profile and the main profile differ from each other only in that in the simple profile bidirectionally predicted pictures are not permissible. Therefore the encoder and decoder are not so costly; however, the loss in coding efficiency is so high that the simple profile most probably will not be used in practice. For the time being, scalability is not possible in either of the profiles, and the chrominance format has been fixed at 4:2:0. The SNR-scalable profile permits the utilisation of SNR scalability, while the spatially scalable profile is added for the spatial scalability. It is the high profile which additionally permits the 4:2:2 chrominance format.

The table as a whole is composed to indicate downward-compatibility, which means that a decoder which can be allocated to a particular box in the table must also be able to decode the data streams of all the profiles and levels situated to the left and below that box. Correspondingly, each MPEG-2 decoder must be able to decode all profiles and all levels of an MPEG-1 data stream. The standard also defines particular characteristics which do not appear in any profile/level combination, such as the 4:4:4 coding. Possible future extensions of the table will be reserved for these. The boxes which have been defined up to now, however, may not be changed. The rationale for defining this matrix of profiles and levels is to make available a practical subset from the multiple tools of the MPEG-2 standard for certain uses. This obviates the necessity to equip all decoders for all eventualities, thus accelerating the implementation of the algorithm in the respective hardware and saving costs.

4.3 Summary

Chapter 4 discusses the video coding in accordance with JPEG, MPEG-1 and MPEG-2. The JPEG standard, which uses the DCT together with subsequent quantisation and redundancy reduction, was primarily developed as a standard for compressing still pictures. However, due to the early availability of efficient and reasonably priced IC solutions it was soon used for coding moving pictures. This so-called Motion JPEG varies, though, from manufacturer to manufacturer, and as a consequence the various solutions are incompatible with one another. In the domain of video editing systems JPEG has secured itself an established place.

Right from the beginning MPEG-1 was conceived for lower-quality applications in computers/multimedia, and the standard is now firmly established in this domain. Because of the attraction of achievable cost reduction by means of data reduction this standard was used beyond its design objectives for the transmission of television signals with a higher quality and data rate before the completion of MPEG-2. Just like JPEG, MPEG-1 uses transformation coding, but supplements this by exploiting the temporal similarities in pictures and by controlling the constancy of the data rate.

Finally, MPEG-2 provides all the necessary tools for the coding of television signals of very different qualities, from SIF (similar to VHS) to HDTV, thus offering the Digital Video Broadcasting (DVB) Project a toolbox for video coding. DVB has compiled the "Implementation guidelines for the use of MPEG-2 systems, video and audio" [ETR 154] which incorporate the list of restrictions that need to be placed on MPEG-2 parameters in oder to ease the implementation of MPEG-2 in DVB. ETR 154 thus warrants the compatibility between equipment from different manufacturers. This has made DVB the first major application of MPEG-2 video coding.

Symbols in Chapter 4

A	signal value of a DPCM pixel
$A(f_x, f_y)$	auxiliary function in the spatial-frequency domain
B	signal value of a DPCM pixel; bidirectionally predicted picture
C	signal value of a DPCM pixel
$C(f_x, f_y)$	scaling constants in the spatial-frequency domain
C_B	digital colour difference signal: blue
C_R	digital colour difference signal: red
f_x	spatial frequency in x-direction
f_y	spatial frequency in y-direction
$G(f_x, f_y)$	spectral coefficients of DCT, transformed from $g(x,y)$
$G_Q(f_x, f_y)$	quantised spectral coefficients of DCT
$g(x,y)$	image signal in x,y spatial domain
I	intraframe-coded image
N, n	power of 2
P	predicted image (unidirectional)
p	multiplier (integer)
$Q(f_x, f_y)$	quantisation step size for DCT coefficients
Q_F	quantisation factor for data-rate control
round	mathematical rounding to integers
x	spatial co-ordinate
Y	digital luminance signal
y	spatial co-ordinate
τ	delay in the DPCM loop

5 MPEG-2 Systems and Multiplexing

Apart from the coding of audio and video signals (see chapters 3 and 4) the MPEG-2 standard [ISO 13818] also defines the multiplexing of audio, video and auxiliary data into one single bit stream, the so-called MPEG-2 System ([TEICHNER, KNOLL]). The joining together, however, is not the only task of the multiplex. It must also provide transmission capacity for information about the current programme or broadcast and about the transmission path, as well as other information required either for the technical servicing or as a navigational aid through the maze of programmes offered to the viewer. This information was defined by the DVB Project ([ETS 468, ETR 211]). Further functions covered by the MPEG-2 System are provisions for clock recovery in the decoder, the synchronisation of video and audio in order to retain the synchronism of the lip movements, and the provision of transmission capacity for conditional-access data (cf. chapter 8). These functions are described in the first part of the MPEG-2 standard entitled "Systems". They cannot be portrayed here in all their dimensions, but the following sections should serve to give an insight into what are sometimes very complex interconnections.

5.1 Differences between Programme Multiplex and Transport Multiplex

The operation of multiplexing is shown as a rough block diagram in figure 5.1. First of all the video, audio and auxiliary data are packetised, i.e. they are divided into relatively large units ("packets") and furnished with controlling information. It is only after this that the data streams are combined into a single one, in which process these so-called "packetised elementary streams" (PES) are divided into smaller packets which are then multiplexed. This step can lead either to a "program stream" (PS) with a single unified time basis or to a "transport stream" (TS) with the possibility of transmitting several different time bases and therefore several programmes in one channel. The main differences can be seen below:

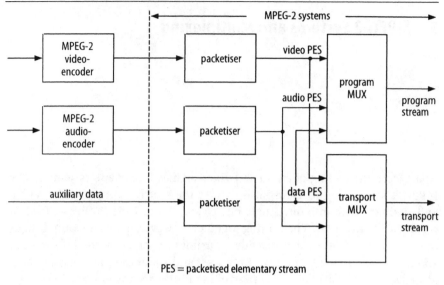

Fig. 5.1. Program multiplex and transport multiplex

Program multiplex:
- all elementary streams have one common time basis
- suitable for use in (relatively) error-free transmission channels (e.g. recording on hard disk)
- packets may be of variable length

Transport multiplex:
- several different time bases are possible
- suitable for use in error-prone channels (i.e. satellite transmission)
- fixed packet length of 188 bytes

The DVB Project has opted for the use of the transport multiplex for television broadcasting in Europe over satellite, cable and terrestrial transmitters, because this is the only one suitable for transmission on error-prone channels. The program multiplex will therefore not be discussed further.

5.2 Positioning of Systems in the ISO/OSI Layer Model

In order to ensure the applicability of the MPEG-2 system bit streams to most existing and future data networks, the Moving Pictures Experts Group has aligned the development of the multiplex with the ISO/OSI layer model ([ISO 7498], ISO = International Standardization Organization, OSI = Open Systems Interconnection). Figure 5.2 is an attempt to show the functionality of

Layer	Definition	Example: written message
7	application layer	reading and understanding
6	presentation layer	arrangement of writing, structuring
5	session layer	alphabet, language
4	transport layer	paper, ink
3	network layer	letterbox
2	data link layer	address on letter
1	physical layer	transportation by the mail company

Fig. 5.2. The ISO/OSI layer model

the model. The example has been kept as instructive as possible; therefore the analogy is not quite flawless.

The seven layers each perform different tasks during the transportation of a message. The seventh layer, the so-called application layer, can be associated with the recipient who reads and understands, as well as with the author who conceives and writes the message. The presentation layer is responsible for structuring the contents of the message. The coding of the text, i.e. the use of the agreed alphabet and the agreed language, takes place in the session layer. In layer four, paper and ink are the media used to make the message ready for despatch. The transition to layer three, i.e. the way to the letterbox, represents the transfer of the completed message to the transport network. The three lowest layers comprise the various functions of the network.

The source coding of the video and audio signals in accordance with MPEG corresponds to the fifth layer, the session layer. The definition of the individual bits in terms of the MPEG video or audio syntax can be seen as an analogy to the language alphabet. The multiplex, i.e. the preparation of the data for transport, corresponds to the manifestation of a written message put on paper and represents the transport layer. The two uppermost layers represent the generating of the programme content by means of camera and microphone or, at the decoder, by playing and viewing the video; hence they do not form part of the MPEG standards. Neither do the three lowest layers, the network layers that provide the transport path for the data, lie within the domain of MPEG. The systems required for these layers were developed by the DVB Project. The service information [ETS 468] required for the completion of the MPEG-2 system documents represents the connection between DVB

ISO/OSI	MPEG-2	
5 session layer	compression layer	– video and audio encoding/decoding
4 transport layer	system layer	**PES packet layer** – synchronisation of bit streams (e.g. video/audio)
		transport packet layer (transport MUX) or pack layer (program MUX) – multiplexing/demultiplexing – buffer management – timing – data transmission/reception

Fig. 5.3. MPEG-2 systems in the ISO/OSI layer model

and MPEG at the level of the transport layer. Just as, in the example of the letter, the precise definition of the interfaces between the layers makes it possible to replace an individual layer or several layers by others that have the same function (in the case of a postal strike, for instance, the letter completed in layer 4 can also be conveyed by a private courier service), so the MPEG-2 transport multiplex can be transmitted via any network capable of a transparent data transmission. That is exactly what was aimed for by specifying the ISO/OSI layer model.

Both layers specified in MPEG-2 are considered more closely in figure 5.3. Whilst layer 5, referred to in the ISO/OSI layer model as the "session layer" and in MPEG-2 usage as the "compression layer", includes the audio and video coding and decoding as described in chapters 3 and 4, the transport layer (or system layer) is divided into two sublayers which perform the various tasks of the multiplex. The PES packet layer is essentially responsible for the synchronisation of the elementary streams. Multiplexing/demultiplexing, buffer management, timing, and the actual transmitting and receiving of the data (which means, the transfer to or from the next-lower layer) take place in the transport packet layer (or, in the case of the program multiplex, in the pack layer).

5.3 End-to-end Synchronisation

A transmission path from the source (i.e. the camera) to the sink (i.e. the display) is shown in figure 5.4. Along the whole path, a constant transmission delay must be ensured so that the individual images of a sequence of moving

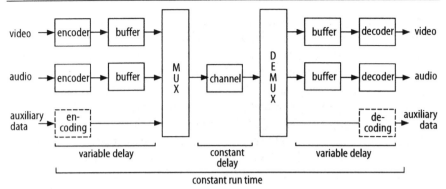

Fig. 5.4. End-to-end synchronisation

pictures can be reproduced correctly at the given frame frequency. In coding and decoding, however, delay times vary; moreover, the image delay is not, as a rule, identical with that of the corresponding audio- and auxiliary information. In order to nevertheless ensure the regular display of the images, all data must be synchronised. This task is fulfilled by the system layer.

A basic prerequisite for decoding is the recovery of the encoding clock in the decoder. An example of the method used for synchronising the decoder is shown by the block diagram in figure 5.5. Following the channel demodulation and the evaluation of the error protection, a transport stream is supplied

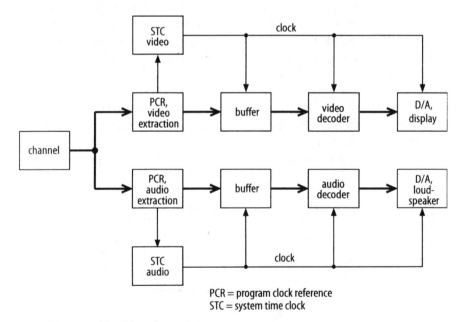

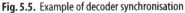

Fig. 5.5. Example of decoder synchronisation

to the decoder and is then fed in its entirety into the separate decoding branches for video and audio. In the case of the video branch, the first step is to eliminate all those data which are not required (e.g. audio data) and, with the aid of the "packet identification" (Packet ID, PID) of the individual packets, to extract only those packets which form part of the image information of the chosen programme. Included in these is the "program clock reference" (PCR). The "system time clock" (STC) compares itself to the reference supplied in the bit stream at least every 0.1 seconds and is corrected if necessary. In this way even the smallest deviations of the crystal-controlled 27-MHz decoder clock are compensated and the required level of precision is achieved.

The structure of the circuit in figure 5.5 with separate clocks for video and audio is an example only. It is, of course, also possible to have only one single extraction block, which generates a common clock and supplies video and audio to two separate outputs. All components belonging to the same programme have the same time reference anyway. Several different time references in one receiver are only necessary when several different programmes have to be decoded simultaneously, for example when there is a "picture in picture" feature or when a video recorder has to be fed separately.

The stabilisation of the decoder clock by the timing reference included in the bit stream is shown in figure 5.6. The "program clock reference" (PCR) is compared with the "system time clock" (STC) in the decoder which was generated by the crystal-controlled 27-MHz system clock. If variations occur, the "numerically controlled oscillator" that generates the system clock is caused to increase or decrease the required clock frequency accordingly. Apart from that, the PCR is loaded into the clock as a supporting value. Between the reference moments, defined by the MPEG-2 standard as having a maximum interval of 0.1 seconds, the clock runs free, a fact which makes a high demand on the precision of the oscillator.

The clock recovery in the decoder is a task to be performed in common for all elementary streams of a programme and is therefore carried out at the transport packet layer, the parent multiplex stage. That is why the PCR from the encoder is not yet inserted at the level of the PES bit streams, but only at

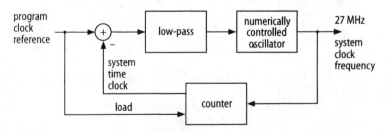

Fig. 5.6. Clock recovery in the decoder

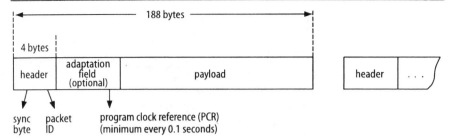

Fig. 5.7. Syntax of the transport stream (TS)

the level of the transport multiplex. A transport packet is shown in figure 5.7. Like every packet, it contains a header whose main task consists in signalling the beginning of the packet through the specified sync byte and in informing the decoder, by means of the packet ID, which type of payload the packet contains. The adaptation field may follow as an optional additional header which, among other things, contains the time reference. The header and the adaptation field contain a great amount of further important signalling information, which cannot be dealt with in detail here. The multiplexer provides for the repetition of the PCR at the prescribed intervals, depending on the data rates for the whole multiplex and the individual data streams. It can be seen that any change in the configuration of the multiplex ("remultiplexing"), for instance in a cable head end, makes a "restamping" of the PCR necessary.

Following the evaluation of the PCR, the decoder clock is in synchronisation with the encoder clock. On this basis the individual processes in the decoder, up to display and loudspeaker, can now be synchronised. This is a problem which has to be solved for each elementary stream individually. Therefore the respective "decoding time stamps" (DTSs) and "presentation time stamps" (PTSs) are to be found at the level of the PESs. Figure 5.8 shows that

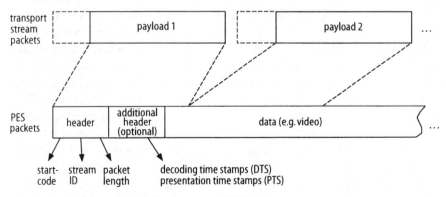

Fig. 5.8. Syntax of the packetised elementary stream (PES)

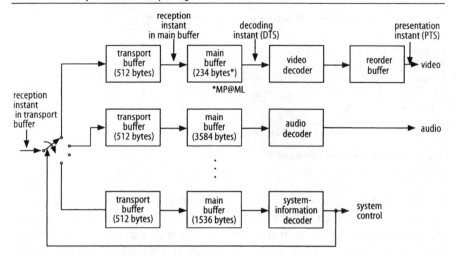

Fig. 5.9. Transport stream system target decoder

the signalling at the PES level is comparable to that at the TS level. The only addition is the indication of the packet length in the header since this length has not been fixed and can vary from packet to packet. Figure 5.9 illustrates the role of DTS and PTS. The PTS indicates at which moment, measured by the 27-MHz clock, the corresponding image should be displayed or the respective audio data be rendered audible to ensure lip sync. The DTS, on the other hand, controls the moment at which the data received must enter the decoder. This is important for the management of the data within the decoding branch.

Apart from the paths for decoding the video and audio information, figure 5.9 shows one more branch, which fulfils the functions that are vital for the system control. For example, the demultiplexing of transport packets, which is indicated by a multiple switch, is controlled by this branch, since the allocation of the packet ID is not predetermined but can be varied by the multiplexer within certain boundaries. In order to make a decoding at all possible under these conditions, the PID o was allocated to the so-called "program association table" (PAT) (figure 5.10). This table contains a breakdown of the multiplex into individual programmes and refers to the associated packet ID. When an MPEG-2 television receiver is switched on, only the packets with the PID o are evaluated to start with, and a list of the available programmes is thus compiled. It is only thereafter that the components of the chosen programme can be fed into the individual decoding branches, that the clock can be regenerated, the decoding carried out, and the video and audio information reproduced.

The MPEG-2 transport multiplex contains some further tables, which will be discussed in the next section.

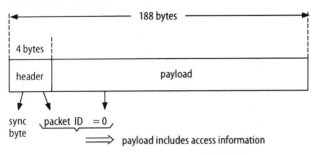

Fig. 5.10. Transmission of program association table (PAT)

5.4 Service Information

As mentioned in the beginning of this chapter, it is necessary to include information in the multiplex which allows the receiver to adjust certain technical parameters and which assists the user in navigating through the program content.

This information can be separated into Programme Specific Information (PSI) and Service Information (SI), both organised in the form of tables. The structure of the tables itself will be discussed in section 5.4.2, but to start with, it is necessary to give a brief overview of the purpose of each table and to explain how these tables are inserted into the Transport Stream. Section 5.4.1 is intended to explain this without looking too closely at the finer details, while section 5.4.2 will give a more detailed insight into the table structures.

5.4.1 PSI/SI Tables and their Insertion into the MPEG-2 Transport Stream

The PSI is defined by the MPEG-2 standard in its Systems part [ISO/IEC 13818-1] and includes 4 tables, as shown in figure 5.11:

- Program Association Table (PAT)
  - contains a list of all programs in the transport multiplex and points to the PIDs of the respective Program Map Tables (PMT)
  - enables the receiver to find the Network Information Table (NIT)
  - is transmitted in packets with the PID-value 0x0000
  - transmission in the actual TS is mandatory
- Program Map Table (PMT)
  - points to the individual PIDs of the respective program and in particular to the packets with the PCR
  - may contain copyright information
  - transmission in the actual TS is mandatory
- Network Information Table (NIT)

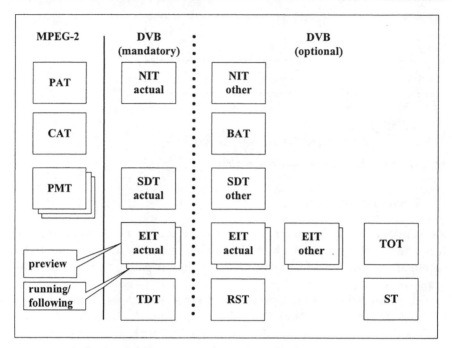

Fig. 5.11. Overview of the PSI/SI tables

- contains private data (i.e. not defined by MPEG but by the user) on the transmission system (e.g. orbit position, transponder number,...)
- content was defined by DVB
- transmission in the actual TS is mandatory, for other TS optional
- Conditional Access Table (CAT)
  - contains private data for conditional access

While the PAT and PMT are necessary to decode the Transport Stream and thus were well defined within the MPEG-2 standard, this standard does not explicitly define the content of the NIT and the CAT. This definition was made by DVB directly, in order to describe the broadcast networks.

Apart from this, the MPEG-2 standard leaves room for the definition of further tables with regards to particular applications. The DVB Project took advantage of this possibility in order to provide an automatic tuning of the Integrated Receiver Decoder (IRD) and to support a convenient user interface in view of the multiplicity of programmes. It specified seven additional tables [ETS 468], [ETR 211] which contain the so-called Service Information (SI). These tables and their contents include:

- Bouquet Association Table (BAT)
  - provides information about the bouquet, such as the bouquet provider, included services, etc.

- services in other TS, even from other networks can belong to one bouquet
- transmission is optional
• Service Description Table (SDT)
  - includes information about the services within the network, i.e. service name, service provider name, text information associated with the service and at least all services included in the actual TS
  - transmission for the actual TS is mandatory
• Event Information Table (EIT)
  - contains the name of the event, start time, duration, information about the present, the following and optionally about future events
  - each service has its own EIT sub-table
  - transmission for the actual TS is mandatory
• Running Status Table (RST)
  - indicates the status (e.g. running/not running) of an event. Information is permanently updated and allows to switch to an event when it starts
• Time And Date Table (TDT)
  - denotes the actual date and time (in UTC)
  - transmission is in a separate table to facilitate frequent updating
• Time Offset Table (TOT)
  - shows the difference between local time and UTC
• Stuffing Table (ST)
  - may invalidate any existing sections at a delivery system boundary e.g. at a cable head-end

Each table can have several sub-tables. The function of the sub-tables varies with the function of the table. Each service in the network has its own EIT sub-table, each TS has its own SDT sub-table and each bouquet in the network its own BAT.

Transmission of all these tables is also done within the MPEG-2 transport stream. But instead of using PES packets, like for the transmission of video, audio and auxiliary data (figure 5.8) the tables are split up into smaller segments – so called sections – and then inserted into the payload of the MPEG-2 transport stream.

These sections can have varying length, but are limited to a maximum length of 1024 bytes for PSI tables (i.e. PAT, PMT and CAT) and 4096 bytes for private sections (i.e. sections which are not defined in their content by MPEG-2 but by the user, e.g. DVB), which are used to transmit the SI tables. Each section has a header of three bytes, in which the first byte denotes the table ID and the next two bytes give additional information like the length of the section and the used syntax. Two different types of syntax are used, the long and the short syntax, as shown in figure 5.12. Only the long syntax supports the description of sub-tables within a header extension. This extension

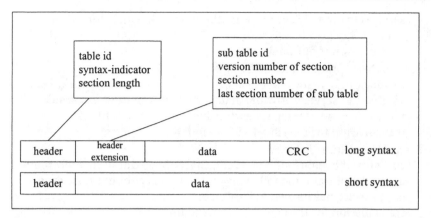

Fig. 5.12. Syntax of Sections

also includes the number of the section, the number of the last section and a version number, all referring to the sub-table which is identified by table ID and table ID extension. A section with long syntax ends with a 32 bit long CRC word, for error detection. The short syntax only mandates the three bytes of the header, the rest of the section may take any form the user determines.

The insertion of the sections into the payload of the TS packets is shown in figure 5.13. Whenever a new section starts within a TS packet, the payload unit start-indicator in the header is set. This indicator signals that the first byte of the payload field includes a pointer towards the position at which the new section starts. If a section is shorter than a TS packet, the rest of the packet can either be filled with stuffing bits or a new section can be started within this TS packet. Longer sections can be distributed over several TS packets.

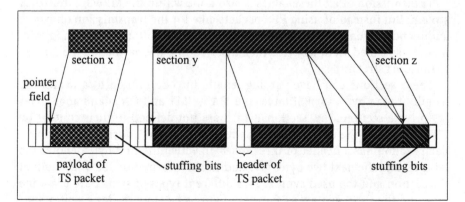

Fig. 5.13. Insertion of sections into TS packets

5.4.2 Section and Table Structure

After introducing the principle of sections and their insertion into the TS the structure of these sections and of the tables which they assemble can be shown.

Figure 5.14 presents the general structure of a section. Each section is defined by a combination of the following fields:

- **table_id:** names the table to which the section belongs (e.g. 0x4A for BAT).
- **section_syntax_indicator:** shows whether a sub-table structure including a CRC-checksum is used or not.
- **section_length:** denotes the length of the section in bytes.
- **subtable_id:** describes which sub-table is described here, e.g. for the BAT it gives the bouquet_id which makes it possible to recognize which bouquet is described in this section of the BAT.
- **version_number:** signals changes of the TS information, i.e. a higher version_number shows the receiver that the content of the section has changed.
- **current_next_indicator:** each section can be marked as „valid now" or „valid in future". This indicator enables transmission of sections which are valid in the future.
- **section_number:** makes sure that the sections are combined into tables in the right order. Numbering of the sections starts with 0x00 so that the maximum number of sections belonging to a table is 256.
- **last_section_number:** denotes the last section of a table so that the receiver knows when a table is received completely.

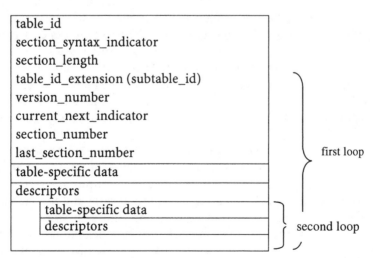

Fig. 5.14. General structure of SI sections

A TS packet carrying a PSI/SI section is marked by a specific packet identifier (PID, see section 5.3). The value of this PID depends on the table to which the section belongs. Values of PIDs and Table Ids for PSI/SI tables are shown in Table 5.1, where the 'x' in the second position of the values denotes that the values are given in hexadecimal notation.

Table 5.1. Identifier values for PSI/SI tables

Table Name	PID	Table ID
Program Association Table (PAT)	0x0000	0x00
Conditional Access Table (CAT)	0x0001	0x01
Program Map Table (PMT)	varying	0x02
Network Information Table, actual network (NIT actual)	0x0010	0x40
Network Information Table, other network (NIT other)	0x0010	0x41
Service Description Table, actual network (SDT actual)	0x0011	0x42
Service Description Table, other network (SDT other)	0x0011	0x46
Bouquet Association Table (BAT)	0x0011	0x4A
Event Information Table (EIT)	0x0012	0x4E – 0x6F
Running Status Table (RST)	0x0013	0x71
Time and Date Table (TDT)	0x0014	0x70
Time Offset Table (TOT)	0x0014	0x73

For the description of the chronological order of events contained within each service, the EIT can take on four classifications (0x4E, 0x4F, 0x5F and 0x6F).

Subsequent to the mandatory elements of the section – table_id, section syntax indicator and section length – one or two loops are included (figure 5.14), depending on the used section syntax. In the case of long syntax, which is used when PSI/SI tables are transmitted, the first loop denotes the different sub-tables described above. The second loop contains descriptors which are used to give information about the elements of the sub-tables. Each of these loops has a specific purpose, depending on the kind of table. In the case of a NIT, the first loop gives information about the network itself, while the second loop describes the respective transport streams within this network.

Description of the elements of the loops is provided by a number of descriptors, i.e. data fields which make it possible to insert information into the tables. For each purpose there are different descriptors, for example the service_name_descriptor giving the name of a service and the name of the provider. By inserting and transmitting only those descriptors that are necessary, the organization of a table becomes very flexible and provides the possibility for future extensions. All descriptors must have the syntax shown in figure 5.15.

descriptor name _descriptor
descriptor_tag
descriptor_length
<descriptor-specific data area>

Fig. 5.15. General structure of descriptors

The descriptor_tag is a value identifying the descriptor (e.g. 0x4E for extended_event_descriptor). Descriptor_length shows the number of bytes following this entry. What then follows is an area that is specific for each descriptor, depending on the elements described by it.

The appearance of a descriptor within a specific table defines the meaning of the description. For example the linkage_descriptor, when inserted into the first loop of the NIT gives a link to another service or to the TS attached to the network operator. It is allowed to appear more than once in this table, in order to be able to link to several other services. In the descriptor-specific data area a linkage_type can be found which gives a further definition of the kind of service to which this descriptor points (e.g. service containing information about the network, an Electronic Program Guide (EPG) for the network or comprehensive Service Information related to the network). Appearance of the linkage_descriptor in the first loop of the BAT on the other hand is used to give a link to another service or TS attached to the bouquet provider. It is also allowed to be used more than once here and also includes a linkage_type for specifying the service that it is linked to. Here the possible usage is to point at services containing information about the bouquet, an EPG for the bouquet or comprehensive SI related to the bouquet.

After the general description of the tables and table elements we will now describe two complete examples of these tables.

5.4.3 Examples of Table Usage: NIT and SDT

The following section shows the structure of the Network Information Table (NIT, figure 5.16) and the Service Description Table (SDT, figure 5.17). In the tables, abbreviations of the names of identifiers are used, having the following meanings:

uimsbf	unsigned integer, most significant bit first
bslbf	bit stream, left bit first
rpchof	remainder polynomial coefficients, highest order first

It is clearly visible, that the first elements (not the values) of both tables are the same, following the structure shown in figure 5.14. Up to the „last_section_number" all the elements in both tables are the same, except the table_id_extension. In the case of the NIT, this extension is named „network_id" and serves as a label to differentiate the described delivery system from any other delivery system while in the case of the SDT, it is named „transport_stream_id" and labels the TS described in the SDT. The loops that follow this more general part of the tables include data that is table-specific.

5.4.3.1 Network Information Table (NIT)

The NIT generally includes sub-tables for each network that has to be described but in the case of the NIT, only one actual network is listed and therefore the table contains only one sub-table and thus only one loop. In contrast the other NIT describes all other networks in the vicinity of the actual network and thus often shows more than one sub-table.

The part of the NIT following the general table information includes in the first loop information about the network. It has to be noted that in the beginning of each loop an overall decriptor_length is given. In conjunction with the individual length of each descriptor following the descriptor_tag (s. figure 5.15), it is possible to separate the different descriptors. Descriptors that are allowed to appear in this loop of the NIT are the following:

- **Network_name_descriptor:** used to indicate the name of a physical network, e.g. „ASTRA"
- **Linkage_descriptor:** appearance of the linkage_descriptor in the first loop means that it links to a service containing additional information (e.g. about the network, an EPG-Application or comprehensive service information.)
- **Multilingual_network_name_descriptor:** may be used to convey the name of the network in additional languages.

In the second loop information about each TS in the network is provided. Transport_stream_id and original_network_id together provide a unique reference to a TS, even if that was transferred to another delivery system from that of the delivery system where it originated. For each TS within the second loop the following descriptors may be given:

- **Delivery_system_descriptors:**
  i.e. satellite_delivery_system_descriptor, cable_delivery_system_descriptor, terrestrial_delivery_system_descriptor, used to transmit the physical parameters of each transport multiplex in the network

	Syntax	No of bits	Identifier
	network_information_section(){		
	table_id	8	uimsbf
	section_syntax_indicator	1	bslbf
	reserved_future_use	1	bslbf
	reserved	2	bslbf
	section_length	12	uimsbf
	network_id	16	uimsbf
	reserved	2	bslbf
	version_number	5	uimsbf
	current_next_indicator	1	bslbf
	section_number	8	uimsbf
	last_section_number	8	uimsbf
	reserved_future_use	4	bslbf
	network_descriptor_length	12	uimsbf
network information sections	descriptors of the first loop: for (i=0; i<N; i++){ descriptor() }		
	reserved_future_use	4	bslbf
	transport_stream_loop_length	12	uimsbf
	2. loop: transport streams: for (i=0; i<N; i++){		
	transport_stream_id	16	uimsbf
	original_network_id	16	uimsbf
	reserved_future_use	4	bslbf
	descr. of the second loop: Transport_descr_length	12	uimsbf
	for (j=0; j<N; j++){ descriptor() }		
	}		
	CRC_32	32	rpchof
	}		

Fig. 5.16. Structure of the Network Information Table (NIT)

- **Service_list_descriptor:** lists the services, identified by ***service_id (MPEG-2 program_number)***. The transport_stream_id and original_network_id, which are necessary to identify a DVB service uniquely, are given at the beginning of the descriptor loop
- **Frequency_list_descriptor:** contains all further frequencies – apart from the frequency given in the delivery_system_descriptor – used in the network
- **Cell_list_descriptor:** lists the geographical location and extension of each cell in a terrestrial network

- **Cell_frequency_link_descriptor:** lists cell_ids together with the frequencies used in these cells. Thus this descriptor is only used in terrestrial networks

5.4.3.2 Service Description Table (SDT)

The SDT has a syntax similar to that of the NIT, but it refers to transport streams and services. The SDT actual, similar to the NIT actual, describes only the presently received transport stream, so that it incorporates only one single sub-table in the first loop.

Syntax				No bits	Identifier
service_description_section(){					
	table_id			8	uimsbf
	section_syntax_indicator			1	bslbf
	reserved_future_use			1	bslbf
	reserved			2	bslbf
	section_length			12	uimsbf
	transport_stream_id			16	uimsbf
	reserved			2	bslbf
	version_number			5	uimsbf
	current_next_indicator			1	bslbf
	section_number			8	uimsbf
	last_section_number			8	uimsbf
	original_network_id			16	uimsbf
	reserved_future_use			8	bslbf
service description sections		for (i=0; i<N; i++){			
		service_id		16	uimsbf
		reserved_future_use		6	bslbf
		EIT_schedule_flag		1	bslbf
		EIT_present_follow_flag		1	bslbf
	2. loop: service	running_status		3	uimsbf
		free_CA_mode		1	bslbf
		descriptors_loop_length		12	uimsbf
		descr. loop	for (j=0; j<N; j++){		
			descriptor()		
			}		
		}			
	CRC_32			32	rpchof
	}				

Fig. 5.17. Structure of the Service Description Table (SDT)

The structure of the SDT is different from that of the NIT because it only includes one descriptor loop in the description of the services. The most important descriptors here are:

- **Bouquet_name_descriptor:** used to transmit the name of the bouquet the service is allocated to. This is allowed more than once in the loop because a service may belong to more than one bouquet. Transmission is optional.
- **CA_identifier_descriptor:** may be used to transmit data of a conditional access (CA) system. It is not part of any CA control function but rather gives an indication whether a service is scrambled. Thus the main purpose of this descriptor is to avoid frustration of users which could be caused by services being displayed for selection but can not be decoded. Usage of it is optional.
- **Country_availability_descriptor:** indicates whether a service is available in the specified country. Usage is optional.
- **Data_broadcast_descriptor:** identifies data broadcast services in the DVB framework.
- **Linkage_descriptor:** appearance of the linkage_descriptor points to a service being attached to the service described in the SDT and containing additional information (e.g. about the actual service, an EPG-Application for the actual service or a replacement service)

A receiver that is switched on can gather all the information needed for decoding a specific program from the PSI/SI tables. From the NIT it gets the data on the network, from the PAT and PMT it derives information on where to find a service.

Information about the content is broken down into more and more detailed layers by using the BAT (data on the bouquet), the SDT (data on the service) and the EIT (on each event within a service). Therefore the receiver can collect the necessary information step by step.

6 Forward Error Correction (FEC) in Digital Television Transmission

In MPEG source coding (see chapters 3 and 4) redundancy and irrelevance are eliminated from the digital signal. However, this means that the signal is more vulnerable to disturbances in the transmission path. If a digital television image which has not been source-coded is transmitted at a data rate of, for example, 270 Mbit/s and one bit has been altered during transmission, only one pixel of the image will be affected in most cases. However, if an MPEG-coded image is transmitted, a bit error can result in at least one erroneous macroblock, if not in several.

A bit error can occur, for example, if because of too much noise a detection threshold in the demodulator is crossed and the value received is thus assigned to a different, incorrect symbol (see chapter 7).

Therefore channel coding is required which provides for forward error correction (FEC). This enables the receiver to correct the errors which have occurred in the transmission path.

In chapter 6 the two relevant methods of error protection in the transmission of digital television will be introduced. For the sake of simplicity, examples will be shown which, although their parameters are not those of the European DVB standard (cf. chapters 9 to 11), will nevertheless enable the reader to easily understand the principle of error correction. It is relatively easy to transfer these parameters to those actually used in the DVB standard (see section 6.4.3).

6.1 Basic Observations

In this section there will be a short introduction to the basics of error correction, without going too deeply into the theory. A more comprehensive insight into the theoretical and mathematical aspects can be obtained from [SWEENEY].

The principle of the transmission with error-protection coding is shown in figure 6.1.

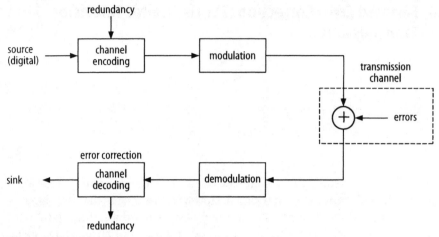

Fig. 6.1. Transmission with error-protection coding

At the transmitter, redundancy is added to the source-coded digital signal in the channel encoder. This is to enable the channel decoder in the receiver to correct any errors. The addition of the redundancy (resulting in the transmission of additional data not included in the source signal) leads to an increase in the data volume to be transmitted. This must be taken into account when choosing the type of modulation in order not to exceed the maximum possible channel capacity.

The digitally modulated signal is overlaid with errors within the transmission channel which are caused by the invalidation of one or more bits. A '1' becomes a '0' and vice versa. The task of channel coding in the receiver is to find the position of the incorrect bits by the evaluation of the redundancy, which is possibly also affected by transmission errors, and to invert these. The added redundancy is then removed.

Figure 6.2 outlines the various types of error which can occur.

Errors occurring singly within a data stream are called bit errors. An n-bit burst error is defined by a block of the length of n bits in which at least the first bit and the last bit are erroneous. Individual bits within this block, however, are not necessarily erroneous. Finally, a symbol error denotes one erroneous symbol, which for example in figure 6.2 has a length of 8 bits (1 byte). Within this symbol up to 8 bit errors can occur in a random constellation.

With the knowledge of the actual types of error occurring in the channel, various codes can be constructed which will successfully correct the most common error types, but less successfully the less common error types. Figure 6.3 gives an overview of the most common classes of codes. It is neither complete nor is every code dealt with individually here (see [CLARK]).

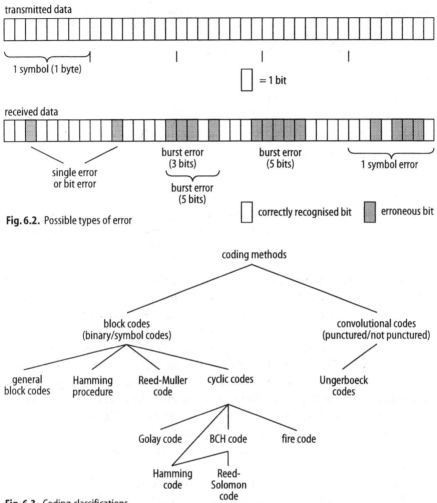

Fig. 6.2. Possible types of error

Fig. 6.3. Coding classifications

The most important criterion is the distinction between the block code and the convolutional code. In the case of block codes the input data stream is divided into blocks of the fixed length m, where m denotes the number of symbols. Such symbols can either be comprised of one bit or, as in the present case, of several bits. In the first case the codes are binary codes and in the second case, symbol codes. A symbol-oriented code is particularly well suited to the correction of symbol errors, concerning which it is of no significance which bit of a symbol is erroneous. The error correction must then not only find the erroneous symbol but also determine the original value of the symbol, whereas with binary codes only the erroneous bit needs to be inverted.

(n,m) block code

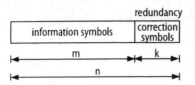

code rate: $R = \dfrac{m}{n} = 1 - \dfrac{k}{n}$

convolutional code

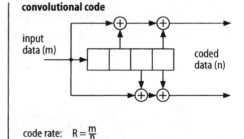

code rate: $R = \dfrac{m}{n}$

Fig. 6.4. Block codes and convolutional codes

In the block codes the calculated redundancy is appended to the actual m information positions in the k correction positions so that finally a block with a length of $n = m + k$ is transmitted. The code rate indicates the ratio between the information m and the transmission symbols n.

The convolutional code differs from the block code in that a binary shift register is always used. There is no division into predetermined segments of the input data stream, but the input information is spread over several output data. This takes place through storage of the input data in a shift register and generation of the output data through a combination of various taps at the shift register. It is because of this type of coding that the convolutional codes are destined for the correction of individual bit errors. The code rate is here defined as the ratio of the number of input bits to the number of output bits which are generated at the same step. As in the case of the block codes, the code rate is always <1, as otherwise no redundancy would have been added. Figure 6.4 shows a comparison of the principles of the block codes with those of the convolutional codes.

Finally, the "bit-error rate" (BER) should be introduced at this point. This is the term used to denote the proportion of erroneously decoded bits to the total number of bits received. Accordingly, the term "symbol-error rate" (SER) denotes the proportion of erroneously decoded symbols to the total number of symbols received.

6.2 Reed-Solomon Codes

Reed-Solomon codes are symbol-oriented block codes. That means, the error correction must not only recognise which symbol of the block of length n is erroneous, but also compute the value of the original symbol. As a rule, one symbol contains 8 bits. To make the following section easier to understand, examples will use 3 bits per symbol.

The Reed-Solomon encoding and decoding require a considerable amount of computation. As we are moving within a finite range of numbers which has 256 elements in the case of 8 bits per symbol or, correspondingly, 8 elements in the case of 3 bits per symbol, the arithmetic to be employed must guarantee that the result of an operation using elements from this number space will again be an element from the same number space. A prerequisite is that all arithmetical operations, i.e. addition, multiplication, subtraction, division, etc. must be valid. It is the arithmetic of the Galois field which is used for the Reed-Solomon code. Therefore it is necessary to understand the system of the Galois field in order to understand the way the Reed-Solomon code works. That is why the following subsection will be an introduction to the arithmetic of the Galois field, in which great care has been taken to choose examples that are easy to understand. Those readers who require more detail and who are interested in the mathematical proofs of the matters to be discussed should consult [SWEENEY], [CLARK]. The examples of the Reed-Solomon code shown here are based on an encoding and a decoding within the frequency domain. This procedure, although corresponding exactly to the definition of the Reed-Solomon code, is not actually used in practice. However, it is considerably easier to understand than the respective procedure with encoding and decoding in the time domain, so that it is more suitable for an introduction to the functional principal of the Reed-Solomon code. The transfer to the procedure which works in the time domain is relatively easy (see section 6.2.5 or [CLARK]). But for this, too, the most important precondition is to understand the Galois field.

6.2.1 Introduction to the Arithmetic of the Galois Field

The Galois field is defined as follows:

> A Galois field, short form GF(q) (here always $q = 2^w$, $w = 3$), is a finite field which contains a set of q different elements. The following rules govern the elements of this field:
> - The arithmetical operations of addition and multiplication are so defined that the combination of two field elements will always result in a field element.
> - The neutral element of addition (o) as well as the neutral element of multiplication (1) are contained within the field.
> - For each element β an additive inverse element $(-\beta)$ and a multiplicative inverse element β^{-1} exist, so that $\beta + (-\beta) = o$ and $\beta \cdot \beta^{-1} = 1$. The subtraction and division of field elements are defined in this way.
> - The associative and commutative laws apply.

At this point a short definition of the modulo operation is appropriate as this is not known to all readers. The modulo operation of two numbers results in the residual quantity of the whole-number division of the two numbers. For example, if the number 7 is divided by 3, the result is 2 and the residual quantity is 1: $7 = 2 \cdot 3 + 1$, or $7 \bmod 3 = 1$.

The result of a modulo operation is always smaller than the argument of the modulo operation, for example a modulo-2 operation can only have 1 or 0 as a result.

In the same way as the modulo operation can be carried out with whole numbers it can be applied to polynomials. The result of a modulo operation with two polynomials is the residual quantity (a polynomial!) which results from a polynomial division. The degree of this (residual) polynomial is therefore, by definition, smaller than the degree of the polynomial through which it was divided.

On the basis of the Galois field as defined above and with the knowledge of the modulo operation we can now assign the following properties to the $GF(2^w)$:

(1) The elements of the $GF(2^w)$ are polynomials of degree $<w$.
(2) The coefficients of the polynomials are 0 or 1.[1]
(3) The addition of two elements is the modulo-2 addition, that is, the EXCLUSIVE-OR linkage (XOR) of the polynomial coefficients which correspond to each other. Therefore an element is equal to its inverse element and the following is valid: $\beta + (-\beta) = \beta + \beta = 0$.
(4) The multiplication of two elements is the multiplication of the polynomials (taking property 3 into account) and the subsequent modulo operation with the generator polynomial $g(x)$ of the Galois field $GF(2^w)$.
(5) The generator polynomial $g(x)$ of the Galois field $GF(2^w)$, comprised of the coefficients 0 or 1, can be freely selected; however, it must be of degree w and irreducible, i.e. incapable of being factorised.[2]

These properties can be illustrated by the following example:

Let the two polynomials

$$\beta_1 = x^2 + x + 1 \quad \in GF(2^3), \quad \text{degree } (\beta_1) = 2 < w = 3 \tag{6.1a}$$

$$\beta_2 = x^2 + 1 \quad \quad \in GF(2^3), \quad \text{degree } (\beta_2) = 2 < w = 3 \tag{6.1b}$$

be defined. Their degree is 2 and their coefficients are 0 or 1. Herewith properties 1 and 2 are satisfied. The addition is now the modulo-2 addition of the corresponding coefficients. Therefore the following is valid:

---

[1] The coefficients are elements of the $GF(2)$.
[2] Even the generator polynomial has coefficients from the $GF(2)$. 'Irreducible' means that the polynomial cannot be broken down into a product of further polynomials unless these have coefficients which are not derived from the $GF(2)$.

$$\begin{aligned}
\beta_1 + \beta_2 &= (x^2 + x + 1) + (x^2 + 1) \\
&= x^2 + x^2 + x + 1 + 1 \\
&= x
\end{aligned} \tag{6.2}$$

The coefficient of x^2 is equal to 1 in both β_1 and β_2. The result of the modulo-2 addition (or XOR operation) is therefore 0. The same is the case for x^0. For the multiplication, a generator polynomial must be defined which has the properties as defined under (5). For this there are tables, for example in [FURRER]. The following polynomial was selected:

$$g(x) = x^3 + x + 1 \quad \text{irreducible, degree } [g(x)] = w = 3 \tag{6.3}$$

The polynomials β_1 and β_2 are now multiplied with each other while still satisfying property 3. The result is then subjected to the modulo operation with the generator polynomial $g(x)$ (once again taking property 3 into account). The result of this modulo-$g(x)$ operation represents the multipliers of β_1 and β_2 in the Galois field.

$$\begin{aligned}
\beta_1 \cdot \beta_2 &= [(x^2 + x + 1)(x^2 + 1)] \bmod (x^3 + x + 1) \\
&= (x^4 + x^3 + x^2 + x^2 + x + 1) \bmod (x^3 + x + 1) \\
&= (x^4 + x^3 + x + 1) \bmod (x^3 + x + 1) \\
&= x^2 + x
\end{aligned} \tag{6.4}$$

As expected, the result is yet again an element of $GF(2^w)$. It has the properties 1 and 2. The modulo operation with the generator polynomial of degree w ensures that the degree of the resulting polynomial is $<w$. One of the properties of all Galois fields is that for each field a so-called primitive field element α exists, the exponents of which can be used to generate or represent all other elements of the field. The Galois field can be completely and uniquely defined by this primitive field element and the generator polynomial as defined by property 5. In most cases the primitive field element α corresponds to the element of the Galois field which is described by the polynomial $\alpha = x$.

On the assumption that $\alpha = x$ is the primitive field element and $g(x) = x^3 + x + 1$ is the generator polynomial, we are listing below – by way of an example – the elements of the respective Galois field $GF(2^3)$.

The calculation of the field elements is as follows: the first element is zero, which in accordance with the definition must be included in the Galois field. All further elements can be represented by the involution of α, for example, α^0, α^1 and α^2. When commencing with the exponent α^3 one has to bear in mind that the degree of the resulting polynomial must be <3. Here the modulo operation with $g(x)$ must be used and this will lead to the result $\alpha^3 = x + 1$.

Table 6.1. Calculation and binary representation
of the elements in a Galois field

Calculation	Binary representation
0	000
$\alpha^0 = 1$	001
$\alpha^1 = x$	010
$\alpha^2 = x^2$	100
$\alpha^3 = x^3 = x+1$	011
$\alpha^4 = x^2+x$	110
$\alpha^5 = x^3+x^2 = x^2+x+1$	111

The calculation of α^4, α^5 and α^6 is done in the same way. When calculating α^7 it can be seen that $\alpha^7 = \alpha^0$, which means that the Galois field is cyclic. $\alpha^x = \alpha^{x \bmod 7}$ or, more generally, $\alpha^x = \alpha^{x \bmod (2^w - 1)}$. The binary notation of the elements of the Galois field can be found in the second column of table 6.1. One element of the Galois field is represented here by 3 bits, one bit each for one coefficient of the polynomial corresponding to the field element. The element which was generated by α^6 is taken as an example. After computing α^6 (by applying the modulo-$g(x)$ operation) the result is $\alpha^6 = x^2 + 1$. The coefficient of x^2 is 1, that of x^1 is 0, and that of x^0 is 1. Therefore the binary notation of the field element is '101'.[3]

On the basis of table 6.1 it is easy to perform the operations of addition and multiplication:

- The addition of two elements corresponds to the EXCLUSIVE-OR linkage (XOR) of the elements in the binary notation.
- The multiplication is the addition and the subsequent modulo-7 operation of the exponents of the powers of α.[4]
- To convert the binary mode to the exponential mode and vice versa, it is necessary to use a table like the one reproduced above.

It is now necessary to reiterate why all calculations are made in the Galois field rather than in a real or natural space of numbers: the aim is to define an error-correcting code which operates on the basis of symbols. Each symbol can take on a certain number of 2^w values. Each calculation or arithmetical link of two such symbols must guarantee that the result of this operation lies within the original range of values.

Provided that the above requirement is taken into account, a discrete Fourier transform (DFT) can now be defined in the Galois field. The mapping of a set into another set, for example the transfer from the time domain to the frequency domain, must now also be effected in a way which ensures that again

---

[3] Also other binary representations are possible. For this see [SWEENEY].
[4] In general a modulo-(2^w-1) operation should be carried out in $GF(2^w)$.

only the elements of the Galois field are present in the image domain. The corresponding formulae for the DFT and its inverse function (IDFT) will only be mentioned here, without giving the mathematical derivations. These can be found, for example, in [SWEENEY]. Here again the Galois field is assumed as being $GF(2^w)$, and its primitive element as being α.

DFT in the Galois field $GF(2^w)$:

$$A_l = \sum_{i=0}^{N-1} a_i \alpha^{+il} \qquad N = 2^w - 1 \qquad l = 0 \ldots N-1 \tag{6.5}$$

IDFT in the Galois field $GF(2^w)$:

$$a_l = \sum_{i=0}^{N-1} A_i \alpha^{-il} \qquad N = 2^w - 1 \qquad l = 0 \ldots N-1 \tag{6.6}$$

When, in the following, the expressions 'time domain' and 'frequency domain' are used, it is in order to differentiate between the two domains of transformation and inverse transformation.

The definition of the Reed-Solomon code is derived from another characteristic of the Galois field which is explained by the following 'theorem of roots and spectral components' and will be illustrated by an example in which the previously defined DFT is used.

The polynomial

$$a(X) = a_{l-1}X^{l-1} + \ldots + a_i X^i + \ldots + a_2 X^2 + a_1 X + a_0 \tag{6.7}$$

from $GF(2^w)$ has a root α^i if and only if the spectral component A_i, equals zero.

A prerequisite is that the coefficients $a_0 \ldots a_{l-1}$ are elements of the Galois field $GF(2^w)$ and that X is the argument of the polynomial $a(X)$, i.e. for X, too, an element of the Galois field $GF(2^w)$ must be substituted. So if α^i is substituted for X and if $a(X) = a(\alpha^i) = 0$ results, it follows that the spectral component A_i, i.e. the coefficient A_i of the transformed polynomial, equals zero.

Strictly speaking, this theorem again describes (6.5), in which a sum is formed for a fixed α^l which corresponds exactly to (6.7). If that sum is zero, i.e. if α^l is a root, then and only then is A_l also equal to zero.

For the inverse transformation the following is valid:

The transformed polynomial

$$A(Z) = A_{l-1}Z^{l-1} + \ldots + A_i Z^i + \ldots + A_2 Z^2 + A_1 Z + A_0 \tag{6.8}$$

has a root α^{-i} if and only if the component a_i of the inverse transform, i.e. the coefficient a_i of the polynomial $a(X)$, equals zero.

Here the coefficients $A_0 \ldots A_{l-1}$ are again elements of the Galois field $GF(2^w)$ and Z is the argument of the polynomial $A(Z)$. If $Z = \alpha^{-i}$ is substituted and results in $a(Z) = A(\alpha^{-i}) = 0$, the coefficient a_i of the inverse transformed polynomial $a(X)$ equals zero.

The specimen calculation below will again be in $GF(2^w)$. For convenience, let us define the polynomial

$$a(X) = \alpha^2 X^6 + \alpha^3 X^5 + \alpha^6 X^4 + \alpha^5 X^2 + \alpha^4 . \tag{6.9}$$

It can be seen that $a_3 = a_1 = 0$. In accordance with the above theorem the transformed polynomial $A(Z)$ must show zero roots for $Z = \alpha^{-3}$ and $Z = \alpha^{-1}$. In order to check this, first the spectral components $A_0 \ldots A_6$ are computed in accordance with (6.5). For the addition and transformation of the two representations of the elements table 6.1 can again be consulted.

$$
\begin{aligned}
A_0 &= \alpha^4 + 0 + \alpha^5 + 0 + \alpha^6 + \alpha^3 + \alpha^2 \\
&= \alpha^4 + \alpha^5 + \alpha^6 + \alpha^3 + \alpha^2 = \alpha^3 \\
A_1 &= \alpha^4 + 0\alpha + \alpha^5\alpha^2 + 0\alpha^3 + \alpha^6\alpha^4 + \alpha^3\alpha^5 + \alpha^2\alpha^6 \\
&= \alpha^4 + \alpha^0 + \alpha^3 + \alpha^1 + \alpha^1 = \alpha^2 \\
A_2 &= \alpha^4 + 0\alpha^2 + \alpha^5\alpha^4 + 0\alpha^6 + \alpha^6\alpha^8 + \alpha^3\alpha^{10} + \alpha^2\alpha^{12} \\
&= \alpha^4 + \alpha^2 + \alpha^0 + \alpha^6 + \alpha^0 = \alpha^5 \\
A_3 &= \alpha^4 + 0\alpha^3 + \alpha^5\alpha^6 + 0\alpha^9 + \alpha^6\alpha^{12} + \alpha^3\alpha^{15} + \alpha^2\alpha^{18} \\
&= \alpha^4 + \alpha^4 + \alpha^4 + \alpha^4 + \alpha^6 = \alpha^6 \\
A_4 &= \alpha^4 + 0\alpha^4 + \alpha^5\alpha^8 + 0\alpha^{12} + \alpha^6\alpha^{16} + \alpha^3\alpha^{20} + \alpha^2\alpha^{24} \\
&= \alpha^4 + \alpha^6 + \alpha^1 + \alpha^2 + \alpha^5 = \alpha^1 \\
A_5 &= \alpha^4 + 0\alpha^5 + \alpha^5\alpha^{10} + 0\alpha^{15} + \alpha^6\alpha^{20} + \alpha^3\alpha^{25} + \alpha^2\alpha^{30} \\
&= \alpha^4 + \alpha^1 + \alpha^5 + \alpha^0 + \alpha^4 = \alpha^2 \\
A_6 &= \alpha^4 + 0\alpha^6 + \alpha^5\alpha^{12} + 0\alpha^{18} + \alpha^6\alpha^{24} + \alpha^3\alpha^{30} + \alpha^2\alpha^{36} \\
&= \alpha^4 + \alpha^3 + \alpha^2 + \alpha^5 + \alpha^3 = \alpha^6
\end{aligned}
\tag{6.10}
$$

The transformed polynomial therefore gives

$$A(Z) = \alpha^6 Z^6 + \alpha^2 Z^5 + \alpha Z^4 + \alpha^6 Z^3 + \alpha^5 Z^2 + \alpha^2 Z + \alpha^3 . \tag{6.11}$$

The substitution of the zero roots to be checked results in

$$A(\alpha^{-1}) = \alpha^6\alpha^{-6} + \alpha^2\alpha^{-5} + \alpha\alpha^{-4} + \alpha^6\alpha^{-3} + \alpha^5\alpha^{-2} + \alpha^2\alpha^{-1} + \alpha^3$$

$$= \alpha^0 + \alpha^{-3} + \alpha^{-3} + \alpha^3 + \alpha^3 + \alpha^1 + \alpha^3$$

$$= \alpha^0 + \alpha^1 + \alpha^3 = 0 \tag{6.12}$$

$$A(\alpha^{-3}) = \alpha^6\alpha^{-18} + \alpha^2\alpha^{-15} + \alpha\alpha^{-12} + \alpha^6\alpha^{-9} + \alpha^5\alpha^{-6} + \alpha^2\alpha^{-3} + \alpha^3$$

$$= \alpha^{-5} + \alpha^{-6} + \alpha^{-4} + \alpha^{-3} + \alpha^{-1} + \alpha^{-1} + \alpha^3$$

$$= \alpha^2 + \alpha^1 + \alpha^3 + \alpha^4 + \alpha^3 = 0$$

QED.

6.2.2 Definition of the RS Code and the Encoding/Decoding in the Frequency Domain

A t-error-correcting RS code, with $q = 2^w$ symbol values and a block length $n = q - 1$, is the set of all words whose spectrum within GF(q) is zero in $k = 2t$ consecutive components.

An RS code vector can now be generated using the 'theorem of roots and spectral components' as given in the previous section.

- First of all, a particular Galois field $GF(2^w)$ is chosen, which is defined by its generator polynomial $g(x)$ and its primitive field element α.
- With this, the block length n ($n = q - 1$) is determined.
- After choosing the error-correction possibility t of the code, k – and therefore also m – is known (remember: $n = m + k$).
- Starting from the spectrum, the k positions $C_0 \ldots C_{2t-1}$ of the block with length n are set to zero. Therefore there are $2t$ consecutive zeros.
- The information symbols are placed in the m positions $C_{2t} \ldots C_{n-1}$.
- If, finally, an IDFT is performed in accordance with (6.6), this operation results in a valid transmittable RS code vector $c(X)$ which corresponds to the above definition.

This procedure is shown again clearly in figure 6.5.

In the transmitting channel the (polynomial) code vector $c(X)$ transmitted is overlaid with a (polynomial) error vector $e(X)$ which, however, is unknown to the receiver.

At the receiving end, which may have received an erroneous vector $r(X)$, the signal spectrum $R(Z)$ can be computed using a DFT. At this point it is already possible to decide whether an error occurred during the transmission. If the positions $E_0 \ldots E_{2t-1}$ are zero, a valid RS code vector has been received (see the above definition of the RS code) and the positions $R_{n-1} \ldots R_{2t}$ contain the original information symbols. Hence, the transmission was error-free. If the positions $E_0 \ldots E_{2t-1}$, or even only one of them, are not zero, the spectrum

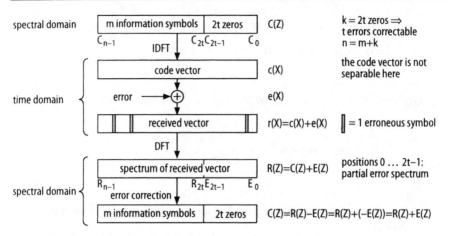

Fig. 6.5. Encoding and decoding of an RS code vector in the frequency domain

$R(Z)$ of the received vector has been additively overlaid with the spectrum $E(Z)$ of the error vector. It is the task of the subsequent error correction to define the spectrum of the error vector, so that

$$R(Z) = C(Z) + E(Z) \tag{6.13}$$

again results in

$$C(Z) = R(Z) - E(Z) = R(Z) + [-E(Z)] = R(Z) + E(Z) \tag{6.14}$$

if $E(Z)$ is known.[5] To understand this clearly we must remember that the positions $R_0 \dots R_{2t-1}$ correspond to $E_0 \dots E_{2t-1}$, as here there were originally zeros. Since the spectrum of the error-vector polynomial additively overlays the zeros, a number of the spectral coefficients of the error vector (namely $E_0 \dots E_{2t-1}$) are available in the receiver immediately after the DFT, ready to be used by the subsequent error correction.

6.2.3 Error Correction Using the RS Code

The task of the error correction is to compute all positions of $E(Z)$ from the known positions $E_0 \dots E_{2t-1}$ of the spectrum $E(Z)$ of the error vector $e(X)$, so that the only thing left to do is to additively combine $E(Z)$ with $R(Z)$ in order to receive the original information $C(Z)$ (see (6.14)). To solve this problem an error-locator polynomial $\lambda(X)$ is introduced to begin with. This is so defined that at the position at which the error-vector polynomial $e(X)$ has a coeffi-

---

[5] In this case the addition of two polynomials corresponds of course again to the modulo-2 linkage of the coefficients which correspond to each other.

cient not equal to zero, i.e. at the position at which an error has occurred, the corresponding coefficient λ_i of $\lambda(X)$ is zero. The precise knowledge of this position, however, is not required for the time being. The transformed error-locator polynomial $\lambda(X)$ is denoted as $\Lambda(Z)$. Therefore the product of λ_i and e_i, as well as its DFT, is:

$$\lambda_i e_i = 0 \quad \xrightarrow{\;DFT\;} \quad \sum_{i=0}^{n-1} \Lambda_i E_{l-i} = 0 \tag{6.15}[6]$$

The following simplifications can be used in general without restriction:

- $\Lambda_0 = 1$
- $\Lambda_i = 0$ for $i > t$: a maximum of t errors should be able to be corrected, therefore there are a maximum of t positions at which $\lambda_i = 0$, thus $\Lambda(Z)$ is the maximum degree of t.[7]

$\Rightarrow i = 0 \ldots t$

$\Rightarrow l = t \ldots 2t - 1$, when at first only $E_0 \ldots E_{2t-1}$ are known.

Using these simplifications, (6.15) can be rewritten as the following set of equations:

$$
\begin{aligned}
\Lambda_0 E_t &+ \Lambda_1 E_{t-1} &+ \ldots + \quad \Lambda_t E_0 &= 0 \\
\Lambda_0 E_{t+1} &+ \Lambda_1 E_t &+ \ldots + \quad \Lambda_t E_1 &= 0 \\
&&\vdots& \\
\Lambda_0 E_{2t-1} &+ \Lambda_1 E_{2t-2} &+ \ldots + \quad \Lambda_t E_{t-1} &= 0
\end{aligned}
\tag{6.16}
$$

In this so-called key equation for $\Lambda(Z)$ both $E_0 \ldots E_{2t-1}$ and Λ_0 are known. It comprises t equations with t unknowns and can therefore be solved.

In the following the above simplifications shall apply. Moreover it can be assumed that $\Lambda_0 \ldots \Lambda_t$ are known after the solution of the set of equations. Then (6.15) can be once again rewritten as

$$\sum_{i=0}^{n-1} \Lambda_i E_{l-i} = \Lambda_0 E_1 + \sum_{i=1}^{n-1} \Lambda_i E_{l-i} = 0 \quad \Rightarrow \quad E_1 = \sum_{i=1}^{t} \Lambda_i E_{l-i} . \tag{6.17}[8]$$

On the basis of this equation each E_l can be recursively computed from $E_{l-1} \ldots E_{l-t}$. This directly leads to a feedback shift register implementation (recursive extension) in accordance with figure 6.6.

First, the registers are loaded with $E_t \ldots E_{2t-1}$ in accordance with $R_t \ldots R_{2t-1}$. With every cycle, the following E_l is now to be found at the output. All coeffi-

---

[6] In the space of real numbers the summation shown here is also known as discrete convolution.

[7] See 'theorem of roots and spectral components'.

[8] Taking into account $\Lambda_0 = 1$, $-E_i = E_i$ and $i = 1 \ldots t$.

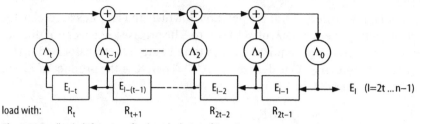

Fig. 6.6. Feedback shift register for the calculation of $E_{2t} \ldots E_{n-1}$

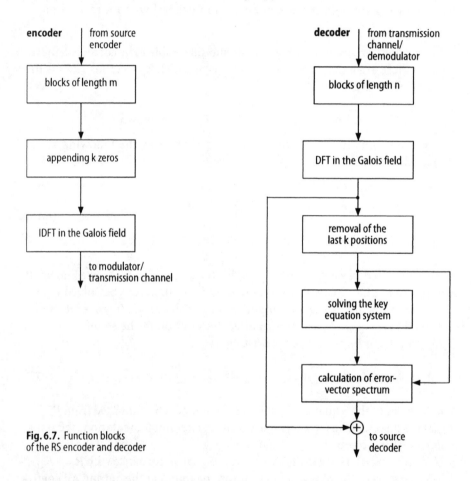

Fig. 6.7. Function blocks of the RS encoder and decoder

cients of $E(Z)$ can be calculated in this way. By adding $R(Z)$ and $E(Z)$ one finally arrives at $C(Z)$ (see figure 6.5).

Figure 6.7 summarises the procedure of encoding and decoding Reed-Solomon code vectors in the frequency domain.

6.2.4 Examples of Encoding/Decoding in the Frequency Domain

We shall now explain, by using a simple example, how the Reed-Solomon code works. For a better understanding of the process the reader is referred to figure 6.5. We are dealing with a $(7,3)$-RS code which is defined in $GF(2^3)$. $\alpha = x$ is defined as the primitive field element, the generator polynomial of the Galois field is $g(x) = x^3 + x + 1$. The code can correct two errors and thus transmit 3 information symbols. The three information symbols α^5, α^2 and α^3 which are to be transmitted are written in the positions $C_6 \ldots C_4$. The last four positions $C_3 \ldots C_0$ are set to zero.

$$\left.\begin{array}{l} C_0 \ldots C_3 = 0 \\[4pt] C_4 = \alpha^3 \\[4pt] C_5 = \alpha^2 \\[4pt] C_6 = \alpha^5 \end{array}\right\} \Rightarrow C(Z) = \alpha^5 Z^6 + \alpha^2 Z^5 + \alpha^3 Z^4 \tag{6.18}$$

By an IDFT in accordance with (6.6) the transmitting code vector $c(X)$ is obtained.

$$\left.\begin{array}{l} c_0 = \alpha^3 + \alpha^2 + \alpha^5 = 0 \\[4pt] c_1 = \alpha^3\alpha^{-4} + \alpha^2\alpha^{-5} + \alpha^5\alpha^{-6} = \alpha^4 \\[4pt] c_2 = \alpha^3\alpha^{-8} + \alpha^2\alpha^{-10} + \alpha^5\alpha^{-12} = 0 \\[4pt] c_3 = \alpha^3\alpha^{-12} + \alpha^2\alpha^{-15} + \alpha^5\alpha^{-18} = \alpha^5 \\[4pt] c_4 = \alpha^3\alpha^{-16} + \alpha^2\alpha^{-20} + \alpha^5\alpha^{-24} = \alpha^6 \\[4pt] c_5 = \alpha^3\alpha^{-20} + \alpha^2\alpha^{-25} + \alpha^5\alpha^{-30} = \alpha \\[4pt] c_6 = \alpha^3\alpha^{-24} + \alpha^2\alpha^{-30} + \alpha^5\alpha^{-36} = \alpha^4 \end{array}\right\} \Rightarrow c(X)$$

$$c(X) = \alpha^4 X^6 + \alpha X^5 + \alpha^6 X^4 + \alpha^5 X^3 + \alpha^4 X \tag{6.19}$$

In the channel, the error $e(X)$ then overlays the code vector, and the result is the received vector $r(X)$:

$$e(X) = \alpha^2 X^5 + \alpha X \tag{6.20}^9$$

$$\Rightarrow r(X) = c(X) + e(X) = \alpha^4 X^6 + \alpha^4 X^5 + \alpha^6 X^4 + \alpha^5 X^3 + \alpha^2 X.$$

The spectrum $R(Z)$ of the received vector is calculated in accordance with (6.5) as

---

[9] $e(X)$ has here been chosen at random.

$$\left.\begin{array}{l}
R_0 = \alpha^2 + \alpha^5 + \alpha^6 + \alpha^4 + \alpha^4 = \alpha^4 \\
R_1 = \alpha^2\alpha + \alpha^5\alpha^3 + \alpha^6\alpha^4 + \alpha^4\alpha^5 + \alpha^4\alpha^6 = \alpha^6 \\
R_2 = \alpha^2\alpha^2 + \alpha^5\alpha^6 + \alpha^6\alpha^8 + \alpha^4\alpha^{10} + \alpha^4\alpha^{12} = \alpha^2 \\
R_3 = \alpha^2\alpha^3 + \alpha^5\alpha^9 + \alpha^6\alpha^{12} + \alpha^4\alpha^{15} + \alpha^4\alpha^{18} = \alpha^6 \\
R_4 = \alpha^2\alpha^4 + \alpha^5\alpha^{12} + \alpha^6\alpha^{16} + \alpha^4\alpha^{20} + \alpha^4\alpha^{24} = \alpha^4 \\
R_5 = \alpha^2\alpha^5 + \alpha^5\alpha^{15} + \alpha^6\alpha^{20} + \alpha^4\alpha^{25} + \alpha^4\alpha^{30} = \alpha^2 \\
R_6 = \alpha^2\alpha^6 + \alpha^5\alpha^{18} + \alpha^6\alpha^{24} + \alpha^4\alpha^{30} + \alpha^4\alpha^{36} = 0
\end{array}\right\} \Rightarrow R(Z)$$

$$R(Z) = \alpha^2 Z^5 + \alpha^4 Z^4 + \alpha^6 Z^3 + \alpha^2 Z^2 + \alpha^6 Z + \alpha^4 \qquad (6.21)$$

As the last four positions $R_0 \dots R_3$ of the spectrum $R(Z)$ are not all zero, they are bound to be overlaid with an error vector whose positions $E_0 \dots E_3$ directly correspond to the received and transformed positions $R_0 \dots R_3$. Using $E_0 \dots E_3$ we can state and solve the key equation.

$$\left.\begin{array}{l}
\Lambda_0\alpha^2 + \Lambda_1\alpha^6 + \Lambda_2\alpha^4 = 0 \\
\Lambda_0\alpha^6 + \Lambda_1\alpha^2 + \Lambda_2\alpha^6 = 0
\end{array}\right\} \Rightarrow \Lambda_0 = 1, \ \Lambda_1 = \alpha^6, \ \Lambda_2 = \alpha^6 \qquad (6.22)$$

The values of $\Lambda_0 \dots \Lambda_2$ determine the coefficients in the feedback branches of the shift register (see figure 6.8).

If the latter is loaded with E_3 and E_2, the spectral coefficients at the output are $E_4 \dots E_6$ and thus part of the error vector. They result in $E_4 = \alpha^6$, $E_5 = 0$ and $E_6 = \alpha^5$. Finally, the original information can be reconstructed if $E(Z)$ and $R(Z)$ are known:

$$E(Z) = \alpha^5 Z^6 + 0Z^5 + \alpha^6 Z^4 + \alpha^6 Z^3 + \alpha^2 Z^2 + \alpha^6 Z + \alpha^4$$

$$R(Z) = 0Z^6 + \alpha^2 Z^5 + \alpha^4 Z^4 + \alpha^6 Z^3 + \alpha^2 Z^2 + \alpha^6 Z + \alpha^4 \qquad (6.23)$$

$$C(Z) = E(Z) + R(Z) = \alpha^5 Z^6 + \alpha^2 Z^5 + \alpha^3 Z^4$$

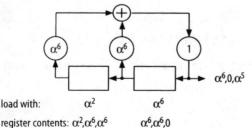

load with: α^2 α^6

register contents: $\alpha^2, \alpha^6, \alpha^6$ $\alpha^6, \alpha^6, 0$

Fig. 6.8. Example of the structure of a feedback shift register for the definition of $E_4 \dots E_6$

6.2.5 Encoding and Decoding in the Time Domain

The procedure introduced so far performs the RS encoding and decoding in the frequency domain. The advantage of this procedure is that it can be easily understood. It was chosen to demonstrate the basic modus operandi of the RS code to the reader. The disadvantage lies in its considerable implementation requirement, as in both the transmitter and the receiver a transformation has to be effected. Further, the original information is not identifiable in the transmitting code vector, i.e. the code vector is not separable into information and checking symbols.

These disadvantages are avoided if the encoding and the decoding, which are only outlined here, are performed in the time domain (a more detailed description can be found, for example, in [CLARK]):

– The arithmetical operations in the Galois field, the generator polynomial of the Galois field, and the primitive field element α are the same as in the encoding and decoding in the frequency domain.
– As a first step, the information symbols are placed in the first m positions of the code vector to be transmitted.
– The encoding takes place with the aid of a code-specific generator polynomial which is constructed in such a way that its zero positions are bound to produce $2t$ consecutive zeros in the spectrum.
– If the previously generated polynomial, in which the information is stored in the first m positions, is divided modulo the generator polynomial and if the result is transferred to the last k positions of the code vector, there is a guarantee that the code vector can be divided by the generator polynomial without a remainder and that it therefore has the same roots. Thus it is a valid code vector in accordance with the definition in section 6.2.2 (its spectrum containing $2t$ consecutive zeros).
– During the decoding a so-called syndrome is initially calculated with the aid of which the key equation for $\Lambda(Z)$ is set up.
– One can solve this key equation by using Euclid's or Berlekamp's algorithms. In addition, a so-called error-evaluator polynomial $\Omega(Z)$ is generated.
– By substituting all powers of α for $\Lambda(Z)$ and $\Omega(Z)$ and by evaluating the results, one obtains the information needed about place and value of the error in the received vector.

Although the theory is somewhat more difficult to understand than that of the encoding/decoding in the frequency domain, which is mainly due to the introduction of another polynomial and the evaluation of the polynomials $\Lambda(Z)$ and $\Omega(Z)$, the implementation is much easier to realise. At the transmitting end no transformation is any longer required, and at the receiving end

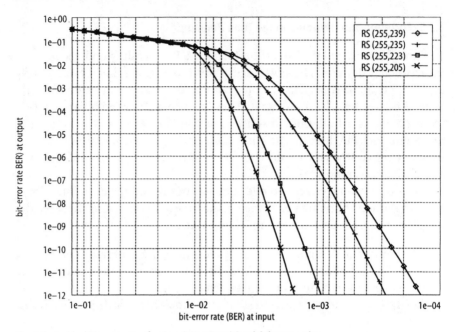

Fig. 6.9. Residual bit-error rate of various (255, 255-2t) Reed-Solomon codes

the only transformation to be done is that of the $2t$ characters of the syndrome.

6.2.6 Efficiency of the RS Code

Independently of whether the encoding or decoding takes place in the time or the frequency domain, the efficiency of both procedures with regard to their correctability is the same. Figure 6.9 shows the curve of the residual bit-error rate, i.e. the bit-error probability at the decoder output, using various RS codes as defined in $GF(2^8)$. Although the RS codes are symbol-oriented codes, the analysis of the efficiency takes bit errors into account. On the assumption that the bit errors are evenly distributed a statement can be made about the symbol-error rate on the basis of which the efficiency of an RS code can be analysed.

The efficiency of the code increases, of course, with an increase in the number of test symbols. In this way, at an input bit-error rate of $2 \cdot 10^{-3}$ the residual bit-error rate of the RS(255, 205) code is approx. $1 \cdot 10^{-10}$ – the coding gain is thus more than 10 to the power of 7 –, whereas in the case of the RS(255, 239) code at the same input bit-error rate of $2 \cdot 10^{-3}$ the output bit-error rate is $9 \cdot 10^{-4}$, the coding gain thus being only slightly greater than 0,5. The correction limit is reached when the bit-error rate at the output is greater than, or the same as, the input bit-error rate. It is in the order of $7 \cdot 10^{-3}$ at RS(255, 205) and in the order of $2 \cdot 10^{-3}$ at RS(255, 239). From this limit onwards more errors occur than the code can correct. As a consequence, additional errors occur which are caused by the corrective algorithm in the output data stream, which leads to the bit-error rate at the output of the decoder being greater than at the input.

For all DVB transmission standards a modified (shortened) RS(255, 239) code is used which makes it possible to guarantee a residual bit-error rate of approx. $1 \cdot 10^{-11}$ at an input bit-error rate of $2 \cdot 10^{-4}$ while correcting up to 8 symbol errors per block (see chapters 9 to 11).

6.3 Convolutional Codes

6.3.1 Basics of the Convolutional Codes

As already explained in section 6.1, the convolutional codes are binary codes in which the information is spread over several transmitting symbols. A convolutional code is therefore always bit-oriented. The information, consisting of individual bits, is fed into a shift register in order to be encoded. The transmitting signal is obtained by combining various taps at the shift register. Figure 6.10 shall serve as an explanation of the encoding and the technical terms used here.

Using the relatively simple convolutional encoder, as shown in figure 6.10, for each individual bit fed into the shift register two bits are generated as output symbols. The number of input lines is $m = 1$ bit, the number of output lines is $n = 2$ bits, and the code rate $R = m/n = 1/2$. The memory, that is, the storage depth of the encoder, is defined as the number of the preceding bits which contribute to the encoding of the current bit. In this example this would be $S \cdot m = 4$, with $S =$ length of the shift register. The constraint length K

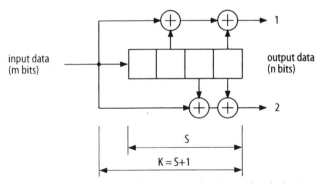

Fig. 6.10. Model construction of a convolutional code to explain the basic terms

describes the total number of bits contributing to the coding process. In the example $K = (S + 1) \cdot m = 5$. If the convolutional encoder is conceived as an automaton, as known from informatics, it can be characterised by the number of possible internal states. This number can be determined from the number of memory elements and the corresponding number of possible i/o combinations. The convolutional encoder shown therefore has $2^{S \cdot m} = 16$ possible states. Finally, a convolutional encoder is characterised by the number and positioning of the taps at the shift register, which are indicated by generator polynomials G the coefficients of which are 0 or 1, according to whether there was a tap at the respective position or not.[10] It is common practice to combine the coefficients as octal numbers. Polynomials or octal numbers must be indicated individually for each output branch. The characteristic parameters of this example of a convolutional encoder can be summarised as follows:

Number of input lines	m	$= 1$
Number of output lines	n	$= 2$
Code rate	R	$= m/n = 1 / 2$
Memory	$S \cdot m$	$= 4$
Possible states	$2^{S \cdot m}$	$= 16$
Constraint length	K	$= (S+1) \cdot m = 5$
Generator polynomial 1	G_1	$= 1+X^2+X^4$ (25_{OCT})
Generator polynomial 2	G_2	$= 1+X^3+X^4$ (31_{OCT})

Figure 6.11 shows a further example of a convolutional encoder which, with a number of input lines of $m = 2$, uses two shift registers with a length of $S = 3$.
The characteristic parameters of this convolutional encoder are:

Number of input lines	m	$= 2$
Number of output lines	n	$= 3$
Code rate	R	$= m/n = 2/3$
Memory	$S \cdot m$	$= 6$
Possible states	$2^{S \cdot m}$	$= 64$
Constraint length	K	$= (S+1) \cdot m = 8$
Generator polynomial 1	G_1	$= 1+X^2+X^3$ (15_{OCT})
Generator polynomial 2	G_2	$= 1+X^1+X^2$ (07_{OCT})
Generator polynomial 3	G_3	$= 1+X^1+X^3$ (13_{OCT})

---

[10] The LSB of the generator polynomial describes the input of the shift-register cell, whereas the MSB describes the tap at the most delayed shift-register cell. See [PROAKIS].

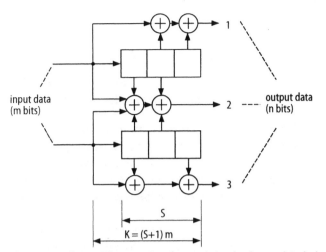

Fig. 6.11. Further model construction of a convolutional code to explain the basic terms

6.3.2 Examples of Convolutional Encoding and Decoding

6.3.2.1 Construction of a Model Encoder

In this section a very simple convolutional encoder is introduced, by means of which an example of encoding and decoding will be carried out in section 6.3.2.3. Figure 6.12 shows the construction of this encoder. It is to be noted that the input data are fed in from the right, as opposed to those in figures 6.10 and 6.11. In this way the state value of the shift register can be used, for example, to represent a state diagram (see next section).

The characteristics of this convolutional encoder are:

Number of input lines m $= 1$
Number of output lines n $= 2$
Code rate R $= m/n = 1/2$
Memory $S \cdot m$ $= 2$
Possible states $2^{S \cdot m}$ $= 4$

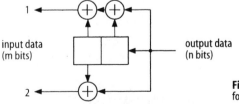

Fig. 6.12. Convolutional encoder as a basis for the following encoding and decoding example

Constraint length K $= (S+1) \cdot m = 3$
Generator polynomial 1 G_1 $= 1+X^1+X^2$ (7_{OCT})
Generator polynomial 2 G_2 $= 1+X^2$ (5_{OCT})

6.3.2.2 State Diagram and Trellis Diagram of the Model Encoder

In section 6.3.1 it was briefly mentioned that a convolutional encoder can also be considered an automaton. First of all, an automaton is characterised by the number of its internal states. As a function of the current input symbol and of its current internal state the automaton outputs one or more symbols – here bits – and changes to a new state. This can usually be represented by a state diagram as shown in figure 6.13.

The 1/0 combinations in the circles describe the state of the automata or the actual contents of the shift register. Two transition arrows diverge from each state and a further two converge to it. The first bit of the 1/0 combination at the transition arrow describes the input bit. The last two bits are the data of the outputs 1 and 2.

Commencing with the state "00" (i.e. the shift register contains "0" in each cell), let us feed in a "1" at the input. In accordance with figure 6.12, the output signals are "11". This process is represented in the state diagram by the left-hand arrow pointing from the bottom to "01", the designation of which is, logically, "1/11", between the states "00" and "01". The encoder is now in the state "01".

Another way to document the same combinations is the trellis diagram. Here the states are plotted below one another and time-sequentially next to each other (see figure 6.14).

From every state two paths lead to a new state each, depending on whether a "1" (thin line) or a "0" (bold line) has been input. The output data is registered at the transition lines. If, once again, we take state "00" as the starting

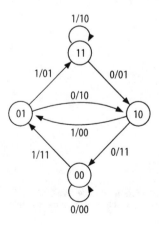

Fig. 6.13. State diagram of the model encoder

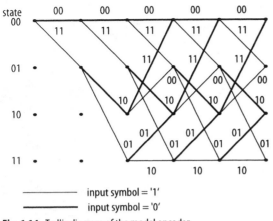

Fig. 6.14. Trellis diagram of the model encoder

position and read in a "1", we obtain state "01" (moving along the thin line) while reading out a "11". This corresponds to the state diagram represented in figure 6.13.

The following examples will be explained on the basis of the trellis diagram and the state diagram.

6.3.2.3 Example of Encoding with Subsequent (Viterbi) Decoding

Let the information sequence to be coded and transmitted be "1011000". In the transmission channel, let two positions be overlaid with an error, leading to the inversion of the respective bits. The transmission process can thus be summarised as follows:

Information (assumed):

1	0	1	1	0	0	0

Coded bit sequence (to be transmitted):

11	10	00	01	01	11	00

Error vector (assumed):

01	00	10	00	00	00	00

Received sequence:

10	10	10	01	01	11	00

The total Viterbi decoding is explained on the basis of the trellis diagrams in figure 6.15 together with the state diagram in figure 6.13. As opposed to the encoding, in the decoding process the input and the output of the state diagram now have to be exchanged.

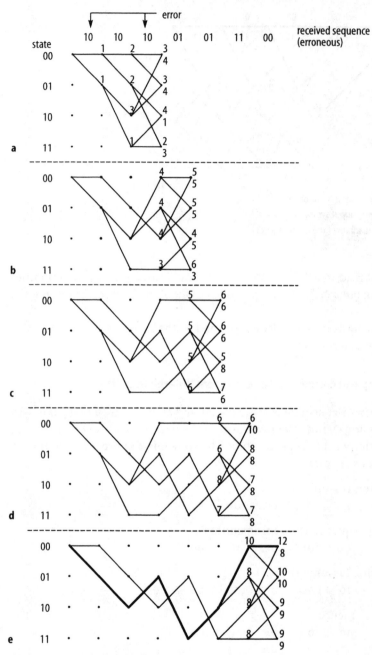

Fig. 6.15. Trellis diagrams for Viterbi decoding

The decoding takes place as follows:

(1) The decoder is in the state "00" and receives the bit sequence "10". As shown in the state diagram, the encoder could not have generated this bit sequence, because commencing with the state "00" there are only two alternative possibilities:
- Sending a "00", and keeping the state "00" (in the trellis diagram in 6.15a the top left-hand horizontal transition line). The decoder 'knows' that in this case only one correct bit was received. As the sum of the correct bits a "1" is recorded in the transition path.
- Sending a "11", and transition to state "01" (in the trellis diagram in 6.15a the diagonal line from the top left-hand side). The decoder 'knows' also in this case that only one correct bit was received. A "1" is again recorded as the sum of the correct bits.

(2) The decoder again receives the bit sequence "10".
- Starting from state "00" when decoding the second input bit sequence, the reception of a "00" and the retention of state "00" is again expected or, alternatively, the reception of a "11" and a transition to state "01". This step, too, would produce one incorrect bit. In each of the two cases 2 correct bits have now been recognised (out of 4 bits received in the meantime).
- Starting from state "01", the reception of a "01" with a transition to state "11" and one correct bit still in the sum, or the reception of a "10" with a transition to state "10" and hence a total of 3 correct bits is expected.

At this point we can introduce the term 'metric'. In the present case the metric Δ denotes the sum of the correctly received bits, while following a chosen path through the trellis diagram. The larger the metric, the higher the probability that the path through the trellis diagram will correspond to the path that has been followed through the encoder. For a mathematical description of the metric see section 6.3.3, (6.24) to (6.26).

(3) After the third bit sequence "10" has been received (see figure 6.15a), all possible transitions between the states have been analysed, and the input bit sequence has been compared with the expected reception values, the case occurs that in each state position two transitions converge. The principle of the Viterbi decoder (see [VITERBI]) is now to choose precisely that transition out of two which has the larger metric or, in other words, to delete the transition with the lower metric and, as the case may require, the preceding transitions as well, since this is the less probable path through the trellis diagram. The result can be seen in the trellis diagram shown in figure 6.15b. Should two transitions with the same metric converge, then a transition can be chosen at random, as from this position backwards there

is no unambiguous decision possible. For a proof of the above the reader is referred to [VITERBI].

(4)–(6) By processing the following received bit sequences "01", "01" and "11" and successively deleting the transitions with lower metrics a path through the trellis diagram emerges whose metric is larger than that of the other paths (figures 6.15b to 6.15d).

(7) The last bit sequence received is "00". The transitions to be chosen can be seen in figure 6.15e.

The path of the metric $\Delta = 12$ through the trellis diagram is now the most probable (drawn boldly in figure 6.15e). By retracing, one finally obtains the most probable state sequence, and with the aid of the state diagram one obtains the original data sequence:

Most probable state sequence:

00　　01　　10　　01　　11　　10　　00　　00

Corrected received sequence:

11　　10　　00　　01　　01　　11　　00

Decoded information sequence:

1　　0　　1　　1　　0　　0　　0

The errors in the received sequence have thus been corrected. In addition, by comparing the highest metric with the number of bits received, a statement can be made about the number of errors which occurred (in this case: 14 – 12 = 2 errors) and therefore also about the actual state of the transmission channel.

6.3.3 Hard Decision and Soft Decision

Figure 6.16 shows the generalised probability density of an originally binary received signal which has been transmitted over a disturbed channel. Either the symbol $x_i = 0$ or the symbol $x_i = 1$ has been transmitted. Due to noise in the transmission channel no signal with discrete states has been received but,

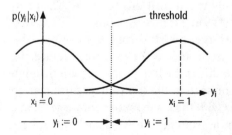

Fig. 6.16. Probability density of the received signal and hard-decision threshold

instead, a signal y_i with a wide range of values and with the conditional probability density function $p(y_i | x_i)$.

In hard-decision decoding, the range of y_i is divided by a threshold. For all reception values of y_i which are above the threshold, $y_i := 1$ is assumed. For all others $y_i := 0$ is valid. The metric δ_i for the comparison of two bits in hard-decision decoding is therefore

$$\delta_i = \begin{cases} 1 & \text{for } x_i = y_i \\ 0 & \text{for } x_i \neq y_i \end{cases}. \tag{6.24}$$

In soft-decision decoding, on the other hand, several interim states for y_i are evaluated. The range of values for y_i is divided by several thresholds and quantised as represented, for example, in figure 6.17 which depicts a 3-bit soft decision.

The range of the metric for the comparison of two bits x_i and y_i in soft decision extends from 0 to 1 in 8 steps and is identified by the parameter d_i (cf. figure 6.17):

$$\delta_i = \begin{cases} d_i & \text{for } x_i = 1 \\ 1 - d_i & \text{for } x_i = 0 \end{cases}. \tag{6.25}$$

The metric Δ of a path through the trellis diagram results in both cases (hard decision and soft decision) from the sum of the metrics δ_i of the bit comparisons, as conveyed in the description of the decoding.

$$\Delta = \sum_i \delta_i \tag{6.26}$$

We can see that in soft-decision decoding non-integer metrics are also possible. This leads to a far more accurate estimation of the probability that a chosen path through the trellis diagram is correct. The typical coding advantage, which is to be gained by the use of soft decision in the decoder, is in the range of 2 dB (cf. figure 9.11, chapter 9).

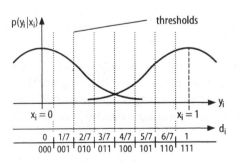

Fig. 6.17. Probability density of the received signal and soft-decision thresholds

6.3.4 Puncturing of Convolutional Codes

One of the disadvantages of convolutional codes is the low code rate, which is $R = 1/2$ in the encoder used here. This means that there are twice as many bits transferred than the actual information would require, or to state it differently: the data stream contains 50% redundancy. By puncturing, the code rate can be increased, which of course increases the correction requirement. As an example, we will again revert to the information sequence from section 6.3.2.3. If the model encoder is used, this information sequence results in the coded data of rate $R = 1/2$. In that now every third bit of the coded data sequence is not transmitted, or is punctured, the code rate increases to $R = 3/4$. For the purpose of explaining the puncturing, an error-free transmission is here assumed. The decoder tries to reconstruct the original coded data of rate $R = 1/2$ by assuming an X (= "don't care") after each second symbol received. When calculating the metric for the Viterbi decoding each symbol received is used in accordance with the rules of soft decision in section 6.3.3, whilst for "don't care" a metric of $\delta_x = 0{,}5$ is assumed. The principle of encoding, including puncturing and decoding, is recapitulated in the following table.

Information sequence (assumption):

1	0	1	1	0	0	0

Coded data, rate 1/2:

11	10	00	01	01	11	00

Puncturing to rate 3/4:

11	$_1$0	^{0}0	01	0^1	1$_1$	00

Data transferred (here error-free):

11	00		01	11		00

Reconstruction for decoding:

11	Xo	oX	01	X1	1X	00

Metric δ_i with soft decision:

$\delta_1 \, \delta_2$	$0{,}5 \, \delta_4$	$\delta_5 \, 0{,}5$	$\delta_7 \, \delta_8$	$0{,}5 \, \delta_{10}$	$\delta_{11} \, 0{,}5$	$\delta_{13} \, \delta_{14}$

6.3.5 Performance of Convolutional Codes

In figure 6.18 the residual bit-error rate of convolutional codes of rate $R = 1/2$ is plotted as a function of E_b/N_0. E_b/N_0 is defined as the energy E_b, which is transmitted per bit, divided by the noise-power density N_0 of the white Gaussian noise with which the signal on the transmission channel is overlaid.[11]

---

[11] For further explanations see chapter 7.

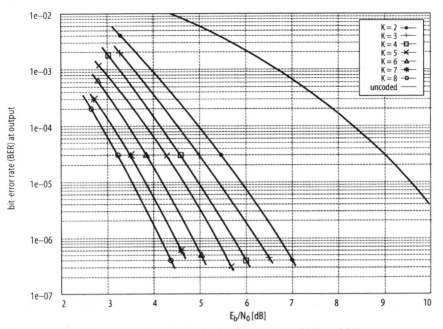

Fig. 6.18. Residual bit-error rate of convolutional codes of rate R = 1/2 in QSPK modulation

Furthermore, we assume a transmission with QPSK modulation (see chapter 7), the parameter K describing the constraint length of the code used.

The performance of the error correction increases – as expected – with increased constraint length of the code used. The more preceding information bits are used for the encoding of an output symbol, i.e. the wider the input information is "smeared", the more reliable the data are, as a result of the decoding. So, for example, with an E_b/N_0 of 5 dB at the input of a decoder, the residual bit-error rate at a constraint length of $K = 2$ is less than two powers of ten better than in the uncoded case, while for $K = 6$ a residual bit-error rate can be achieved which is more than four powers of ten better than in the uncoded case.

For the DVB standard a convolutional code of rate $R = 1/2$ with a constraint length of $K = 7$ is used. In this way it is possible, while having an E_b/N_0 of only 3.2 dB[12], to achieve a bit-error rate of less than $2 \cdot 10^{-4}$ at the output of the decoder, this ratio corresponding to the maximum permissible bit-error rate at the input of the RS decoder, so that finally a bit-error rate at the output of the RS decoder of less than $1 \cdot 10^{-11}$ is obtained (cf. section 6.2.6).

---

[12] This value corresponds to the one that was theoretically determined. No allowance for an implementation margin has been made here (cf. section 9.5.2).

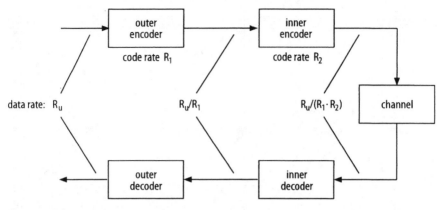

Fig. 6.19. Code concatenation

6.4 Code Concatenation

By the concatenation of various codes it is possible to increase the performance of the error correction; however, this also increases the gross transmission data rate. Figure 6.19 shows the concatenation of two codes.

Let the information data rate to be transmitted be R_u and the first-used code have the code rate R_1. This code is called outer code, as it is either right at the beginning of the transmission path or right at the end. The second code has the code rate R_2 and is called the inner code. Therefore the data rate to be transmitted is $R_u/(R_1 \cdot R_2)$. The advantage of code concatenation is explained in the following section by means of the concatenation of two block codes.

6.4.1 Block-Code Concatenation

The simplest example of a concatenation of different codes is the concatenation of block codes (see figure 6.20).

The outer code here is an RS code, which adds k redundant symbols to the m information symbols. The frame length of the outer code is therefore $n = m + k$. Each individual symbol is comprised of m' bits. The inner code, in this case a Hamming code, now adds to the m' bits of each symbol of the outer code k' redundant bits for error protection. An individual bit error will now be corrected by the inner code in the decoder so that the corresponding symbol is correct for the outer code, i.e. the outer decoder is not burdened by bit errors. On the assumption that the inner code can only correct one bit error, up to n bit errors can occur per frame of the outer code without the outer decoder having to correct any symbol error. A prerequisite for this is that each of the n bit errors fits perfectly within a frame of the inner code vector. Without

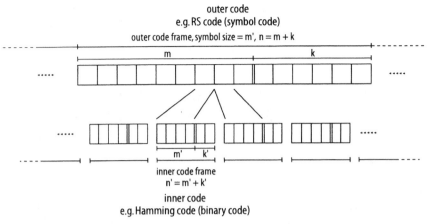

Fig. 6.20. Block-code concatenation

the inner code there would have been a decoding breakdown in the outer decoder. Should there be more than one bit error per inner code frame, then these errors can no longer be corrected by the inner code. In the outer code there is, however, only one erroneous symbol in each code vector which has to be corrected. A code concatenation, as shown in figure 6.20, thus combines the advantages of the individual codes, viz, reliable correctability of individual bit errors in the inner code and reliable correctability of short burst errors in the outer code.

6.4.2 Interleaving

In order to correct long burst errors in addition to bit errors and short burst errors, an interleaver is inserted between the outer and the inner code (see figure 6.21).

The interleaver supplies no additional error correction code, it merely initiates a rearrangement of the symbols generated by the outer code. The principle is shown in figure 6.22.

The symbols generated by the outer encoder (here 1 symbol = 1 byte) are read into the storage matrix of the block interleaver line by line. Thereafter the matrix is read out column by column and the symbols are fed individually to the inner encoder. After interleaving, two adjacent symbols generated by the outer encoder are separated from each other by exactly the number of

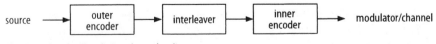

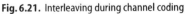

Fig. 6.21. Interleaving during channel coding

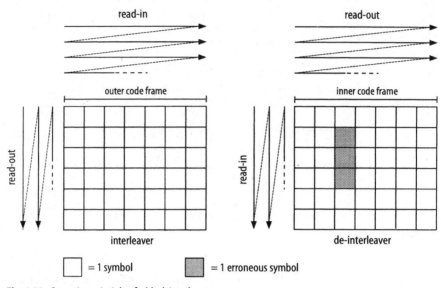

Fig. 6.22. Operating principle of a block interleaver

symbols that can be found in a column of the matrix. This number is the main parameter of an interleaver and is called interleaving depth I. In the example of figure 6.22 I equals 6.

At the receiver the output symbols of the inner decoder – which may be overlaid by errors – are deposited column by column in the matrix of the de-interleaver. A lengthy burst error, which could not be corrected by the inner decoder, will show in one of the columns (shaded in figure 6.22). Due to the line-by-line read-out of the matrix, there will only be one symbol error per frame of the outer code. Without an interleaver/de-interleaver all symbol errors of a burst would be within one frame of the outer code. A decoding breakdown would be the result.

The disadvantages of the block interleaver are that periodic disturbances which invalidate the same symbol in each column would result in a decoding breakdown of the outer decoder. This would then place the erroneous symbols in one line of the de-interleaver. Apart from this, the block interleaver has a relatively high requirement for storage capacity, which is also a disadvantage; another disadvantage being the two-dimensional synchronisation, i.e. not only the beginnings of the outer code vectors have to be found, but also the first outer code frame which stands in the first line of the de-interleaver. These disadvantages are avoided by the utilisation of the convolutional interleaver, according to [FORNEY], as shown in figure 6.23.

The convolutional interleaver consists of $(I - 1)$ shift registers with the length $M, 2M, \ldots, (I - 1)M$ and corresponding multiplexers and demultiplex-

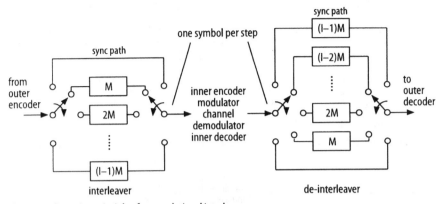

Fig. 6.23. Operating principle of a convolutional interleaver

ers (represented here as switches), each of which connects a shift register with an input or an output. As before, I represents the interleaving depth and M the so-called base delay. $M = n/I$ applies if n, as before, represents the frame length of the outer code. With each step the multiplexer and the demultiplexer switch on to the next respective input or output. The next symbol is read into the shift register currently connected to the input, and a another symbol is picked up from the output of that shift register. When the top path in the interleaver, the sync path, happens to be active, the input is connected directly to the output. This ensures that, between adjacent symbols at the input, $M \cdot I$ further symbols are transmitted.

The de-interleaver is constructed in such a way as to ensure that the non-delayed symbols of the interleaver are delayed to a maximum. Hence, the total delay is $M \cdot (I-1) \cdot I$ for all symbols. At the beginning of each outer frame all multiplexers and demultiplexers (in this case switches) must be in the initial position, i.e. only a synchronisation of one level is required. The interleaving depth I must therefore be a whole-number divisor of the frame length n of the outer code (see above).

6.4.3 Error Correction in DVB

For the transmission of digital TV over satellite in accordance with [ETS 421] (see chapter 9) and via terrestrial transmission networks in accordance with [ETS 744] (see chapter 11) an RS code is concatenated with a convolutional code by means of an interleaver (see figure 6.24).

The RS code is based on the Galois field $GF(2^8)$ and therefore has a symbol size of 8 bits. An RS code (255, 239) was chosen which processes a data block of 239 symbols and can correct up to 8 symbol errors by calculating 16 redundant correction symbols. As an MPEG-2 transport packet is 188 bytes long

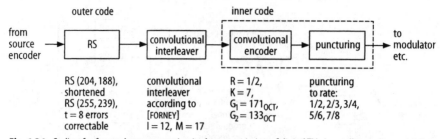

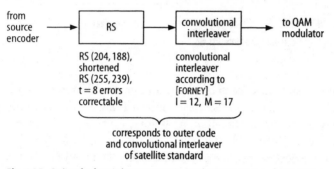

Fig. 6.24. Coding for forward error correction in the transmission of digital TV via satellite and terrestrial network

Fig. 6.25. Coding for forward error correction in the transmission of digital TV via cable

(see chapter 5), the code was shortened, i.e. the first 51 information bytes were set to zero and not transmitted at all. In this way an RS code (204,188) is generated.

After the outer code a convolutional interleaver according to [FORNEY] is used. The interleaving depth is $I = 12$. From the frame length of the outer code with $n = 204$ the base delay results as $M = n/I = 17$.

Finally, a convolutional code is applied to the interleaved symbols. Its rate is $R = 1/2$, the constraint length is $K = 7$. The taps at the shift register are described by the two generators $G_1 = 171_{OCT}$ and $G_2 = 133_{OCT}$. Optionally, a puncturing to the rates $R = 2/3, 3/4, 5/6,$ and $7/8$ is possible. Coding for error correction by transmission over cable in accordance with [ETS 429] is similar to the aforementioned coding (see figure 6.25), only the convolutional code is not required as the signal-to-noise performance in the cable channel is very much better than in the satellite channel.

A detailed description of the systems in their entirety as well as their performance can be found in chapters 9, 10 and 11.

6.5 Further Reading

To those who wish to read more extensively in the theory of codes for error correction and error recognition and who intend to consolidate their understanding of the concepts introduced here, the following works are recommended for further reading. A very basic and comprehensible overview of block and convolutional codes is given in [SWEENEY], whilst the block codes, from the basics to the deeper theory, are treated in [FURRER]. There is a standard work by G.C. Clark and J.B. Cain [CLARK], which, among other things, discusses aspects of implementation, compares various algorithms and contains performance graphs (residual bit-error rate versus signal-to-noise power). The original publication by S. Reed and G. Solomon on the code of the same name is to be found in [REED], that by A.J. Viterbi in [VITERBI]. [FORNEY] discusses the convolutional interleaver for the channel with overlaid burst errors, while [HAGENAUER] concentrates on the theory of the puncturing of convolutional codes.

Symbols in Chapter 6

$A(Z)$	polynomial of the Galois field in the frequency domain
A_i, A_l	i^{th}, l^{th} coefficient of $A(Z)$
$a(X)$	polynomial of the Galois field in the time domain
a_i, a_l	i^{th}, l^{th} coefficient of $a(X)$
$C(Z)$	code vector (polynomial) transformed from $c(X)$
$C_i,$	i^{th} coefficient of $C(Z)$
$c(X)$	code vector (polynomial) to be transmitted
c_i	i^{th} coefficient of $c(X)$
d_i	quantised input signal for soft decision of ith bit comparison
$E(Z)$	error vector (polynomial) transformed from $e(X)$
E_b	energy/bit
E_i	i^{th} coefficient of $E(Z)$
$e(X)$	error-vector polynomial
e_i	i^{th} coefficient of $e(X)$
$GF(q)$	Galois field of size q
G_i	generator polynomial at output i of a convolutional encoder
$g(x)$	generator polynomial of Galois field
I	interleaving depth
i	running variable, integer
K	constraint length of a convolutional code
k	number of correction symbols of a block code/outer block code
k'	number of correction symbols of an inner block code
l	running variable, integer
M	base delay of the convolutional interleaver
m	number of information symbols in a block code/outer block code or input bits in the convolutional code
m'	number of information symbols in the inner block code
N	$2^w - 1$
N_o	noise-power density

n	number of code symbols in a block code/outer block code or input bits in the convolutional code
n'	number of code symbols in an inner block code
$p(y_i \mid x_i)$	conditional probability of the received symbol y_i provided that the symbol x_i is sent
q	size (number of elements) of a Galois field
R	code rate
$R(Z)$	received vector (polynomial) transformed from $r(X)$
R_1,R_2	rate of the outer code, rate of the inner code
R_i	i^{th} coefficient of $R(Z)$
R_u	useful data rate
$r(X)$	received vector (polynomial)
S	length of a shift register
t	number of correctable errors in RS coding
w	exponent of 2 in the definition of q
X	argument of polynomials defined in time domain of GF
x	polynomial argument in GF
x_i	i^{th} symbol sent
y_i	i^{th} reference symbol, to be compared to the symbol received
Z	argument of polynomials defined in frequency domain of GF
α	primitive element of Galois field
α, β_i	elements of the Galois field
Δ	metric of a trellis path
δ_i	metric of the ith bit comparison
$\Lambda(Z)$	transformed polynomial from $\lambda(X)$
Λ_i	i^{th} coefficient of $\Lambda(Z)$
$\lambda(X)$	error-locator polynomial
λ_i	i^{th} coefficient of $\lambda(X)$
$\Omega(Z)$	error-evaluator polynomial

7 Digital Modulation Techniques

MPEG source coding, which achieves a data reduction in audio and video signals, has been discussed in chapters 3 and 4. As explained in chapter 5, the various elementary streams are combined in the MPEG transport multiplexer to form a single data stream. This is followed by a coding of the data stream in which redundant signal portions are inserted (see chapter 6). The entire processing of the baseband signals can take place in a computer (or a digital circuit) in which the data are available as a sequence of numerical values. For these values to be transmitted on a channel they have to be converted into genuine data signals. The signals are output sequentially by an interface, synchronisation being provided by an internal processing clock. Each information bit possesses finite energy, which will be referred to as bit energy E_b. The methods by which data signals can be adapted to the respective transmission channel are the subject of the following section.

7.1 NRZ Baseband Signal

The physical shape of the signal which is mainly used for the processing of digital signals in the baseband is called non-return-to-zero (NRZ). An ideal signal is usually described as a sequence of weighted Dirac pulses [LÜKE 1]. The data are read out at the output interface of the computer, where they adhere to a rigid time-slot pattern nT_B. Their shaping into an NRZ pulse is achieved by a hold unit, which holds the value of the information for the period of the timing pulse T_B so that each ideal Dirac pulse is overlaid with a rectangular pulse. Figure 7.1 depicts an NRZ signal with the period T_B. The amplitude assumes the value of A when state 1 is to be transmitted, and the value of 0 when state 0 is to be transmitted.

If the signals, having assumed this form, are to be used for data transmission, it is important to know their spectral properties. The power spectral density (PSD or spectrum, for short) of the NRZ signal can be computed on the assumption that the signal states consist of a statistically independent sequence of values occurring with equal frequency. The PSD consists of one discrete portion PSD_{dis} and one continuous portion PSD_{con} [MORGENST]:

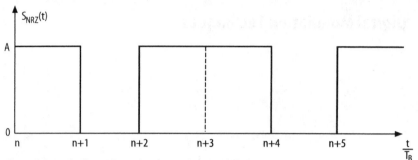

Fig. 7.1. Example of an NRZ signal with a symbol period of T_B

$$PSD_{NRZ}(\omega) = PSD_{dis} + PSD_{con} = \frac{A^2}{4}\delta(\omega) + \frac{A^2}{4}T_B \cdot si^2\left(\frac{\omega T_B}{2}\right) \qquad (7.1)$$

The discrete component in the first part of (7.1) is limited to a single Dirac pulse at a radian frequency $\omega = 0$. It represents the continuous portion of the signal, which can also be seen in fig 7.1. The rectangular pulse shape of the time-domain signal generates the continuous portion in the PSD. This rapidly diminishes with an increase in frequency. Ideally, in the case of infinitely steep signal edges there would result an infinitely wide PSD. This is plotted in fig. 7.2.

The post-filter effect of the rectangular pulse shape is conspicuous [SCHÖNFD 1]. However, as a rule this does not suffice for a real transmission. In most cases the signal spectrum has to be limited to a finite value by an additional filter to account for the finite bandwidth of the channel. The suppression of the high-frequency portions of the signal results in a widening of the impulses in the time domain and causes temporal crosstalk between the individual impulses. This behaviour is called "intersymbol interference (ISI)". In accordance with the first Nyquist criterion it is sufficient for the signal to have no ISI at the sampling points nT_B. The course of a single impulse traversing the sampling points nT_B, should have the following sampling values:

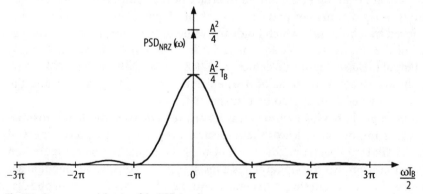

Fig. 7.2. Power spectral density of the NRZ signal

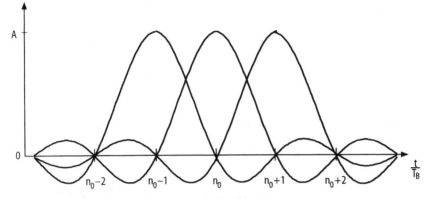

Fig. 7.3. Interference-free transmission due to ideal band limitation

$$s_{NRZ}(t - nT_B) = A \quad \text{for} \quad n = n_0$$
$$s_{NRZ}(t - nT_B) = 0 \quad \text{for} \quad n \neq n_0 \tag{7.2}$$

This first Nyquist criterion is satisfied, among others, by the si-shaped course of an impulse, which can be generated by an ideal band limitation at half the clock frequency. The cut-off frequency is referred to as the Nyquist frequency f_N. Figure 7.3 shows the curve represented by three consecutive si-impulses. The resulting signal is interference-free at the sampling points because the si-functions of all consecutive impulses are zero at these points.

In real systems, however, the ideal sampling accuracy required cannot be maintained and, moreover, the filters used always have finitely steep filter slopes. Nyquist proved in his second theorem that, with a cosine-wave frequency response $H(f)$, the ISI on both consecutive impulses is limited. For each of the individual impulses, an additional requirement – apart from the first Nyquist criterion – is that the amplitude values which are exactly midway between the two consecutive samples have half the basic impulse amplitude (in this case 0.5 A) [NYQUIST]:

$$s_{NRZ}(t - nT_B) = \frac{A}{2} \quad n = n_0 \pm \frac{1}{2}, \tag{7.3}$$

A cosine-wave frequency response is shown in figure 7.4 a. Figure 7.4 b shows the corresponding eye diagram. The amplitude values can be clearly seen from (7.2) and (7.3). The disadvantage resulting from adherence to the second Nyquist criterion is due to the doubling of the channel bandwidth required for the transmission. For this reason the cosine-wave drop in frequency response is generally reduced to a narrower transition band which is symmetrical to the Nyquist frequency. The resulting frequency responses correspond to (7.4), with the roll-off factor indicating the steepness of the filter slope and the ensuing extension of the frequency range. The roll-off factor can assume a

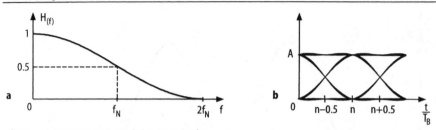

Fig. 7.4. a Frequency response, **b** Eye diagrams for $\alpha = 1$

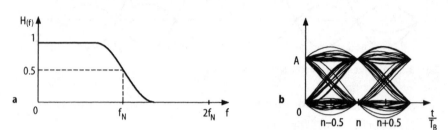

Fig. 7.5. a Frequency response, **b** Eye diagrams for $\alpha = 0{,}35$ (e.g. DVB satellite)

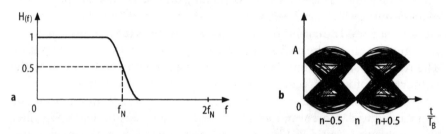

Fig. 7.6. a Frequency response, **b** Eye diagrams for $\alpha = 0{,}15$ (e.g. DVB cable)

value between 0 and 1. The frequency responses have a so-called "raised co-sine" characteristic. Two examples with differing roll-off factors are shown in figures 7.5 a and 7.6 a, which are used for the transmission of DVB signals over satellite (see chapter 9) and cable (see chapter 10). It can be clearly seen by the eye diagrams in figures 7.5 b and 7.6 b that the first Nyquist criterion is satisfied. The second Nyquist criterion, however, is infringed, as not all individual impulses have the required amplitude values at $(n \pm 0.5)T_B$ according to (7.3).

$$H(f) = 1 \qquad \text{for } |f| < f_N (1-\alpha)$$

$$H(f) = \tfrac{1}{2} + \tfrac{1}{2} \sin\left[\frac{\pi}{2f_N} \left(\frac{f_N - |f|}{\alpha} \right) \right] \qquad \text{for } f_N (1-\alpha) \le |f| \le f_N (1+\alpha) \qquad (7.4)$$

$$H(f) = 0 \qquad \text{for } |f| > f_N (1+\alpha)$$

In practice the required Nyquist signal frequency response is generally obtained by two subfilters, one connected after the other. The first subfilter is at the transmitter output. It limits the signal spectrum to $f_N(1 + \alpha)$ and shapes the impulse. In the following it will be assumed that the data signal is disturbed by additive white Gaussian noise (AWGN). The subfilter at the receiver input has the task of minimising the noise power which was added during the transmission and of maximising the signal-to-noise ratio S/N at the sampling point. An arrangement in which the impulse response of the filter is identical to the impulse response of the signal transmitted is called a matched filter [NORTH]. The effect of the noise signal no longer depends on the wave-form of the useful signal, but only on the average energy E_b which is required per bit. If the useful signal has no redundancy (e.g. no error-protection portion), this results in a signal-to-noise ratio of

$$\frac{S}{N} = \frac{E_b}{N_0} \tag{7.5}$$

[LÜKE 1, KAMMEYER] at the optimal sampling point at the filter output.

The change in the digital signal brought about by the AWGN signal causes the decoder to make a false decision if the noise amplitude exceeds a given decision threshold. In accordance with the above assumption the noise signal has a Gaussian amplitude-density distribution [GAUSS], also called "normal distribution"

$$f(a) = \frac{1}{\sigma_N \sqrt{2\pi}} \cdot e^{-\frac{(a-\mu)^2}{2\sigma_N^2}} \tag{7.6}$$

with

amplitude	a,	
mean value	$\mu = 0$	for average-free noise,
variance	$\sigma_N^2 = N$.	

(7.6a)

The curve of this function has been tabulated in various publications (e.g. [BRONST]). Figure 7.7 shows the superposition of the two possible values (A and o) in a symmetrical channel, each having the same Gaussian amplitude-density function. The optimal decision threshold is marked by a broken line and is to be found at the amplitude $\alpha_{thresh} = 0.5 \, A$.

The probability of a detection error can be calculated for each state from the area below that part of the curve which lies beyond the decision threshold. The primitive of this function necessary for this purpose is not available in closed form. Hence, the so-called error function was introduced for the path integral:

$$erf(x) = \frac{2}{\sqrt{\pi}} \int_0^x e^{-z^2} dz. \tag{7.7}$$

The probability of a false decision is given by the so-called complementary error function (erfc):

$$erfc(x) = 1 - erf(x).$$
(7.8)

A connection between the maximum amplitude of data signal A and noise power N results from the substitution of the variable z from (7.7) and by making use of (7.6.a):

$$z = \frac{a - a_{thresh}}{\sqrt{2\sigma_N^2}} \Rightarrow x = \frac{A - \frac{A}{2}}{\sqrt{2N}} = \frac{A}{\sqrt{8N}}.$$
(7.9)

In various publications (e.g. [LÜKE 1]) an instantaneous power $S_a = A^2$ is defined which is used as a reference quantity for computing the bit-error rate. In this section an average signal power S shall serve as reference quantity (as, for instance, in [KAMMEYER]). Assuming that the transmitted states and the corresponding amplitude values A and 0 occur, on average, with equal frequency, the average signal power S can be computed as follows:

$$S = \tfrac{1}{2}(A^2 + 0) = \frac{A^2}{2} \Rightarrow A = \sqrt{2S}.$$
(7.10)

Substituting (7.10) into (7.9) and making use of (7.5) we obtain:

$$x = \sqrt{\frac{E_b}{4N_0}}.$$
(7.11)

As the signal in accordance with figure 7.7 can only be invalidated in one direction, the complementary error function must be multiplied by the value 0.5. Thus the bit-error rate (BER) for the baseband transmission of a unipolar NRZ signal results in the required dependency of E_b/N_0:

$$BER_{unipol} = \tfrac{1}{2} erfc \left(\sqrt{\frac{E_b}{4N_0}} \right).$$
(7.12)

In a bipolar baseband transmission, state 0 is mapped onto the value $-A$ by a negative excursion of the amplitude, which causes the mean signal power to increase to

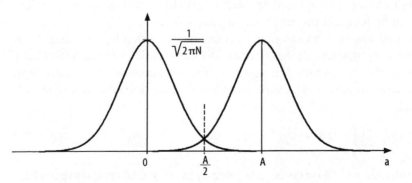

Fig. 7.7. Overlay of amplitudes of NRZ signal with Gaussian amplitude-density function of noise

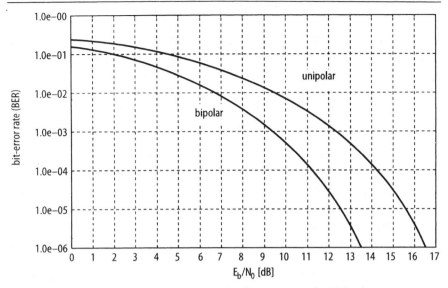

Fig. 7.8. Bit-error rate of a unipolar and a bipolar baseband transmission with NRZ signals

$$S = \tfrac{1}{2}(A^2 + A^2) = A^2 \Rightarrow A = \sqrt{S}.$$ (7.13)

The decision threshold α_{thresh} from (7.9) is shifted to the DC potential. If these changes are taken into account, the result is the following bit-error rate for a bipolar baseband transmission:

$$BER_{bipol} = \tfrac{1}{2}erfc\left(\sqrt{\frac{E_b}{2N_0}}\right).$$ (7.14)

Both bit-error rates are shown in figure 7.8 as functions of E_b/N_0.

Although in bipolar transmission, the distance of the amplitude range between the decision threshold and the two possible amplitude values has doubled as opposed to the unipolar transmission, this only results in a noise-ratio improvement by 3 dB owing to the duplication of the average signal power S.

In those publications which define the above-mentioned instantaneous power S_a as reference quantity, this gain increases to 6 dB.

7.2 Principles of the Digital Modulation of a Sinusoidal Carrier Signal

There are various transmission systems for the distribution of broadcast signals. These include satellite systems, cable networks and terrestrial channels, which are generally organised as frequency multiplex systems. For the transmission of user data the available frequency ranges are divided into system-specific channel spacings, a fact which requires an adaptation of the baseband signals to the particular channel conditions. This takes place by the modulation of a carrier signal. Heed has to be taken that the power of the data signal be concentrated as much as possible at the required position in the frequency spectrum. For this reason sinusoidal carrier signals are chosen:

$$s_{cos}(t) = \text{amplitude} \cdot \cos(2\pi \cdot \text{frequency} \cdot t + \text{phase}). \tag{7.15}$$

Before the signals are converted, band limitation and impulse pre-shaping are usually performed at the transmitter (see section 7.1). After transmission in the high-frequency range, the signal must be down-converted to the baseband and filtered by the matched filter at the receiver. Usually each signal which passes through one of the above-mentioned transmission media would be processed in accordance with section 7.1 prior to modulation and subsequent to demodulation. The previous considerations concerning the baseband transmission with regard to the filters used in the transmitter and the receiver therefore retain their validity for the transmission in a real band-pass channel. In this case, too, the useful signal is overlaid with an AWGN signal with noise-power density N_0 which is constant over the whole frequency range. An optimal receiver synchronously demodulates the signal and then calculates the real part of the baseband signal [KAMMEYER]. After the signal has passed through the matched filter, sampling takes place with the double Nyquist frequency $2f_N$, which provides discrete amplitude values A' as a result.

The above findings can be translated into the high-frequency range before demodulation. On account of the band-pass character of the modulated signals we obtain, for each transmitted state, a defined amplitude and a reference state (or rather, a reference frequency for frequency shift keying) of the carrier signal used. If these states are plotted in a complex plane, the Euclidean distance d is a measure of the resistance of this method against disturbances. This complex amplitude plane is described in the literature as a constellation diagram. Moreover, to facilitate a comparison between the various modulation techniques it is common practice to scale to the value of 1, resp. $\sqrt{2}$, the smallest signal amplitude which differs from zero.

By analogy with what has been said about baseband transmission, the probability of an erroneous band-pass transmission can be defined by substituting into (7.9) the Euclidean distance d for the transmitted amplitude A. On this assumption the noise power N is equal to the real effective noise power. Various publications, such as [LÜKE 1], describe the modulation techniques on the basis of this approach. However, in our subsequent discussion the total power which is added to the noise signal during transmission over a real transmission channel will be referred to as N in accordance with [KAMMEYER]. This means that on account of the above-mentioned calculation of the real part of the signal in the receiver only half the noise power N is effective, so that, in addition, 0.5 N needs to be substituted for σ_N^2 in (7.9). Therefore the result for the argument of the complementary error function is:

$$x = \frac{d}{\sqrt{4N}}.$$
(7.16)

According to (7.15) it is possible to modulate the information to be transmitted onto the amplitude (amplitude shift keying – ASK), the frequency (frequency shift keying – FSK), the phase (phase shift keying – PSK), or onto a combination (e.g. quadrature amplitude modulation – QAM). In this way symbols are generated whose feature-carrying information (e.g. the amplitude in ASK) can assume different states. As an introduction, the principles of the digital modulation techniques will be explained by means of the three basic methods. This will be followed by a discussion of the techniques which are intended for the transmission of DVB signals.

7.2.1 Amplitude Shift Keying (2-ASK)

In 2-ASK the unipolar NRZ signal, discussed in section 7.1, will be fed to a modulator. After its connection with the carrier, the NRZ signal switches the modulator on when its state is 1, and switches it off when a 0 is to be transmitted. Because of this keying mode the 2-ASK is also referred to as "on-off keying". It is very easy to implement by means of a switch or, more generally, by a multiplier. The block diagram of the modulator and the wave-forms of the time-domain signals are shown in figure 7.9.

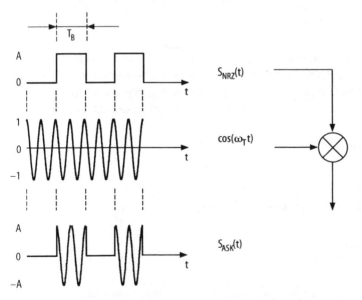

Fig. 7.9. Block diagram of the 2-ASK modulator and time graph of signals

It can be clearly seen that the information transmitted is contained in the envelope of the signal. This phenomenon is typical of amplitude modulation and also known from analogue technology.

Under the same conditions as in section 7.1 (the same occurrence probability and statistical independence of the two states) the temporal multiplication of the unipolar NRZ signal by the sinusoidal carrier results in a PSD which arises from a convolution of the PSDs of the two individual signals and consists of one discrete portion and one continuous portion [JOHANN].

$$PSD_{2-ASK}(\omega) = PSD_{dis}(\omega) + PSD_{con}(\omega)$$

$$= \frac{A^2}{16}\left[\delta(\omega_T + \omega) + \delta(\omega_T - \omega)\right] \tag{7.17}$$

$$+ \frac{A^2}{16} T_B\left[si^2\left((\omega_T + \omega)\frac{T_B}{2}\right) + si^2\left((\omega_T - \omega)\frac{T_B}{2}\right)\right]$$

Because of the discrete signal portion of the carrier frequency ω_T a synchronous demodulation in the receiver is possible without additional measures being taken.

Figure 7.10 shows the one-dimensional constellation of 2-ASK. Also plotted is the square root of the mean signal power S so that the relationship between S and the Euclidean distance d can be read off. This results in:

$$d = \sqrt{2S}\,. \tag{7.18}$$

From (7.18), (7.16) and (7.5), we obtain the bit-error rate

$$BER_{2-ASK} = \tfrac{1}{2}erfc\left(\sqrt{\frac{E_b}{2N_o}}\right). \tag{7.19}$$

With 2-ASK it is not imperative to perform a synchronous demodulation. Figure 7.10 depicts the decision threshold by means of a broken circular line with a radius of half the amplitude. As can be seen, it is also possible to perform a demodulation of the envelope, independently of the phase angle transmitted.

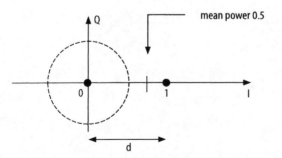

Fig. 7.10. Constellation diagram of 2-ASK and of the amplitude value which corresponds to the mean ASK power

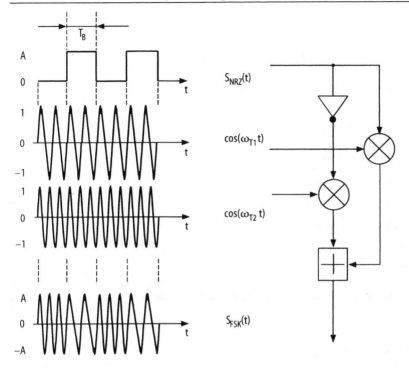

Fig. 7.11. Block diagram of the 2-FSK modulator and time graph of signals

7.2.2 Frequency Shift Keying (2-FSK)

The findings from section 7.2.1 can be used to explain the 2-FSK technique if the 2-FSK modulator is conceived as two 2-ASK modulators connected in parallel (cf. figure 7.9). The two multiplier outputs are interconnected by means of an adder. The unipolar NRZ signal discussed in section 7.1 is read into one of the multiplier inputs. The inverted unipolar NRZ signal is fed into the other multiplier as an input signal. The carrier signals used for the modulation have two different frequencies. They are sampled by the input signals whose wave-forms are in inverse relation to each other, so that the symbol occurring at the output of the adder always differs from zero. This process results in the wave-form of a time-domain signal with a constant envelope. The block diagram of the 2-FSK modulator and the wave-forms are shown in figure 7.11.

The PSD of the 2-FSK technique results from the addition of two $PSD_{2\text{-}ASK}$, which, owing to the different carrier frequencies, are spectrally offset against each other.

A more thorough treatment of the 2-FSK technique is given, for example, in [JOHANN].

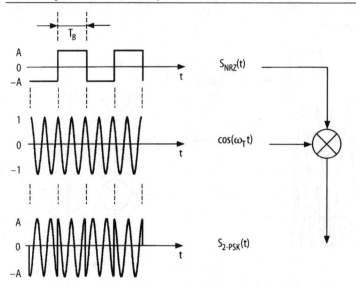

Fig. 7.12. Block diagram of the 2-PSK modulator and time graph of signals

7.2.3 Phase Shift Keying (2-PSK)

A further possibility of transmitting information is to change the phase angle of the carrier signal by 180°. By means of the relation

$$\cos(\omega t + \pi) = -\cos(\omega t) \tag{7.20}$$

the 2-PSK can also be interpreted as an ASK, although the amplitude is not switched on and off, as is the case with amplitude keying, but is inverted. When using a 2-ASK modulator, as shown in figure 7.9, this inversion of the carrier can be achieved very simply by feeding a bipolar NRZ signal to the input instead of a unipolar NRZ signal. The block diagram of the modulator and the corresponding wave-forms of the time-domain signals can be seen in figure 7.12.

The constant envelope of the 2-PSK can be clearly seen in the lower part of figure 7.12. Furthermore, a comparison with the wave-form of 2-ASK (see bottom part of figure 7.9) shows that intervals during which the signal is zero, such as occur in 2-ASK, can be filled by inversion or by a 180° phase shift of the carrier. This causes the power of the 2-PSK signal to double compared with the mean power of a 2-ASK signal. On the assumption that the two symbol states are statistically independently distributed and occur, on average, with equal frequency, the carrier, temporally averaged out over numerous symbol states, must be absent from the spectrum. Hence this is a modulation technique with carrier suppression. The PSD of 2-PSK can be obtained from the

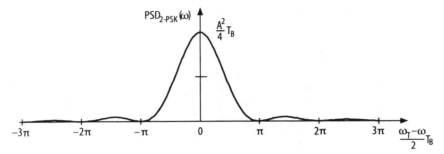

Fig. 7.13. Power spectral density of 2-PSK for the positive frequency range in the neighbourhood of $\omega = \omega_T$

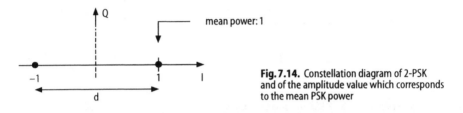

Fig. 7.14. Constellation diagram of 2-PSK and of the amplitude value which corresponds to the mean PSK power

PSD of 2-ASK (see (7.17)) by suppressing the discrete component and, on the basis of the above-mentioned power balance, quadrupling the continuous component. According to [JOHANN]:

$$PSD_{2-PSK}(\omega) = \frac{A^2}{4} T_B \left[si^2\left((\omega_T + \omega)\frac{T_B}{2} \right) + si^2\left((\omega_T - \omega)\frac{T_B}{2} \right) \right]. \tag{7.21}$$

The PSD in figure 7.13 represents the positive frequency range with $\omega = \omega_T$ if $T \gg 2\pi/T_B$.

The one-dimensional constellation of 2-PSK is shown in figure 7.14. The decision threshold, whose course is orthogonal to the transmitted in-phase component, passes exactly through the origin of the co-ordinates. The relation between the Euclidean distance d and the mean signal power S is calculated as

$$d = 2\sqrt{S}. \tag{7.22}$$

As the signal has the same power in both states a symbol decision can only be carried out after its synchronous demodulation. As opposed to 2-ASK, the required carrier has to be recovered by non-linear signal processing. The Costas loop is a method which is often used for the synchronous demodulation of the PSK signals [COSTAS].

The bit-error rate for 2-PSK can be calculated from (7.22), (7.16) and (7.5):

$$BER_{2-PSK} = \frac{1}{2} erfc\left(\sqrt{\frac{E_b}{N_o}} \right). \tag{7.23}$$

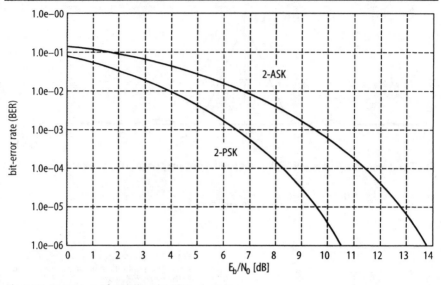

Fig. 7.15. Bit-error rates for 2-PSK and 2-ASK transmission

The bit-error rate BER for 2-ASK and 2-PSK is given in figure 7.15 as a function of E_b/N_o. A doubling of the mean signal power results in an improvement in the signal-to-noise ratio of 3 dB for 2-PSK as compared to 2-ASK. Several publications do not relate the bit-error rate of 2-ASK to the total signal power, but take the AC component of the signal power as a reference value [MÄUSL]. In this way the bit-error rate curve of 2-ASK is shifted by 6 dB to higher signal-to-noise ratios.

7.3 Quadrature Phase Shift Keying (QPSK)

The QPSK technique offers the possibility of simultaneously transmitting two bits per symbol. This doubles the spectral efficiency (controlling the number of bits that can be transmitted per second per required bandwidth) as opposed to the techniques dealt with up to now. The serial data stream of the NRZ signal is first split up into two parallel paths by means of a demultiplexer. This process is shown in figure 7.16 for eight consecutive bits. After the data stream has been transformed into two parallel ones, each two bits can be processed simultaneously. These dibits are assigned the function of a complex symbol with a real and an imaginary part. The wave-forms of the signals are referred to as $Re\{s_{Dibit}(t)\}$ or $Im\{s_{Dibit}(t)\}$. The duration which is available for the processing of a complex symbol is referred to as symbol duration T_S. It is double the bit duration T_B in the case of QPSK.

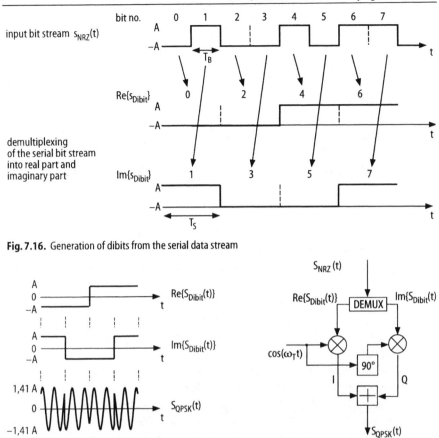

Fig. 7.16. Generation of dibits from the serial data stream

Fig. 7.17. Block diagram of the QPSK modulator and time graph of signals

Figure 7.17 shows the block diagram of the QPSK modulator. The demultiplexer at the input divides the serial data stream into two parallel paths. The scalar components are fed to one multiplier each. Two sinusoidal signals, having the same frequency ω_T and being phase-shifted against each other by 90°, serve as carriers. The two signal parts are then added together. This type of signal processing, as shown in (7.24), can be expressed by a complex multiplication of both quantities with the subsequent calculation of the real part of the signal. Hence the QPSK signal at the output of the modulator can be computed as:

$$s_{QPSK}(t) = Re\left\{ s_{Dibit}(t) \cdot e^{-j\omega_T t} \right\}$$

$$= Re\{s_{Dibit}(t)\} \cdot \cos(\omega_T t) - Im\{s_{Dibit}(t)\} \cdot \sin(\omega_T t). \tag{7.24}$$

From figure 7.17 it is apparent that the QPSK signal results from the addition of two 2-PSK signals. The spectrum of the QPSK signal is the outcome of the

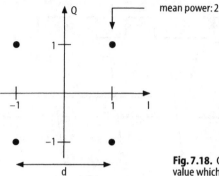

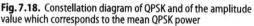

Fig. 7.18. Constellation diagram of QPSK and of the amplitude value which corresponds to the mean QPSK power

addition of the two 2-PSK spectra. The symbol duration T_S must be used in (7.21) instead of the bit duration T_B. Moreover, the signal power is doubled by the addition of the two 2-PSK signals, which leads to the following expression:

$$PSD_{QPSK}(\omega) = \frac{A^2}{2} T_S \left[si^2\left((\omega_T + \omega)\frac{T_S}{2}\right) + si^2\left((\omega_T - \omega)\frac{T_S}{2}\right)\right]. \qquad (7.25)$$

By substituting the relation $T_S = 2T_B$ one realises that the required transmission bandwidth is halved as against that in 2-PSK.

A QPSK receiver separates the input signal into its in-phase and quadrature components, where each can be regarded as an independent 2-PSK signal, as mentioned in the last paragraph. This can be clearly seen once again when we look at the QPSK signal in the now two-dimensional constellation diagram depicted in figure 7.18. Moreover, it can be seen that the signal power S is equally distributed between the two components. Half the signal power $S/2$ is effective for both bit decisions, in the direction of the in-phase component I as well as in the direction of the quadrature component Q. The resulting amplitude value equals half the Euclidean distance, which is illustrated in 7.18.

$$d = 2 \cdot \sqrt{\frac{S}{2}} = \sqrt{2S} \qquad (7.26)$$

The bit-error rate results from the substitution of the Euclidean distance into (7.16). Using (7.5), we have to substitute the bit energy E_b by the symbol energy E_s since the principle adopted in section 7.1 for the optimisation of the signal-to-noise ratio by the matched filter in the receiver no longer refers to individual bits but to the transmitted symbols.

$$\frac{S}{N} = \frac{E_s}{N_0} \qquad (7.27)$$

E_s corresponds to the sum of the energy of the two signal components E_b which are required per bit:

$$E_s = 2E_b \, . \tag{7.28}$$

For the bit-error rate we get:

$$BER_{QPSK} = \tfrac{1}{2} erfc\left(\sqrt{\frac{E_b}{N_0}}\right) . \tag{7.29}$$

As both the signal power S and the noise power N are distributed between the two components I and Q the same conditions apply to each signal component as in 2-PSK. Therefore it is not surprising that the bit-error rate of a QPSK signal is the same as the mean value of the bit-error rates of its two signal parts and that it is thus identical with the bit-error rate of a 2-PSK signal. This statement is valid as long as the bit-error rate is given as a function of E_b/N_0.

Apart from the interpretation of the QPSK signal as a combination of two 2-PSK signals, QPSK can be conceived as a quadrature amplitude modulation (QAM) with 4 states. That both interpretations are valid is shown by a comparison between the bit-error rate curves in figure 7.15 (for 2-PSK) and figure 7.23 (for 4-QAM).

Apart from the basic QPSK variant discussed above, other special forms are discussed in the literature [e.g. KAMMEYER, JOHANN, MÄUSL, PROAKIS] in detail (i.e. Offset-QPSK, DPSK).

QPSK has practical applications in the transmission of digital signals over satellite channels (see chapter 9) and in connection with the OFDM technique (see section 7.6) in terrestrial transmissions (see chapter 11).

7.4 Higher-level Amplitude Shift Keying (ASK) and Vestigial-Sideband Modulation (VSB)

In television engineering, analogue vestigial-sideband modulation is to be found in PAL colour coding as well as in television transmissions. Today's terrestrial transmission channels which are used for radio broadcasting are designed to accommodate VSB, just as are the cable networks, given that the digital variant of the vestigial-sideband modulation only differs from the analogue one in that a finite number of discrete amplitude states have to be transmitted. Therefore there are economic advantages in using the existing systems for digital VSB without any changes having to be made.

The serial data stream at the input of the modulator is first combined to form data words of width m and then allocated to the M possible discrete symbol states by means of a table, where

$$M = 2^m \, . \tag{7.30}$$

The required VSB signal shape can be generated using ASK. As the modulated signal contains no phase information it will suffice for only one of the side-

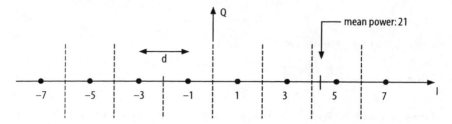

Fig. 7.19. Constellation diagram of 8-level ASK and of the amplitude value which corresponds to the mean ASK power

bands to be transmitted. After the modulation the second sideband is therefore almost completely suppressed by post-filtering. This, however, produces negative side effects. For example, the coherent carrier signal in the receiver which is required for the synchronous demodulation cannot be recovered from the transmitted VSB signal. Hence the modulator must not completely suppress the carrier. The residual carrier becomes noticeable in the one-dimensional constellation of the VSB by an additive DC component and thus increases the mean signal power. The one-dimensional constellation of an eight-level ASK, as shown in figure 7.19, differs from the eight-level VSB only by the missing DC component. As a rule, the individual standardised amplitude levels A_l' are indicated in the literature in accordance with (7.31):

$$A_l' = 2l-1-M \quad \text{for} \quad l=1, 2, \ldots, M.$$ (7.31)

At this point it should be mentioned that the ASK technique for the value $M = 2$, as described in this section, does not correspond with the 2-ASK technique introduced in section 7.2.1, because in higher-level ASK the amplitude values of (7.31) are modulated symmetrically to 0. The ASK with the value $M = 2$ corresponds to the 2-PSK introduced in section 7.2.3.

The mean power S of the ASK signal is computed from the expected value of the squares of all possible signal amplitudes A_l', assuming that all amplitudes occur with the same probability [JOHANN, PROAKIS]. For 8-ASK, the amplitude value corresponding to the mean power S is plotted in figure 7.19.

$$S = E\left\{A_l'^2\right\} = \frac{2}{M} \sum_{l=1}^{M/2} (2l-1)^2 = \frac{M^2-1}{3}$$ (7.32)

The basic relationship between S and the Euclidean distance d follows from figure 7.19, using (7.32):

$$\frac{d}{\sqrt{S}} = \frac{2}{\sqrt{\frac{M^2-1}{3}}} \Rightarrow d = \sqrt{4 \cdot \frac{3}{M^2-1} \cdot S}.$$ (7.33)

For higher-level modulation the bit-error rate (BER) can be estimated via the symbol-error rate (SER). The symbol-error rate indicates the probability of an error occurring in the detection of a transmitted symbol. For the computation of the SER an additional correction factor needs to be taken into account, which can be explained as follows: higher-level ASK techniques, as opposed to the techniques discussed before, have more than two amplitude levels. As a consequence, each level has two direct neighbours and therefore two decision thresholds beyond which it may be disturbed and found to be erroneous. Hence, the error probability increases by the factor of 2. The two outer points in the constellation which indicate the two peaks are the exceptions. Since the SER constitutes the mean value of all error probabilities of the M individual states, the correction factor is computed as follows:

$$c_{cor} = 2 \cdot \frac{M-1}{M}.$$
(7.34)

Hence, the M-level ASK has an SER of

$$SER_{ASK} = \frac{M-1}{M} erfc \left(\sqrt{\frac{3}{M^2-1} \cdot \frac{E_s}{N_o}} \right).$$
(7.35)

The transition from one double-sideband ASK signal to a signal modulated in accordance with the VSB technique is explained by the following simplified model:

(1) As already discussed, for the transmission almost the whole of one sideband is suppressed. The receiver filter, as a result, has only half the noise bandwidth. This results in a correction of the signal-to-noise ratio of 3 dB to the advantage of the VSB signal.

(2) In the case of synchronous demodulation, the correlated amplitudes of the upper and the lower sideband in the sampling point of an ASK signal are vectorially superimposed. As the vector representing the lower sideband is missing in a VSB transmission, the Euclidean distance in the constellation is reduced by half. This means that the signal-to-noise ratio is reduced by 6 dB as compared with the signal-to-noise ratio of an ASK signal.

(3) Furthermore, the disadvantage of additionally required power for the transmission of the coherent carrier signal becomes noticeable. In the U.S., where the Grand Alliance favours VSB as the transmission technique for cable networks and terrestrial channels (see chapter 1), a power increase of 0.3 dB has been recommended [GRALLI].

The total balance corresponds to a deterioration in the signal-to-noise ratio by a factor of approx. 2.14, which corresponds to a logarithmic value of 3.3 dB. For the transition from ASK to VSB a reduction in the required transmission bandwidth is exchanged for a loss in immunity to interference. By the suppression of one sideband the spectral efficiency is increased from the ideal $m/2$ bit/s per Hz for ASK to the ideal m bit/s per Hz for VSB [SCHÖPS].

The symbol-error rate for VSB can be estimated from (7.35) by using the above factor of 2.14:

$$SER_{VSB} = \frac{M-1}{M} erfc\left(\sqrt{\frac{3}{2.14(M^2-1)} \cdot \frac{E_s}{N_0}}\right). \tag{7.36}$$

The dependence on the bit energy E_b is computed using the following relation (see also (7.28) for QPSK):

$$E_s = m \cdot E_b. \tag{7.37}$$

The average bit-error rate can be obtained by taking into account the fact that when a symbol is transmitted correctly all m bits are transmitted error-free as well. On the assumption that, firstly, all m bits are statistically independent of each other and, secondly, that they are erroneously detected with the same probability, the following approximation applies:

$$1-SER \cong (1-BER)^m \Rightarrow BER \cong 1-(1-SER)^{\frac{1}{m}} \tag{7.38}$$

Therefore the bit-error rate of an ASK signal and of a VSB signal can be expressed as follows:

$$BER_{ASK} = 1-\left[1-\frac{M-1}{M} erfc\left(\sqrt{\frac{3}{(M^2-1)} \cdot \frac{m \cdot E_b}{N_0}}\right)\right]^{\frac{1}{m}} \tag{7.39}$$

$$BER_{VSB} = 1-\left[1-\frac{M-1}{M} erfc\left(\sqrt{\frac{3}{2.14(M^2-1)} \cdot \frac{m \cdot E_b}{N_0}}\right)\right]^{\frac{1}{m}}. \tag{7.40}$$

The bit-error rate for various VSB techniques is shown in figure 7.20. The curves for ASK can be easily obtained by shifting the VSB curves by a value of 3.3 dB to lesser E_b/N_0 values.

Different characteristics of VSB are to be found in table 7.1. It can be seen that the mean signal power increases by approx. 6 dB per level towards higher constellations, which results in a reduction of the Euclidean distance d and in

Table 7.1. Characteristics of the VSB techniques

Number of symbol states and bits per symbol	Mean signal power	Ratio between peak and mean signal power	Dynamic range
$M, m = ld\,(M)$	$10 \log\left(\frac{M^2-1}{3}\right) +0.3$ dB	$10 \log\left(3\frac{(M-1)^2}{M^2-1}\right)$	$20 \log\,(M-1)$
2, 1	0.3 dB	0.0 dB	0.0 dB
4, 2	7.3 dB	2.6 dB	9.5 dB
8, 3	13.5 dB	3.7 dB	16.9 dB
16, 4	19.6 dB	4.2 dB	23.5 dB

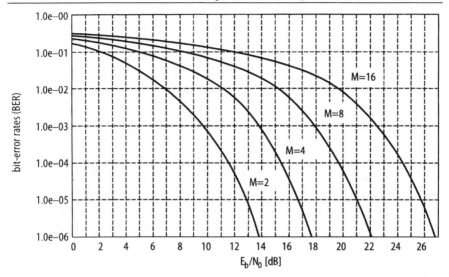

Fig. 7.20. Bit-error rates of the VSB techniques for various M

the correlated reduction of the interference immunity. The ratio of the maximum signal power, which is represented by the outer constellation points, to the mean signal power S is important, if, for example, possible overdrive effects of amplifiers need to be avoided. The minimum signal power is always equal to 1 according to (7.31). A signal dynamics of 0 dB signifies a constant envelope.

7.5 Digital Quadrature Amplitude Modulation (QAM)

In section 7.3 it was shown how by demultiplexing the serial data stream into two parallel branches a complex symbol can be generated. In section 7.4 m bits were combined to a higher-level symbol. By both techniques the spectral efficiency can be increased. Hence it is reasonable to combine the two techniques in order to obtain a further improvement.

In digital QAM m bits are combined and mapped onto a complex symbol word consisting of a real part S_{real} and an imaginary part S_{imag}. In figure 7.21 the mapper is shown at the input of the modulator. In contrast to figure 7.17, the mapper does not necessarily generate binary signals any more but, depending on the desired QAM, also multilevel ones. In QPSK the symbol word represents the dibit. The combination of the m bits results in the symbol duration T_S increasing proportionally to the bit duration. The actual modulation process that follows is identical to the technique described in section 7.3. The real part of the QAM symbol word modulates a cosine-wave carrier signal in the in-phase branch, while the imaginary part of the QAM symbol modulates a sine-wave carrier of the same frequency in the quadrature branch of the modulator. By this process two carrier-frequency oscillations are generated which can be conceived as independent ASK signals. Once more the similarity to the QPSK signal, which is composed of two 2-PSK signals, is easily seen. The required QAM signal results after adding the two components. The wave-forms of the time-domain signals are shown in figure 7.21 by taking 16-QAM as an example. It can be seen that the envelope of the QAM signal is not constant. Therefore, the possibility of carrying out an envelope demodulation in the demodulator cannot be precluded at this stage. Against this, the fact has to be borne in mind that there are various symbol states which have the same amplitude but different phase angles. For this reason only a synchronous demodulation is required. This can be demonstrated by using the two-dimensional constellation shown in figure 7.22 which takes a 64-QAM as an example. The first of the four quadrants contains 16 of the 64 possible amplitude and phase values. The scaling follows the general practice in the literature:

$$A'_{I,k} = 2k - 1 - \sqrt{M} \quad \text{for} \quad k = 1, 2, \ldots, \sqrt{M}$$
$$A'_{Q,l} = 2l - 1 - \sqrt{M} \quad \text{for} \quad l = 1, 2, \ldots, \sqrt{M} .$$

(7.41)

Using (7.41), the amplitude values marked along the negative axes correspond to the amplitude of the in-phase component $A'_{I,k}$ and the amplitude of the quadrature component $A'_{Q,l}$ respectively. The signal powers in the fourth quadrant result from the points of the vectors. If all constellation points occur with the same frequency, the mean power of the transmitted signal is obtained from the expected value of the squares of all amplitude levels, as expressed by the following equation:

$$S = E\left\{A'^2_{k,l}\right\} = \frac{1}{M} \sum_{k=1}^{\sqrt{M}} \sum_{l=1}^{\sqrt{M}} \left(A'^2_{I,k} + Q'^2_{Q,l}\right) = 2\frac{M-1}{3}.$$

(7.42)

Applying the above example to 64-QAM and using $M = 64$, the scaled mean power results in 42. This value is plotted in the fourth quadrant of figure 7.22.

The second quadrant in figure 7.22 shows the phase values for the various states. For convenience, these values have been so arranged as to mirror symmetrically those in the first quadrant, so that the zero phase angle results in the absolute angle of 180°. In a mathematically negative sense of rotation the phase angle increases. The actual values for the second quadrant can be computed by subtracting the given angles from 180°.

The third quadrant contains the optimal decision thresholds.

The four points which are nearest to the origin of the system of coordinates in figure 7.22 correspond to the QPSK values. Therefore QPSK can also be referred to as 4-QAM. The transmission of 16 amplitude and phase

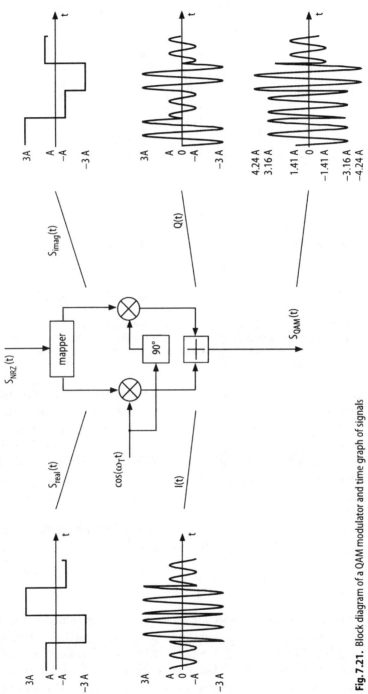

Fig. 7.21. Block diagram of a QAM modulator and time graph of signals

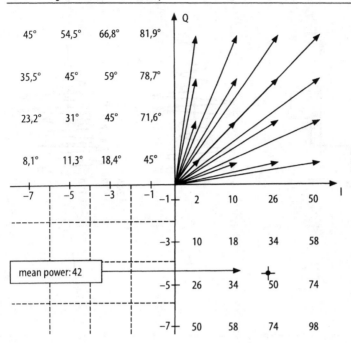

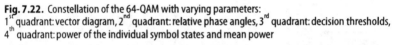

Fig. 7.22. Constellation of the 64-QAM with varying parameters:
1st quadrant: vector diagram, 2nd quadrant: relative phase angles, 3rd quadrant: decision thresholds,
4th quadrant: power of the individual symbol states and mean power

values, in accordance with a 16-QAM, results in a constellation which can be described by the 16 values in figure 7.22 which are nearest to the origin.

The special cases of those QAM constellations which result from the transmission of an odd number of m are not discussed in this section. There are various approaches to arranging the amplitude and phase values in the respective constellation. Some of the more frequently discussed solutions are called cross constellations and are explained, among others, in [LEE].

On the basis of (7.42), the general relation between the Euclidean distance d and S can be deduced from figure 7.22.

$$\frac{d}{\sqrt{S}} = \frac{2}{\sqrt{2\frac{M-1}{3}}} \Rightarrow d = \sqrt{4\frac{3}{2(M-1)}} S \qquad (7.43)$$

Similar to higher-level ASK the bit-error rate is obtained by computing the SER. According to the above findings an M-QAM symbol consists of two separate $\sqrt{M}$-ASK symbols. A complex QAM symbol is only correctly decoded when not only the ASK symbol of the in-phase component but also that of the quadrature component have been correctly detected. As both components are statistically independent of each other, this results in the following SER for the M-QAM technique:

$$SER_{QAM} = 1 - \left(1 - SER_{\sqrt{M}-ASK}\right)^2 = 2SER_{\sqrt{M}-ASK} - SER^2_{\sqrt{M}-ASK} . \quad (7.44)$$

From (7.35), by substituting the number of states M with $\sqrt{M}$, one obtains the error rate for one of two QAM components (in-phase or quadrature). Further, it is to be noted that each component has only half the QAM signal power, so that in the denominator of the argument the factor 2 has to be added:

$$SER_{I,Q} = \frac{\sqrt{M}-1}{\sqrt{M}} erfc\left(\sqrt{\frac{3}{2(M-1)} \cdot \frac{E_s}{N_0}}\right). \quad (7.45)$$

The argument of the complementary error function in (7.45) can be computed by substituting the Euclidean distance d from (7.43) and the matched-filter condition from (7.27) into (7.16).

The expression of the symbol-error rate of an M-QAM symbol follows from (7.44) and (7.45):

$$SER_{QAM} = 2\frac{\sqrt{M}-1}{\sqrt{M}} erfc\left(\sqrt{\frac{3}{2(M-1)} \cdot \frac{E_s}{N_0}}\right)$$

$$- \left[\frac{\sqrt{M}-1}{\sqrt{M}} erfc\left(\sqrt{\frac{3}{2(M-1)} \cdot \frac{E_s}{N_0}}\right)\right]^2 . \quad (7.46)$$

After conversion of the symbol energy E_s into the bit energy E_b (see (7.37)) the bit-error rate results in

$$BER_{QAM} = 1 - \left[1 - \left[2\frac{\sqrt{M}-1}{\sqrt{M}} erfc\left(\sqrt{\frac{3}{2(M-1)} \cdot \frac{m \cdot E_b}{N_0}}\right)\right.\right.$$

$$\left.\left. - \left[\frac{\sqrt{M}-1}{\sqrt{M}} erfc\left(\sqrt{\frac{3}{2(M-1)} \cdot \frac{m \cdot E_b}{N_0}}\right)\right]^2\right]\right]^{\frac{1}{m}} \quad (7.47)$$

in accordance with (7.38).

The bit-error rates for various M are plotted in figure 7.23 as a function of E_b/N_0.

In the representation of the two-dimensional constellations of a 64-QAM signal, which are shown as "snapshots" in figure 7.24, it can be seen how the ideal positions of the transmitted values are invalidated by superimposed noise. Each position represents one sampled complex amplitude value. With $E_b/N_0 = 12$ dB, the points, seemingly uncoordinated, appear to be distributed over the whole constellation. With $E_b/N_0 = 17$ dB the constellation pattern

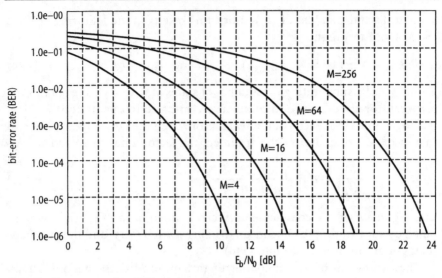

Fig. 7.23. Bit-error rates of the QAM techniques for various M

| $E_b/N_0 = 12$ dB | $E_b/N_0 = 17$ dB | $E_b/N_0 = 22$ dB |

Fig. 7.24. Constellations of a 64-QAM with $E_b/N_0 = 12$ dB, $E_b/N_0 = 17$ dB and $E_b/N_0 = 22$ dB

transmitted can already be clearly distinguished. From figure 7.23 it can be deduced that the signal now has a bit-error rate lower than 10^{-4}. When $E_b/N_0 = 22$ dB the bit-error rate drops below 10^{-6}.

Several characteristics of the QAM technique have been compiled in table 7.2. The mean power increases by about 3 dB per level. A comparison with table 7.1 shows that in the VSB technique, the one-dimensional expansion of the constellation results in an average power increase of approx. 6 dB per level. The ratio between maximum power – represented by the outer points in the constellation – and mean power is identical with the values of the VSB technique, as is the signal dynamics. However, here the values increase, with each level, by the same amount as in the VSB technique, therefore increasing the requirements for the level control of the amplifier used [FRIEDR].

Table 7.2. Characteristics of the QAM techniques

Number of symbol states and bits per symbol	Mean signal power	Ratio between peak and mean signal power	Dynamic range
$M, m = ld\ (M)$	$10 \log \left(2\frac{M-1}{3} \right)$	$10 \log\left(3\frac{(\sqrt{M}-1)^2}{M-1} \right)$	$20 \log (\sqrt{M} - 1)$
4, 2	3.0 dB	0.0 dB	0.0 dB
8, 3	6.7 dB see [PROAKIS]		
16, 4	10.0 dB	2.6 dB	9.5 dB
32, 5	13.2 dB see [PROAKIS]		
64, 6	16.2 dB	3.7 dB	16.9 dB
128, 7	19.2 dB see [PROAKIS]		
256, 8	22.3 dB	4.2 dB	23.5 dB

The QAM techniques will be used for the transmission of DVB signals in cable networks (see chapter 10) and also for terrestrial transmission (see chapter 11) in connection with the OFDM technology (see section 7.6).

7.6 Orthogonal Frequency Division Multiplex (OFDM)

A terrestrial broadcasting channel differs from a satellite transmission link or a cable transmission channel in that it is prone to multipath propagation. Reflections of the transmitted signal from obstacles such as buildings or mountains are superimposed asynchronously on the directly received signal. These reflected signals are, of course, time-delayed and can cause harmful interference. If the delay time of the echo signals approaches the symbol duration of the transmitted signal, this circumstance results in a selective behaviour of the frequency response [KAMMEYER]. Such a scenario is shown in figure 7.25, where it is aggravated by additional interference from a co-channel.

The individual echoes which successively arrive at the receiver, vary in amplitude and delay time. By superimposing themselves on the main signal they cause fluctuations in the complex channel transfer function. A characteristic value of such fading channels is given by the ratio of the directly received signal power to the total of the power of all echo signals.

By using suitable equalisers it is possible to compensate for the distortions in the frequency domain. The time delays of the various echoes, however, often by far exceed the symbol duration. This means that a corresponding number of adjacent symbols affect each other. A filter for ISI reduction must therefore be of a high order, which makes its implementation very expensive.

It is possible to minimise the number of symbols affecting each other by lengthening the duration of the symbol transmitted. This can be done by the

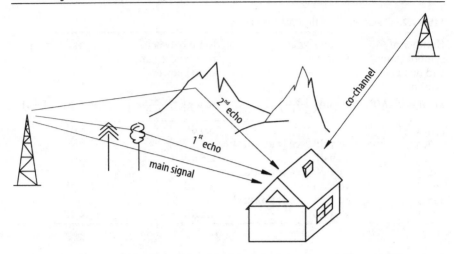

Fig. 7.25. Transmission scenario of a possible multipath reception in one channel with co-channel transmitter

parallel transmission of several symbols. If, for instance, the information to be transmitted is simultaneously modulated onto 1000 symbols of different carrier frequencies, then for each individual symbol (in the following referred to as subsymbol) there is a time slot available, which, before changing to parallel transmission, was allotted to all the 1000 sequentially transmitted symbols together. One of the basic criteria in communication engineering implies that the values of the bandwidth and the transmission time of an information can vary as a function of each other. In this way the frequency range required for the transmission of an individual subsymbol is reduced by the corresponding value. The total bandwidth of all subsymbols remains almost constant as compared with the bandwidth when the single-carrier technique is used. Figure 7.26 shows the block diagram of a multicarrier system at the encoder. As with QAM or QPSK, first of all m bits from the serial data

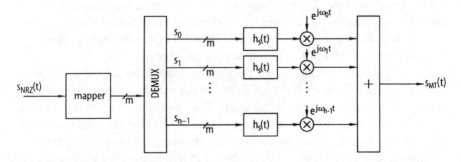

Fig. 7.26. Block diagram of a multicarrier system

stream are combined and mapped into complex subsymbols. Second, after a combination of the subsymbols to obtain the desired number to be transmitted in parallel, a serial-parallel conversion takes place. At the transmitting end, in each of the parallel branches there is a pre-filter (with the impulse response $h_t(t)$) as well as a modulator which modulates the respective subsymbol to the desired frequency position. Thereafter, these signals, which are often referred to as subcarriers, are added to a multicarrier symbol. The signal path which exists in the case of a multicarrier symbol combining n subsymbols s_k (as illustrated in figure 7.26) can be expressed as:

$$s_{MT}(t) = \sum_{k=0}^{n-1} s_k \cdot h_t(t) \cdot e^{j\omega_k t} \ . \tag{7.48}$$

As each individual subsymbol s_k can be modulated in amplitude as well as in phase, the multicarrier technique now also uses the third parameter – the frequency – for transmitting the information.

The complexity of this procedure increases to the extent that the number of parallel branches augments as a consequence of an increase in the number of necessary filters and modulators, so that this kind of implementation can very quickly lead to high costs.

A special case of the multicarrier technique is the OFDM system. For this a prerequisite is that all subcarrier frequencies ω_k be orthogonal to each other:

$$\omega_k \equiv 2\pi k f_o \quad \text{where} \quad k = 0, 1, 2, \dots, n-1$$
$$\text{and} \quad f_o \text{ as base frequency} \ . \tag{7.49}$$

In this case the parallel connection of the modulators in figure 7.27 exactly follows the rule for computing the inverse discrete Fourier transform (IDFT), with a subsequent frequency conversion of the entire signal.

The IDFT is a block-oriented processing algorithm [OPPENHM]. It is necessary for a predetermined number of subsymbols to be available simultaneously at the inputs of the IDFT unit. For this reason the sequentially received data are temporarily stored, until the required number of subsymbols for parallel transmission have accumulated, and are then read out in parallel. Figure 7.28 demonstrates by a simple example the principle of the signal processing within the subsequent IDFT unit. In this example an OFDM symbol is

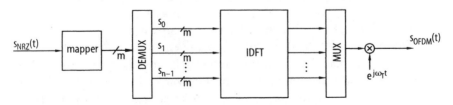

Fig. 7.27. Block diagram of an OFDM modulator

shaped from five consecutive bits. The first diagram represents the serial data stream. After the parallel transformation each bit lies at one of the inputs of the IDFT unit for the duration $T_U = 5\,T_B$ and generates a subsignal. The frequencies of the individual subsignals result in integral multiples of $f_0 = 1/T_U$. They are therefore orthogonal to one another. The total of all five subsignals results in the wave-form of the time-domain signal of an OFDM symbol after IDFT.

$$s_{OFDM}(t) = \sum_{k=0}^{4} s_k \cdot e^{j2\pi k \frac{t}{T_U}} \cdot rect\left(\frac{t - \frac{T_U}{2}}{T_U}\right) \tag{7.50}$$

From (7.50) it can be seen that the impulse response of each subsignal transmitted is rectangular. By applying the Fourier transform the frequency spectrum of an OFDM symbol can be computed.

$$s_{OFDM}(f) = \sum_{k=0}^{4} |s_k| \cdot e^{j\varphi k} \cdot si\left(\pi T_u\left(f - \frac{k}{T_u}\right)\right) \tag{7.51}$$

In the example shown here the third bit has the value 0, so that a notch results at the respective position in the spectrum of the output symbol. This relationship between the sequence of subsymbols before the IDFT in the transmitter and the individual subcarriers facilitates the shaping of the transmitted signal spectrum by way of substituting predefined subsymbols.

The reason for introducing a parallel transmission of numerous subsymbols was less due to the possibility of simple spectrum shaping, than to the fact that this solution satisfied the demand for the longest possible symbol duration. The example given in figure 7.28 results in an extension of the symbol duration T_U by the number of the temporarily stored subsymbols. If this figure is chosen to be very high, the number of adjacent OFDM symbols which contribute to ISI can be reduced considerably. A total avoidance of ISI, however, can only be accomplished by introducing a temporal guard interval whose task is to bridge the transient effects of the transmitted signal in the receiver which are caused by the broadcasting channel. Hence, the duration of the guard interval must be longer than the longest time delay of all echoes. It is therefore adjusted directly to the broadcasting channel.

The effect of the guard interval can easily be explained in the time domain [RUELBG]. The binary data stream from figure 7.28 shall serve again as input signal for the IDFT. For convenience, we here choose its inverted form. This means that all bits, except the third, have the value 0 and therefore do not affect the output signal. The third bit has the value 1. At the output of the IDFT we get a purely sinusoidal signal. It is plotted in figure 7.29 as the main signal. During the period taken up by the guard interval T_g the transmitted OFDM symbol is periodically prolonged in a forward direction. In the two diagrams below it there are two echo signals to be seen. They have the same shape as the

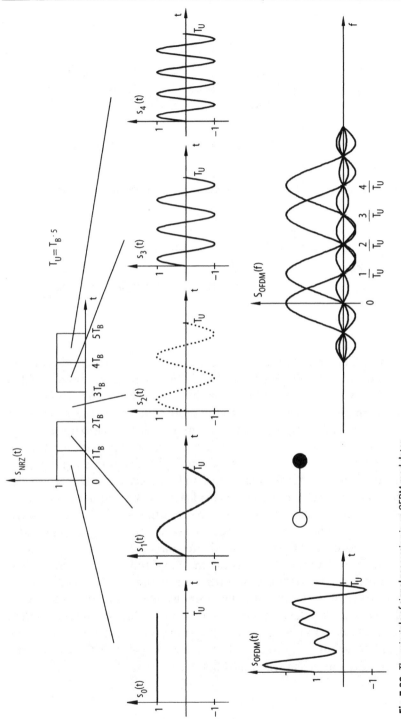

Fig. 7.28. The principle of signal processing in an OFDM modulator

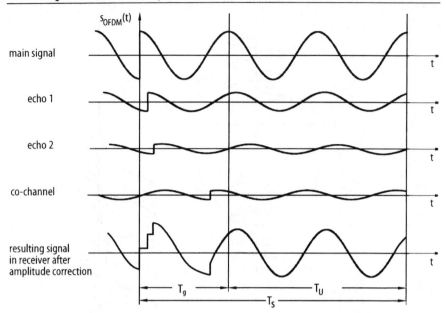

Fig. 7.29. Representation of the guard interval

main signal. Their amplitudes, however, are attenuated and their phases have been shifted due to the longer time delay. The fourth diagram shows a further signal which overlays the main signal on the transmission path. This can occur when a second transmitter sends the same signal on the same channel. The attenuation of the amplitude and the temporal difference to the main signal are relative to the distance between the transmitting stations. The co-channel signal has the same effect as a normal echo. The fifth diagram shows the addition, performed in the receiver, of all four consecutively received signals. The transients occur during the period taken up by the guard interval T_g. After all transient effects have died out, the resulting symbol assumes a stationary state which it retains throughout the entire period T_U. This period is used for the decoding of the transmitted information. In accordance with the function of the channel transmission the amplitude of the symbol has changed and its phase position differs from that of the signal transmitted. It is therefore just as important for the synchronous demodulation of the OFDM symbol to correct the effects of the channel as it is for broadband modulation techniques. The demand for an ISI-free transmission is nevertheless satisfied.

The OFDM technique is used for the transmission of DVB signals over terrestrial broadcasting channels (see chapter 11).

Symbols in Chapter 7

$A; A'$	amplitude in general
a	variable parameter for the amplitude
a_{thresh}	variable parameter for the amplitude boundary
BER_{Ix}	bit-error rate
c_{cor}	correction factor
d	Euclidean distance
$E\{x\}$	expected value of x
E_b	energy per bit
$erf(x)$	error function
$erfc(x)$	complementary error function
E_s	energy per symbol
f	frequency in general
$f(a)$	Gaussian amplitude-density distribution (normal distribution)
f_N	Nyquist frequency
f_o	basic frequency in the OFDM technique
$H(f)$	transfer function
$h_t(t)$	impulse response of the pre-filter at the transmitting end
I	in-phase component
i	running variable (integer)
j	$\sqrt{-1}$
k	running variable (integer)
l	running variable (integer)
M	amplitude states per symbol
m	bits per symbol
N	noise power
N_o	noise-power density in general
n	parameter (integer) or number of OFDM subcarriers
n_o	individual value of the parameter n
PSD_{Ix}	power spectral density
Q	quadrature component
S	mean signal power
$S(f)$	Fourier-transformed function of $s(t)$
S_a	instantaneous power
SER_{Ix}	symbol-error rate
S_{Ix}	signal power
$s_{Ix} s(t)_{Ix}$	signal in time domain
$si(x)$	$(\sin x)/x$
s_k	k subsymbol in OFDM technique
T_B	duration of one bit
T_g	duration of guard interval
T_U	duration of useful interval = $1/f_o$
T_S	duration of one symbol
t	time in general
x	variable in general
z	variable in general
α	roll-off factor
δ	δ (masking) function (Dirac impulse)
ϕ_{sk}	phase shift of the kth OFDM subsymbol
σ_R	standard deviation of a Gaussian noise signal
ω, ω_i	(i^{th}) radian frequency
ω_T	radian carrier frequency
ω_k	k^{th} radian subcarrier frequency in OFDM

8 Conditional Access for Digital Television

"Conditional access" (in the following referred to as CA) is a technique used to protect a programme or a number of programmes from unauthorised viewing. Its implementation requires a variety of technical and commercial system components, which serve the purpose of making the programmes available only to those viewers authorised to receive them (pay TV). Viewers are usually required to pay a monthly or annual fee to gain access to a particular programme channel (pay-per-channel) or, alternatively, a fee for an individual programme (pay-per-view). CA is a technique which originated, and is widely used, in English-language countries, which is why the English expressions are internationally accepted. An overview of a complete CA system is shown in figure 8.1.

The programme signal is processed in a scrambler before transmission. Within the framework of the DVB Project it has been possible to develop a so-called common scrambling system, which is supported by all CA providers. The specification describing this system is not published so that possible "pirates" will have difficulty acquiring the knowledge needed for the construction of illegal descramblers. Although the members of the DVB Project are aware that an absolutely secure scrambling system cannot be found, they are satisfied that the common scrambling system adopted is as secure as possible. As long as the instructions for deciphering are missing in a receiver, it will be impossible to view a scrambled programme.

The DVB Project has taken the initiative to propose anti-piracy laws for Europe and for each individual country. These laws will complement the development of the common scrambling system.

The concept of the common scrambling system is based on the cascading of two ciphering procedures. In the first system, data blocks of 8 bytes, each consisting of 8 bits, are scrambled, and in the second, the resulting data are rescrambled bit by bit [ETR 289].

The procedure used for scrambling is illustrated in figure 8.2. First of all a decision is taken as to which data are to be scrambled. If, for example, scrambling is performed at the level of the transport stream the header cannot be included, because the header is necessary to synchronise the receiver. Furthermore it must be possible for the content provider to scramble only part of the services.

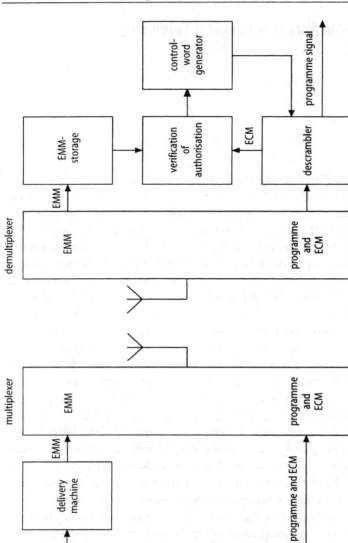

Fig. 8.1. System overview of a conditional-access system

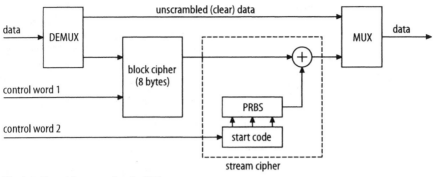

Fig. 8.2. Scrambling procedure for DVB

The first step utilises a block-cipher procedure, a technique based on 8-byte blocks. A first "control word" is required for the ciphering. The data stream coded in that way is then fed into a stream-cipher mechanism which operates with a pseudo-random generator, i.e. it creates a period of pseudo-random data out of another control word. This step can be implemented with the aid of a feedback shift register which, at a given moment, is loaded with a specified initialisation value. The bit stream which is output by this generator is then added modulo-2 to the data to be scrambled.

In the MPEG-2 structure, two levels at which the ciphering can take place are envisaged (see chapter 5): the level of the packetised elementary stream (hereafter referred to as PES) and that of the transport stream (hereafter referred to as TS). Only one of these levels should be used at any one time. The respective header, which is never ciphered, contains special control bits which have the same meaning at both levels (table 8.1): the first shows whether the respective block is coded and the second which cipher (even or uneven) is used for the packet. This differentiation is required for the following reason. The key (i.e. the encrypted representation of the two code words) is changed from time to time. The new key is transmitted with the MPEG-2 data stream, and the second scrambling-control bit then shows that everything following the block to which the header belongs is subject to the changed cipher.

There are some restrictions for scrambling at PES level [ETR 289] with regard to the mapping of the PES onto the TS (see figure 8.3):

Table 8.1. Control bits for scrambling

Bit values	Meaning in TS and PES header respectively
00	no scrambling
01	(not used at the moment)
10	scrambled with even code word
11	scrambled with uneven code word

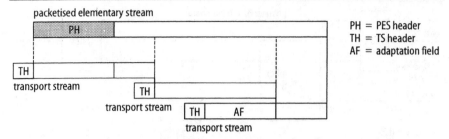

Fig. 8.3. Mapping of a PES packet onto the transport stream (TS)

- The PES header should not be larger than 184 bytes, i.e. it must fit into the useful data range of a TS packet.
- From the beginning of its header, the PES packet is divided into segments of 184 bytes each, which are then mapped onto a TS packet. These TS packets can therefore have no adaptation fields.
- If the last segment is smaller than 184 bytes, it will be preceded within the TS packet by an adaptation field of the appropriate length.
- Should an adaptation field become necessary during the transmission of a scrambled PES packet, a separate TS packet has to be inserted which contains only this adaptation field.

The aim of these limitations is to minimise the storage requirement at the receiver by simplifying the deciphering.

The two control words are required to enable the receiver to descramble the programme information. They are subjected to a separate encryption procedure at the transmission end, which transforms them into entitlement control messages (hereafter referred to as ECMs). The first ECM is used to inform the descrambler in the receiver how the blockwise scrambling can be revoked. The second ECM enables the descrambling of the bit-by-bit scrambling.

In order to transmit the information required for the descrambling in the receiver a conditional-access table has been specified as part of the so-called service information (SI) within the framework of the DVB Project. The entitlement management messages (hereafter referred to as EMMs) can be transmitted in this table as well as the ECMs. The EMMs originate from the customer administration of the CA provider, i.e. from the subscriber management system (SMS). They are attached to the programme stream via a delivery machine which, for example, ensures that each subscriber is supplied hourly or daily with the new, individually valid EMM. If the pay-per-view procedure is to be applied to a programme, then the EMM has to be made available at short notice for each programme as soon as access has been ordered by the subscriber.

The EMM is stored in the receiver. In order to enable an unambiguous allocation of the EMM to a particular receiver (i.e. to the subscriber) each receiver has to have an individual identity number, which, for example, is made available by a memory card. When the receiver has thus been authorised by EMM and memory card to receive a specific channel or programme, the control words required for the deciphering are generated from the ECMs and loaded into the descrambler, and the descrambling can commence.

The development of the common scrambling system, the use of which, as mentioned before, has been agreed within the DVB Project, constitutes a unique step towards a common user terminal for all scrambled services. This system has even been introduced into a European directive by the European Commission. All user terminals, on condition that they have the right ECMs and the correct EMM, will be in a position to descramble all scrambled programmes.

The introduction of pay programmes using common receiver hardware has thus been reduced to the question of whether the providers of scrambled programmes are willing and able to accept a further degree of standardisation. The options for further agreements are as follows:

- All providers of pay programmes agree on a uniform CA system. This option has proved to be completely unrealistic for commercial reasons.
- If every receiver were to have a common interface (CI) for an exchangeable plug-in module, or something similar, then the provider-specific processing of ECMs and EMM as well as all further steps in the CA procedure could be integrated into such a module. All providers could then work with their own CA systems without consideration of one another. This "multicrypt" possibility is generally regarded as viable. As a consequence it was possible to finalise the specification for a common interface which was standardised by CENELEC as [EN 50221]. The reason why the definition of this interface proved to be very problematic was that the knowledge of the specification, if it described an interface right inside a CA system at the receiver, would make it considerably easier for "pirates" to build illegal hardware. Moreover, the costs of such hardware might be so much reduced by the utilisation of the common interface that the commercial threshold for potential "pirates" would also be much lower. For this reason it was decided that the common interface should be placed at the signal level of the MPEG-2 transport multiplex. This entails that the exchangeable hardware must contain the whole CA system.
- If all providers of pay programmes were to agree that, although they each want to market their own receiver which is only suitable for their own CA system XY, they are prepared, for a reasonable fee, to offer every other provider the opportunity of generating their own ECMs and EMMs by means

of this CA system XY, then the objective of a uniform hardware for the subscribers could be achieved. As a result of such an agreement it would also be possible for potential providers to decide freely when, where and for which group of subscribers the marketing of receivers and CA services should commence. This option, named "simulcrypt", was the subject of considerable discussion within the DVB Project. Following lengthy mediation, a code of conduct has emerged which describes the contractual basis for simulcrypt. It is expected that this code will become an integral part of European law. The technicalities of simulcrypt are described in DVB specification [TS 101 197-1].

If, in the future, a network operator wishes to offer his customers – who may, for example, be cable subscribers – a uniform CA concept, then, before feeding scrambled programmes into his network, he must replace the ECMs and EMMs contained, for instance, in the satellite-transmitted pay programmes by data of his own CA system. This type of transformation has been designated as "transcontrol". After holding yet more discussions, the members of the DVB Project have agreed to accept the principle of transcontrol.

9 The Satellite Standard and Its Decoding Technique

Apart from the distribution of broadcast signals by terrestrial stations and cable networks, the transmission via satellite has gained considerable importance over the past few years. For DVB, satellite transmission is also of great significance. The satellite standard developed by the DVB Project was implemented by ETSI as European standard ETS 300 421 and came into force on 1-1-1995 [ETS 421, REIMERS 3].

This chapter will first discuss signal transmission via satellite in general, then look at the encoding and decoding techniques for the DVB satellite standard and, finally, introduce the most important characteristics of the system.

9.1 The Basics of Satellite Transmission

9.1.1 Transmission Distance

Satellites for the distribution of broadcast signals are located in geostationary orbit, i.e. an orbit of about 36,000 km above the equator. The orbit duration is exactly one star day, which means that, seen from the earth, the satellites appear to be stationary. Deviations from the orbit position, due to the solar wind or due to drifting caused by the uneven gravitational pull of the earth, can be corrected by the use of steering jets. The fuel supply for these steering jets is usually the component which determines the life of the satellite. Solar cells provide the energy supply for the operation of the electronic equipment. These do deliver permanent energy (during "daylight hours"), but the power is relatively low. Therefore the satellite channel, at least the downlink, i.e. the transmission from satellite to earth, must be regarded as a channel with very limited power.

The consequences resulting from this limitation will be discussed in detail at a later point.

One of the resources which is available in great abundance is the bandwidth. Communication satellites for broadcasting transmit in a frequency range between 10.7 and 12.75 GHz and in future the band from 21.4 to 22.0

GHz will possibly also be available. Satellites, therefore, have far more bandwidth available than terrestrial transmitters. Moreover, the same frequencies can be used in different orbital positions. This is due to the distinctive directional characteristic of the receiving antenna used in satellite transmission.

9.1.2 Processing on Board a Satellite

Signals which pass through a satellite are distributed to individual transponders. A transponder is the transmission channel placed between the receiving and the transmitting antennas of the satellite and consists of a number of functional units. The frequency range assigned to a transponder has a typical transmission bandwidth of 26 to 36 MHz, but the size of the bandwidth may be much larger in some satellites.

The processing of the signals within the satellite (figure 9.1) is briefly described in the following.

The uplink signals received from the earth station by the receiving antenna are first fed to a band-pass filter, which filters out the frequency range for the particular satellite. Following a pre-amplification, the signals are mixed to the downlink frequency range. There then follows a further amplification. This arrangement with two amplifiers which function at different frequency ranges prevents self-oscillation which would occur with a single amplifier [RODDY].ubsequently the broadband signal is divided into the various frequency ranges by band-pass filters, the so-called input multiplex (IMUX) filters. The signal of each of the frequency ranges is then fed into a travelling wave tube amplifier (TWTA) with a characteristic curve which increases almost linearly in the lower range, then becomes flatter and finally reaches a saturation point (figure 9.2). If the entire capacity of the amplifier tube is to be

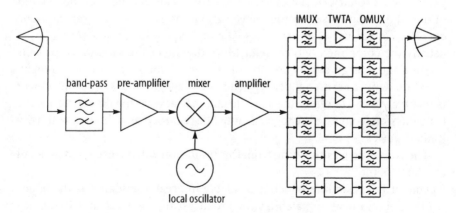

Fig. 9.1. Simplified block diagram of the payload on board a satellite

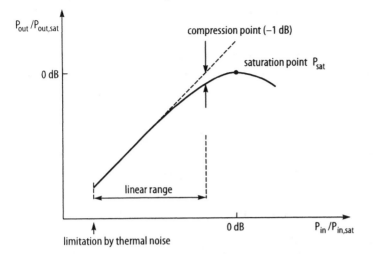

Fig. 9.2. Typical power-transfer characteristic of a travelling wave tube amplifier

used, the non-linearity has to be accepted, which has consequences for the transmission scheme, i.e. the channel coding and modulation:

- With amplitude modulation, the described non-linearity would lead to signal distortion, which means that only frequency- or phase-modulated signals are acceptable. Therefore frequency modulation, due to its immunity to interfering, is chosen for analogue television transmission, whereas for DVB a digital phase modulation (QPSK) is used (see chapter 7).
- If several signals are transmitted by one transponder in the frequency multiplex, intermodulation occurs, i.e. new frequency components are formed by the non-linear components in the amplifier. To avoid these, the operating point must be in the linear part of the TWTA's power-transfer function. This can be achieved by reducing the input power by a few dB, i.e. by applying a so-called input back-off (IBO). As a consequence of this, the output power (output back-off [OBO]) is reduced. Since the maximising of power is the key requirement, the transmitted signal needs to be one single carrier on which the payload is multiplexed in a time-division multiplex. The potential existence of intermodulation products is another reason for the necessity for each transponder to have its own amplifier and to transmit one modulated carrier only.

At the output of the power amplifier the signal from each transponder is filtered once again by an output multiplex filter (OMUX), during which process the harmonics originating in the non-linear amplifier are suppressed. The transponder signals are then recombined and fed to the transmitting antenna, downlinking it to earth.

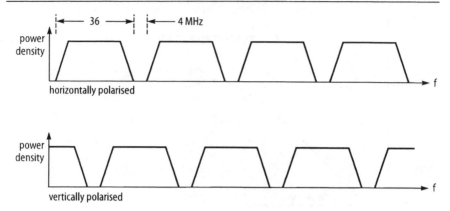

Fig. 9.3. Example of a transponder arrangement with polarisation decoupling

The transmitting antenna has a strong directional characteristic, which causes a relatively high power density in the main direction of the transmission. The power an isotropic radiator needs to have to uniformly generate a maximum of power density in all directions is called "equivalent isotropically radiated power" (EIRP) and constitutes an essential parameter for determining the characteristic features of a satellite.

9.1.3 Polarisation Decoupling

Electromagnetic waves offer the possibility of linear or circular polarisation [RODDY]:

For linear polarisation, a horizontally polarised wave and a vertically polarised wave of the same frequency may be modulated by two independent signals. At the receiver the two signals can then, ideally, be completely separated again (in practice, the crosstalk is <–18 dB). This method is used, for example, by Astra and Eutelsat. A circularly polarised wave consists of a horizontally and a vertically polarised wave of the same amplitude with a phase difference of ±90°. The sign of the phase difference determines whether the polarisation of the signal is left-handed or right-handed. Circular polarisation is used for example for TV-Sat.

Figure 9.3 shows a frequently used scheme in which the horizontally and vertically polarised transponders are staggered with respect to frequency. This has the advantage that the centre frequency of a transponder, which in analogue transmission typically has a maximum power density, faces the gaps in the other polarisation plane. Crosstalk can thus be minimised by this arrangement.

9.1.4 Energy Dispersal

In general, it cannot be assumed that the power density of a digital television signal will be distributed evenly within the transponder bandwidth. In the case of a longer sequence of ones, for instance, as might occur in stuffing packets, the power in a QPSK modulation concentrates on the carrier frequency. Such peak power in the spectrum can cause a distortion in the reception of channels of neighbouring satellites, when these transmit in the same frequency range and the directional characteristic of the receiving antenna on the earth is not narrow enough. Hence the aim is to achieve a power-density spectrum of the modulated signal that is as even as possible.

In analogue television, where these effects on the power-density concentration also occur, the energy dispersal can be achieved by an additional frequency modulation of the carrier with a triangular signal [RODDY]. To avoid interference effects in the image signal, a basic frequency of 25 Hz (50 fields per second) is chosen for PAL transmissions. The carrier oscillates continuously between two frequencies with a separation of one MHz and thereby achieves a somewhat more even energy dispersal.

In DVB the energy dispersal occurs at the very level of the code. A scrambling technique, which is described in detail further down, sees to it that the data stream assumes a seemingly random structure, which results in an almost even energy dispersal.

The energy dispersal is then revoked by the same operation being performed at the receiver.

9.1.5 Signal Reception

At the receiving end the signals transmitted from the satellite in the gigahertz range are directed via a parabolic reflector to the actual antenna (figure 9.4). They are then pre-amplified in the low-noise block (LNB) and downmixed to the first intermediate frequency (IF). This lies in the range of 950 to 2150 MHz and enables the signal to be transferred via a reasonably priced coaxial cable to the receiver. In the receiver there is an arrangement which is similar to that in the satellite: amplifier – mixer – amplifier. The mixing frequency is chosen such that the selected channel within the broadband satellite signal lies in the band pass of the following filter which suppresses all other channels (superheterodyne principle). Subsequently, the signal can be demodulated.

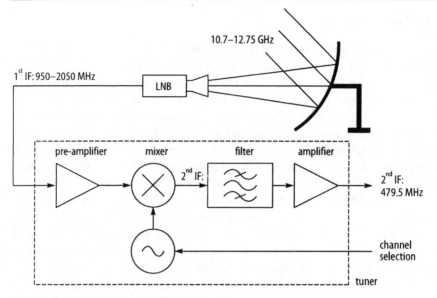

Fig. 9.4. Simplified block diagram of a satellite receiver

9.1.6 Reference Data of a Television Satellite with Astra 1D as an Example

As an example of the order of magnitude of the technical details discussed above, table 9.1 summarises the performance of satellite Astra 1D.

Table 9.1. Data of the Astra 1D satellite

Start:	October 1994
Life expectancy:	13 years
Orbit position:	19.2° east
Total power:	3300 W
Number of transponders:	18
Transponder bandwidths:	26 MHz (–3 dB) for FSS band (10.70–11.70 GHz)
	33 MHz (–3 dB) for BSS band (11.70–12.07 GHz)
Transponder output power:	63 W
EIRP:	52 dBW = 160 kW

9.2 Requirements of the Satellite Standard

The technical requirements of the transmission path on the one hand and the user demands on the other constitute the parameters for the definition of the satellite standard.

The requirements concerning the transmission path can be deduced from the following characteristics of satellite transmission:

- Due to the low-power capacity of the satellite channel the travelling wave tube should run in full saturation. Therefore modulating techniques in which the carrier is also subjected to an amplitude modulation (as for example in the higher-order QAM) cannot be considered.
- To avoid intermodulation in the travelling wave tube, only time-division multiplexing on a single carrier can be used.
- There must be energy dispersal in order to achieve a power density which is distributed as evenly as possible within the transponder bandwidth.
- The low carrier-to-noise ratio at the receiver, which is mainly due to the extremely low received power, makes a high-quality error protection necessary. A quasi error-free (QEF) transmission which is defined by a bit-error rate of less than 10^{-11} is envisaged. In practice, this corresponds to an average of about one false bit per hour per transponder.

The above requirements need to be contrasted with the demands made on the system by the users:

- For television broadcasting – as well as for other digital services – high transmission rates are demanded.
- The available transmission capacities must be used flexibly, i.e. different services with different data rates must be supported.
- The quality of the error protection, too, must be able to be adapted in a flexible way to the various requirements. For example, for a transmission via a high-powered satellite, a lower-quality error protection than that required for a satellite with a lower transmitting power will suffice, so that the net data rate (i.e. the rate of the effectively transmitted uncoded data) is higher although bandwidth and QEF requirement are the same.
- The receiving antenna should have a reflector diameter as small as possible and should be as unobtrusive as possible when mounted on the wall of a house. Furthermore, like the receiver as a whole, it should not be too expensive. However, a small receiving antenna also implies a low received power and therefore a low carrier-to-noise ratio.

Below are listed the basic characteristics of the satellite standard, which was developed taking the above requirements into account [ETS 421]:

- The source coding is carried out in accordance with the MPEG-2 standard, which combines an efficient data compression with a flexible system concept (see chapters 4 and 5).
- As already mentioned, energy dispersal at the encoding end is carried out by scrambling with a pseudo-random sequence.
- A concatenated error protection permitting various code rates is used.
- The QPSK modulation, which guarantees a constant carrier signal amplitude, was chosen for the DVB standard.

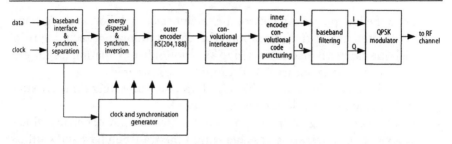

Fig. 9.5. Block diagram of signal processing at the transmitting end

- Only one carrier per transponder is used in order to avoid intermodulation.
- The combination of various services in a "data container" for simultaneous transmission via one transponder takes place in a time-division multiplex. The basis for this is the MPEG-2 transport stream.

9.3 Signal Processing at the Encoder

9.3.1 System Overview

In the following, the satellite standard will be explained on the basis of a block diagram representing the coding and modulation process at the transmitting end (figure 9.5).

The main steps in the adaptation of the transport stream to the satellite transmission link are:

- energy dispersal by scrambling,
- concatenated error protection with interleaving,
- QPSK modulation.

The following two characteristics of the transport stream are of particular significance for the coding and modulation in satellite transmission (cf. chapter 5).

- The MPEG-2 transport stream is composed of individual packets (hereafter referred to as frames) with a length of 188 bytes each. The first four bytes form the header, the first header byte being the synchronisation byte (hereafter referred to as sync byte).
- The "transport-error indicator bit" is also defined in the header. If the packet is no longer decodable due to too many channel errors, then this bit is used to indicate an undecodable erroneous packet for the source decoding.

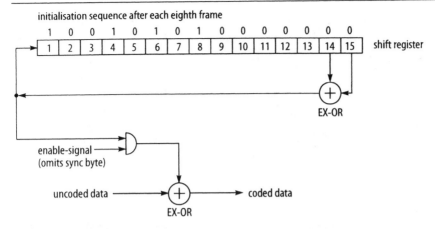

Fig. 9.6. Diagram of the energy dispersal circuit

9.3.2 Energy Dispersal

To generate a channel data stream with the most evenly distributed power-density spectrum possible (see section 9.1.4) the signals are combined bitwise with the output stream of a pseudo-random generator, via an "EXCLUSIVE-OR" operation (modulo-2 addition). This random generator is implemented by a feedback shift register, which is re-initialised at the start of every eighth frame in accordance with a predetermined bit pattern. It is only the sync byte that remains unscrambled, so as to enable synchronisation of the removal of the energy dispersal by the receiver.

When the random generator is re-initialised at the start of every eighth frame, the current sync byte is inverted as a signal for the "energy-dispersal remover" at the receiving end.

9.3.3 Error-protection Coding

The error protection coding follows the energy dispersal. This is a concatenated coding which comprises a block code, an interleaver and a convolutional coder (see chapter 6, [SWEENEY]).

The block code is an RS(204, 188) code, which enlarges each block of 188 bytes (in other words, a complete transport-stream packet) by 16 correction bytes to a gross length of 204 bytes and which can thus correct up to eight erroneous bytes.

The interleaver is designed as a convolutional interleaver with an interleaving depth of $I = 12$. The base delay is 17, and therefore the block length of the interleaver is $n = I \cdot M = 204$ bytes – in other words, a complete RS-coded

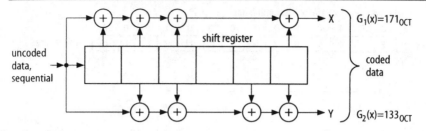

Fig. 9.7. Structure of a convolutional encoder

packet. Bytes which would originally have been contiguous in the data stream are subsequently at least 205 bytes apart.

A convolutional encoder with the basic code rate 1/2 and the generator polynomials $G_1 = 171_{OCT}$ and $G_2 = 133_{OCT}$ follows the interleaver (figure 9.7).

Several possibilities for puncturing are envisaged in order for the coding to be as flexible as possible, to adjust to the existing conditions of the channel and to the requirements of the transmission. According to the DVB specification the following code rates are possible: 1/2 (no puncturing), 2/3, 3/4, 5/6 and 7/8.

Figure 9.8 shows the arrangements which are used for each instance of puncturing. Following the puncturing of some bits from the data streams X and Y delivered by the encoder, the remaining bits must be rearranged so that the original order with two paths can be retained. These two paths are required to supply the modulator with two bits simultaneously (I- and Q-signals).

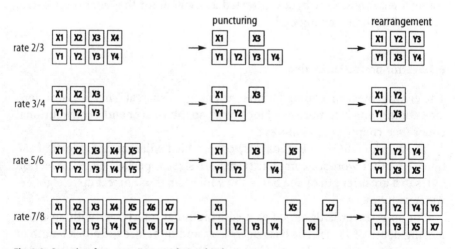

Fig. 9.8. Procedure for puncturing convolutional codes

9.3.4 Filtering

Subsequent to the error-protection coding, the data are pre-filtered and fed into the modulator.

The filtering fulfils two purposes:

- Firstly, the signals must have a band limitation in order to avoid the possibility of crosstalk on adjacent channels.
- Secondly, the filtering serves to shape the signals in accordance with the first Nyquist criterion (see section 7.1).

The Nyquist filter, introduced in section 7.1, with a raised-cosine edge and a roll-off-factor of $\alpha = 0.35$, has been chosen as the filter characteristic for the satellite standard. This filter characteristic is valid for the whole channel, which means that it is the product of all filters in the transmission path. Apart from the somewhat broader-band IMUX- and OMUX-filters in satellites this characteristic applies especially to the pre-filter in the transmitter and to the receiver input filter in the satellite tuner. In accordance with the standard, the filters in the transmitter and in the receiver must have the same transfer function, therefore this must be the square root of the total transfer function; hence these filters are called "square-root raised-cosine" filters. A template for the signal spectrum at the modulator output is given in the appendix to the satellite standard.

9.3.5 Modulation

The filtered signals of both paths are fed into a QPSK modulator [MÄUSL] as in-phase and quadrature components (figure 9.9). The so-called Gray coding allocates the bits to the respective constellation points. If, at demodulation –

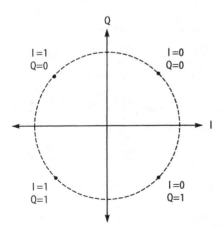

Fig. 9.9. Constellation diagram of the QPSK

owing, for example, to noise in the channel – a phase is decoded which passes beyond one of the two decision thresholds (= axis in figure 9.9), there will only be one erroneous bit.

Before transmission, the IF signal is upconverted to radio frequency.

In this way two bits per symbol are transmitted; hence in ideal circumstances the utilisation of the bandwidth, referred to the gross bit rate, is $B = 2$ bit/(s · Hz). However, owing to the non-ideal filters (finite width of the filter slopes) the bandwidth utilisation is actually smaller. Depending on the configuration of the system, the ratio between bandwidth and symbol rate (BW/R_S) would be in the range of about

$$\frac{BW}{R_s} = 1.27 \frac{\text{Hz}}{\text{symbols}/\text{s}}. \tag{9.1}$$

Hence the bandwidth utilisation would be about

$$B = \frac{\text{bits per symbol}}{BW/R_s} = \frac{2 \text{ bits}/\text{symbol}}{1.27 \text{ Hz}/\frac{\text{symbols}}{\text{s}}} = 1.57 \frac{\text{bit}}{\text{s Hz}}. \tag{9.2}$$

The bandwidth utilisation can also be referred to the net bit rate, i.e. to the number of information bits only, without the error protection. In this case the calculated value, as indicated above, must be multiplied with the code rates of the outer and inner codes.

9.4 Decoding Technique

On reception of the signals transmitted from the satellite, the processing steps that were carried out at the encoder have to be reversed. In particular, the errors which occurred in the channel must be corrected. Moreover, it is necessary to recover the synchronising information, which is required for the channel decoding.

The decoder consists of the following components (figure 9.10):

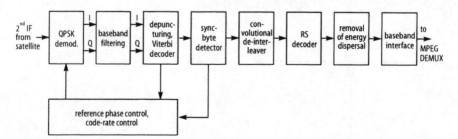

Fig. 9.10. Block diagram of signal processing at the receiving end

- demodulator,
- filter,
- clock recovery,
- sync-byte detector,
- decoder and interleaver for concatenated error protection,
- energy-dispersal removal,
- baseband interface,
- control unit.

9.4.1 Demodulator

The QPSK demodulator must first of all recover the carrier frequency from the input signal. A particular problem is the phase uncertainty of the carrier, which must be resolved in order to demodulate I and Q correctly. The input signal can assume one of four possible phase positions, each of which represents two bits (see figure 9.9) and can serve as a reference for the detection, so that it is possible to have four carrier phases at right angles to each other. As the absolute phase position is unknown to the receiver, it first chooses a phase at random for use in the demodulation.

The demodulator can be implemented together with the carrier recovery by means of a Costas loop [MÄUSL, COSTAS]. The carrier frequency for demodulation is created by a voltage-controlled oscillator (VCO). The VCO is operated by a control signal which is generated, in a control loop, from the two demodulated baseband signals.

The decision as to whether the right phase position is in use can only be made later, in two steps, in the decoder. A phase error of ±90° can be resolved in a first step, and in a second step the residual uncertainty of 180° can be removed. The mechanisms required will be described later in the appropriate sections.

The correction of the absolute carrier phase recognised as wrong generally does not happen at the recovered carrier itself, because this would require considerable expenditure. A correction of 90° can be obtained by exchanging the I and Q components after demodulation and subsequently inverting one of them. A phase error of 180° is compensated by an inversion of the bit stream at the position in the decoder at which it was identified.

9.4.2 Filtering and Clock Recovery

Following demodulation, the baseband signals are subjected to the square-root raised-cosine filtering, as was the case at the transmitting end (see section 9.3.4). Thereafter crosstalk between adjacent symbols at the sampling point is minimal.

The filter can be implemented as a digital filter after an oversampling of the demodulated baseband signal or as an analogue band-pass filter before the demodulator.

The clock of the filtered baseband signal can be recovered with the aid of a PLL-circuit so that the signal can be sampled [LEE].

9.4.3 Viterbi Decoder

The first step in error-correction is the decoding of the convolutional code, performed by the Viterbi decoder, which is preceded by the depuncturing (see section 6.3.4). Soft decision must be made available to the Viterbi decoder by the demodulator. Soft decision improves the error-correction capability and makes possible the correct interpretation of depunctured bits. In most cases 3 bits are implemented, which, in the case of a code rate of 1/2, results in a reduction of the required signal-to-noise ratio by approx. 2 dB for a residual bit-error rate of $2 \cdot 10^{-4}$ (figure 9.11, cf. section 9.5.2). A resolution higher than 3 bits is hardly an improvement and only increases the implementation requirement (see figure 9.11). However, 4-bit soft-decision quantisation can be found in state-of-the-art implementations.

During the processing, the Viterbi decoder counts the number of errors that have been identified and transmits this number to the control unit which has the task of determining the as yet unknown transmission parameters:

- the reference phase for the demodulation (possible uncertainty of 90°),
- the puncturing scheme,
- the synchronisation of the depuncturing. Similar to the puncturing, the depuncturing is a block-oriented periodic process (see figure 9.8). Com-

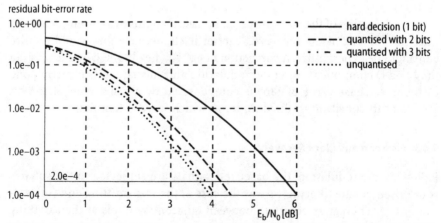

Fig. 9.11. Effect of quantisation on bit-error rate (simulation with DVB convolutional code at code rate 1/2)

mencing at the wrong point, i.e. not at the beginning of a block, will result in the depunctured bits being also inserted at the wrong points.

The determination of these – as yet unknown – parameters is achieved by means of the control unit working through all possible combinations. At the output of the Viterbi decoder the data are then recoded and punctured, in accordance with the parameters determined at the decoding, and subsequently compared with those bits which were originally available at the depuncturing input. In this way the value of the bit-error rate can be computed. Should only one of the above-mentioned parameters be wrong, then this rate would be quite high, due to the fact that the redundancy generated in the encoder would be wrongly interpreted.

It is only when all the parameters are correct that a proper decoding is possible and that the measured bit-error rate will assume a value which is considerably smaller and roughly corresponds to the actual bit-error rate of the channel. The precondition for this is that the signal-to-noise ratio be high enough, so that there is a sufficient distinction between a correct and a false choice of parameters. If the satellite link operates at least close to the standard operating point defined by DVB, the procedure will lead to the desired result. However, if the signal-to-noise ratio is not sufficient, so that the bit-error probability is too high even if the parameters are correct, the decoder will carry on searching. If all possibilities have been exhausted a corresponding signal will be supplied which informs the user that reception is not possible.

A special case in the detection of the missing parameter is the shifting of the reference phase in the demodulator by 180° as against the reference phase in the transmitter. This will result in a bit stream at the receiving end which is inverted compared with the bit stream that was transmitted. The Viterbi decoder will then also emit a bit stream, which (apart from uncorrected errors) is the inverse of the uncoded bit stream from the transmitter. The phase error cannot be detected, hence the uncertainty of 180° remains and is only resolved in the subsequent module, which is the sync-byte detector.

9.4.4 Sync-byte Detector

Further decoding requires that the subdivision of the data stream into the MPEG-2 TS packets as well as the 8-packet structure for the energy dispersal be detected. The sync byte at the beginning of each frame is required for this. At the transmitting end the sync byte remains unchanged by RS encoder, interleaver and energy dispersal and is therefore available at the output of the Viterbi decoder.

The task of the sync-byte detector is to detect the regular occurrence of the sync byte, or of its inversion, every 1632 bits (204 bytes) within the bit stream and from this to derive synchronising signals for the de-interleaver, the RS decoder and the energy-dispersal remover. In the course of this operation the last phase of uncertainty can now be removed from the decoding process: the occurrence of seven inverted sync bytes and one non-inverted sync byte indicates an erroneous phase shift of 180°, in which case the bit stream is simply inverted at the output of the detector.

A byte clock is also derived so that, after a serial-to-parallel conversion, further processing can take place at the byte level.

9.4.5 De-interleaver and RS Decoder

The satellite channel is essentially a Gaussian channel, which means that the signal has been corrupted by additive white Gaussian noise, and that bit errors occur singly rather than in bursts. The convolutional code is well-suited for the correction of such errors. However, in the case of an accumulation of bit errors the Viterbi decoder will fail and supply a wrongly decoded signal, i.e. it will in its turn generate a burst error. To correct this burst error, interleaving and outer error protection (Reed-Solomon code) are applied. If as a result of a decreasing signal-to-noise ratio the number of bit errors in the transmission increases, the bursts become longer and more frequent. Finally, if more than eight erroneous bytes occur in one frame at the input of the RS decoder, the error protection will fail. In most of these cases the decoder recognises that the data can no longer be corrected and transfers an appropriate signal to the subsequent circuits.

Uncorrected errors generally become apparent by distorted blocks appearing in the picture. When the signal-to-noise ratio reaches the critical value at which even the outer coder starts to fail, the number of errors increases very fast in the event of a further drop in the signal-to-noise ratio, until at last even the synchronisation fails and a decoding of the picture is no longer possible. This total breakdown occurs in the range of a signal-to-noise ratio of less than one dB.

9.4.6 Energy-dispersal Remover

Energy dispersal takes place at the transmitting end by means of the modulo-2 addition of a pseudo-random number sequence (see section 9.3.2) and is reversed in the receiver by the same operation. It follows from this that the circuit in the receiver is the same as that in the transmitter. To initiate the pseudo-random generator the energy-dispersal remover requires a signal which indicates the start of the 8-packet sequence and is supplied by the sync-byte detector.

9.4.7 Baseband Interface

After the original MPEG-2 TS has been reconstructed it can be transferred to the TS demultiplexer which feeds the various components of the data stream into the respective source decoders. For the signalling of errors which have occurred in the transmission channel and have been identified by the error protection but which the error protection was unable to correct, MPEG-2 has provided a specific signal: the "transport-error indicator bit", which is the first bit after the sync byte and will be set in such cases. This bit enables the demultiplexer to identify the respective packets as erroneous and take the necessary precautions.

9.5 Performance Characteristics of the Standard

9.5.1 Useful Bit Rates

The useful bit rate R_u can be computed from the symbol rate by

$$R_u = R_S \frac{\text{bit}}{\text{symbol}} R_1 R_2 \tag{9.3}$$

where $R_1 = 188/204 = 0.922$ is the code rate of the Reed-Solomon code and R_2 is the code rate of the convolutional code with or without puncturing.

Table 9.2 [REIMERS 3] lists the possible useful bit rates for a system with $BW/R_S = 1.27$ as a function of the transponder bandwidth and the code rate.

Several members of the Astra and Eutelsat satellite families have commenced transmission of DVB signals, for example via 36 MHz transponders with $BW/R_S = 1.309$ and with a code rate R_2 of 3/4. This results in a symbol rate of 27.5 Mbaud and a useful bit rate of 38.01 Mbit/s.

Table 9.2. Useful bit rate [Mbit/s] in the satellite channel

Channel bandwidth [MHz]	Symbol rate [Mbaud]	Maximum useful bit rate [Mbit/s] at a varying code rate R_2				
		1/2	2/3	3/4	5/6	7/8
54	42.5	39.2	52.2	58.8	65.3	68.5
46	36.2	33.4	44.5	50.0	55.6	58.4
40	31.5	29.0	38.7	43.5	48.4	50.8
36	28.3	26.1	34.8	39.1	43.5	45.6
33	26.0	24.0	31.9	35.9	39.9	41.9
30	23.6	21.7	29.0	32.6	36.2	38.1
27	21.3	19.6	26.2	29.4	32.7	34.4
26	20.5	18.9	25.2	28.3	31.5	33.1

Table 9.3. Internal code rate and required carrier-to-noise ratio

Code rate R_2	Required E_b/N_o [dB]	Corresponding C/N (incl. 0.8 dB) [dB]
1/2	4.5	4.1
2/3	5.0	5.9
3/4	5.5	6.9
5/6	6.0	7.9
7/8	6.4	8.5

9.5.2 Required Carrier-to-noise Ratio in the Transmission Channel

As already explained in section 9.2, it is the aim of the error protection to achieve a bit-error rate of $\leq 10^{-11}$ subsequent to the decoding. For this a bit-error rate of $\leq 2 \cdot 10^{-4}$ is required at the output of the Viterbi decoder. This in turn depends on the code rate and the carrier-to-noise ratio at the receiver input. Table 9.3 [REIMERS 3] shows the corresponding values for the satellite standard which result from a simulation, based on the assumption of an additional loss of 0.8 dB through the practical implementation margin (of modulator, demodulator and TWTA). This takes into account the fact that the hardware components used in practice cannot perform as postulated in theory, that for example the sampling points in the demodulator are slightly changed by the phase noise of the recovered carrier.

The following holds for the above values (see chapter 7):

E_b/N_0 = energy per information bit/noise-power density
C/N = carrier-to-noise ratio.

The "information bits" are defined as the uncoded bits (i.e. before error-protection processing) which are set in relation to the overall energy of the coded bit stream. In this way the fact is taken into account that only part of the transmitted data (all requiring transmission power) are useful data.

With QPSK and concatenated coding at the rates R_1 and R_2, the relation between E_b/N_0 and C/N is

$$C/N = E_b/N_0 \, 2R_1 R_2 \tag{9.4}$$

where factor 2 takes the number of 2 bits into account which are transmitted within one symbol of the QPSK.

The value C/N is dependent on the following conditions for both the uplink and the downlink:

- transmitting power,
- precision of the alignment of the transmission and reception antennas,

Table 9.4. Required antenna diameter for some selected system configurations

Channel bandwidth [MHz]	Data rate [Mbit/s]	Antenna diameter in cm for an average-year service continuity		
		99.7%	99.9%	99.99%
	Code rate 2/3			
54	52.2	50	58	88
33	31.9	39	45	69
26	25.2	35	40	61
	Code rate 5/6			
54	65.3	63	72	111
22	39,9	49	56	86
26	31.5	44	50	77

– diameter of the reception antennas,
– meteorological conditions,
– noise figure of receivers.

9.5.3 Antenna Diameter

Examples of the required antenna diameters are shown in table 9.4 [REIMERS 3]. The data are valid for the hydrometeorological zone E [CCIR 563, CCIR 564] in which Germany is located. Other zones have other parameters due to varying climatic conditions. Further, a downlink frequency of 12 GHz and an EIRP of 51 dBW have been assumed. The service continuity refers to 99.7%, 99.9% and 99.99% respectively, for an average year. The last figure, for example, stands for an average loss of 53 minutes in one year during which time the receiving conditions, for instance due to a thunderstorm, are so poor that the error protection fails completely.

Due to the fact that in the case of a drop in the carrier-to-noise ratio the image quality remains at first unchanged (see section 9.4.5), the breakdown comes as a surprise and could be wrongly interpreted by the viewer as a device error, especially since many or all channels might be affected at the same time, depending on the error protection. For such cases special mechanisms should be installed in the equipment to warn the viewer when the carrier-to-noise ratio drops below the critical level.

9.6 Local Terrestrial Transmission

The satellite standard introduced here is not only conceived for use in satellite transmission, but also for the transmission of MMDS in the frequency band above 10 GHz.

MMDS is a microwave multichannel/multipoint distribution system and operates as a local terrestrial transmission. A microwave transmitter at an elevated point, such as a church spire, covers a relatively small area within a radius of less than 50 km. Reception is possible with a small fixed antenna in line-of-sight with the transmitter. For this the frequency bands envisaged are between 2 and 42.5 GHz, but these differ from country to country.

For transmissions below 10 GHz what matters most is a full utilisation of the available bandwidth, therefore the DVB cable standard (see chapter 10) is used for the transmission (DVB Microwave cable-based – DVB-MC [EN 749]). Above 10 GHz the air causes the attenuation to increase considerably with increasing frequency [RODDY], which calls for a higher robustness of the transmission, as provided by the DVB satellite standard (DVB Microwave satellite-based – DVB-MS [EN 748]).

The fact that the two MMDS standards are based on the DVB-S and DVB-C standards, respectively, ensures that the hardware developed for the latter two can be reused for demodulation and decoding; only the antenna and the tuner need to be adapted.

Symbols in Chapter 9

b_i	bit stream no. i
B	bandwidth utilisation
BW	bandwidth
C/N	carrier-to-noise ratio
E_b	energy per useful bit
f	frequency in general
$G_i(x)$	i^{th} generator polynomial of a convolutional code
I	interleaving depth or in-phase component
i	running variable, integer
IF	intermediate frequency
M	interleaver delay
n	integral control variable, code-word length of a block code
N_o	noise-power density in general
P_{out}	output power
P_{in}	input power
P_{sat}	saturation power
Q	quadrature component
R_i	i^{th} code rate
R_S	symbol rate
R_u	useful bit rate
t	time in general
X	1st output branch of a convolutional encoder
x	argument for a generator polynomial
$X_1, X_2, \ldots$	n^{th} bit of the 1st output branch of a convolutional encoder
Y	2nd output branch of a convolutional encoder
$Y_1, Y_2, \ldots$	n^{th} bit of the 2nd output branch of a convolutional encoder
α	roll-off factor

10 The Cable Standard and Its Decoding Technique

The specification for the transmission of DVB signals in cable networks was developed in the DVB Project between August 1993 and January 1994 [STENGER]. The result was the draft of a baseline system. In the spring of 1994 this was recommended by ETSI as a European transmission standard and in November 1994 it became an ETSI standard, designated as ETS 300 429 [ETS 429]. The specification was also submitted to the International Telecommunication Union (ITU). In June 1995, ITU recommended that it be used as a standard which should not only be valid for Europe.

10.1 Cable Transmission Based on the Example of a German CATV Network

In Germany the cable-based transmission of broadcasting signals from the studio to the user equipment is divided into four sections by the national reference chain. A simplified structure of this chain is shown in figure 10.1.

The supraregional section connects the studios to each other or feeds the signals from a studio to a TV switching station of the public network. In the regional section the signals are distributed to the feeding points of the CATV networks in the cable head-ends. The local section 1 represents the CATV distribution network. The subsequent in-home network and private cable networks are referred to as local section 2.

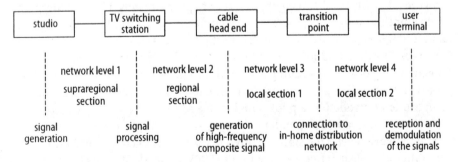

Fig. 10.1. Simplified structure of the reference chain in Germany

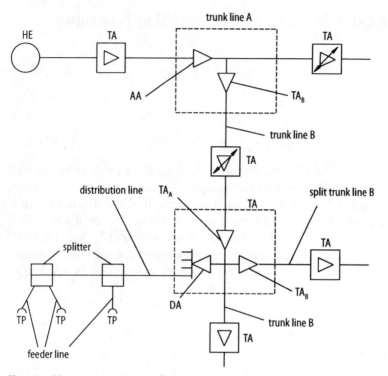

Fig. 10.2. Schematic representation of the CATV network

The CATV network constitutes a tree-and-branch system which is divided into the 4 network levels A, B, C and D [KENTER]. One branch of this structure can be seen in figure 10.2. Network level A starts with its trunk line just after the cable head-end. After a maximum cable length of 412 m an amplifier point (TA) is installed, which basically consists of a cable equaliser and a trunk amplifier. Up to 20 trunk amplifier points can be connected in series at network level A and are all interconnected by a cable route. For the trunk lines of network level A this results in an upper limit of approximately 8 km in length. At each amplifier point it is possible to decouple the signal from network level A and feed it to B (or also directly to C), in which process the signal is attenuated by 14 dB. The first amplifier after the decoupling has the task of compensating for the 14 dB of the decoupling attenuation. The adjacent network level B is constructed similarly to network level A. Each third trunk amplifier point is amplitude-controlled by a pilot signal with a frequency of 80.15 MHz. By this means a temperature-dependent drifting of the amplification is avoided. From network level B the signal is fitted to the C level via a C amplifier. This C amplifier is the last active element in the CATV network. With its high output level it ensures that the signal, subsequent to passing through the passive C

Table 10.1. Index of abbreviations used

Abbreviation	Designation
HE	head-end
TA	trunk amplifier point
TA_A	trunk amplifier of network level A
TA_B	trunk amplifier of network level B
DA	distribution amplifier
TP	transition point

level and also the passive D level, has a guaranteed level at the transition point (TP).

The requirements for the transmission quality of television and radio broadcast FM signals are stated in [FTZ 1] and [FTZ 2] for each section in the national reference chain. No official specifications of requirements for digital signals have so far been established. Therefore we will estimate in the following sections the effect of the overlay of intermodulation and random noise signals on the digital signal.

10.1.1 Intermodulation

In general, transmissions via CATV networks are limited by intermodulation products. In the case of non-linear processing these are caused mainly by the interaction of the image carriers of the analogue TV signals transmitted in different channels. Depending on the frequency position or the spectral distance of the disturbance from the image carrier, the resulting patterns in the picture can produce very coarse structures, which moreover move randomly. These disturbances can easily be perceived by the human eye. This is the reason that the reference chain ensures a signal-to-interference ratio of 60 dB at the user outlets.

However, the digital signal is less sensitive to narrow-band interference. The symbols received are corrupted by an intermodulation signal only if the highest amplitudes of the disturbances exceed the decision thresholds of the symbol decision performed at the receiving end. The power ratio between useful signals and unwanted signals in this case is of course dependent on the type of modulation used. If the peak amplitude of the sinusoidal signal is scaled to value 1, then, according to (7.42), the following expression results if, for instance, a QAM signal is used as the useful signal:

$$\frac{C}{I} = \frac{S_{QAM}}{S_{\sin}} = \frac{2\frac{M-1}{3}}{\frac{1}{2}} = \frac{4}{3}(M-1). \tag{10.1}$$

Hence the following holds for a 64-QAM with $M = 64$ symbol states:

$$10 \log\left(\frac{C}{I}\right)\Bigg|_{64\text{-}QAM} = 10 \log(84) = 19.24 \, \text{dB}. \tag{10.2}$$

If there were no further unwanted signals in the cable network apart from the intermodulation interferers, the signal-to-interference ratio C/I would be allowed to be below 20 dB in the case of a 64-QAM. This also implies that, if the values guaranteed by the FTZ guideline [FTZ 1] were duly observed, the level of the digital signal could be reduced by approximately 40 dB as compared to the level specified for the analogue television signal. Therefore intermodulation disturbances will have no effect on the deterioration of the digital signals.

However, it is to be noted that additional signals fed into the existing networks present an increased burden to the cable amplifiers. Since analogue services will be transmitted over cable networks for many years yet and since the picture quality of these services must not noticeably deteriorate, the total power of the high-frequency composite signal, particularly in the intensely used networks, must not be unduly increased by the introduction of DVB signals. Field tests showed that this condition is satisfied if the DVB signals are fed into networks at a reduced level, which is 10 to 13 dB below the level specified for analogue television signals [SCHAAF].

10.1.2 Thermal Noise

A further deterioration of the useful signal is caused by the superposition of noise. The reference chain defines a signal-to-noise ratio of 44.5 dB at the receiver input for the analogue video signal. The noise figure of this ratio relates to the effective values of a baseband signal which is band-limited at 5 MHz and shaped by a noise-weighting filter. The effect of various factors must be taken into account when converting this data into a carrier-to-noise ratio C/N [VELDERS], during which the correction of the signal power and the noise power can be dealt with separately.

The following characteristics are important for the signal: the carrier attenuation at the point of 50% amplitude response in the filter at the receiving end, the level control of the black-and-white range with 70% of the CVBS signal amplitude, a 90% depth of modulation, and the conversion from peak value to effective value.

The power of the noise signal is changed by the IF filter at the receiving end and by the noise-weighting filter.

The multiplication of the two correction factors results in a factor of approx. −1.5 dB:

$$10 \log\left(\frac{C}{N}\right)\Bigg|_{5 \, \text{MHz}} = 44.5 \, \text{dB} - 1.5 \, \text{dB} = 43 \, \text{dB} \, . \tag{10.3}$$

Let the signal-to-noise ratio of the digital signal apply to an 8-MHz channel. Taking into account the aforementioned 13-dB difference value of the level, the result is a power ratio between DVB signal and noise of:

$$10 \log\left(\frac{C}{N}\right) = 10 \log\left(\frac{C}{N}\right)\Bigg|_{5\,\text{MHz}} - 10 \log\left(\frac{8\,\text{MHz}}{5\,\text{MHz}}\right) - 13\,\text{dB} \cong 28\,\text{dB}. \qquad (10.4)$$

Hence, at the introduction of DVB services, this carrier-frequent signal-to-noise ratio is the least that can be expected in a broadband cable network.

For the estimation of the quality of digital signals one generally uses the bit-error rate of the information bits. This can be computed as a function of the ratio E_b/N_0 as outlined in chapter 7, so that now a connection will be established between the C/N and E_b/N_0 supplied by the transmission channel.

It is the matched filter in the receiver which has a most decisive effect on the connection between these two ratios. The conditions set by (7.27) and (7.37) on the one hand, and the identity of the signal power in transmission channel C and in baseband S on the other hand, result in the following equation:

$$\frac{C}{N} = \frac{E_b}{N_0} \cdot m. \qquad (10.5)$$

As the noise bandwidth of the matched filter in the receiver is twice the Nyquist frequency $2\,f_N$, this results in a reduction of the effective noise power subsequent to passing through the filter. On condition that the spectrum of the transmission channel is optimally utilised this value can also be expressed by the roll-off factor α:

$$\frac{2f_N}{B} = \frac{1}{1+\alpha}. \qquad (10.6)$$

When digital information is transmitted in real systems the modulation is always supported by an error-correction algorithm (see chapter 6). The previously introduced redundancy R is, however, removed again from the signal in the receiver, so that the effective signal power is reduced accordingly. This results in the desired relation:

$$\frac{C}{N} = \frac{E_b}{N_0} \, m \cdot R \, \frac{2f_N}{B}. \qquad (10.7)$$

The following set of parameters which are used in the DVB standard (see section 10.3) shall serve as an example:

$m = 6$ for a 64-QAM,
$R = 188/204$ for a Reed-Solomon error protection with the code rate R,
$2f_N/B = 1/1.15$ for a roll-off factor of 0.15 (see (10.6)).

The CATV network ensures an E_b/N_0 of approx. 21 dB if these parameters are chosen.

10.1.3 Reflections

A cable connection between two network components (e.g. two splitters in a passive network) is not generally ideally terminated. The signal travels along the cable path, for instance from a first splitter to a second. It can be conceived as a forward-travelling wave [UNGER]. At the connection point between the cable and the second splitter a portion of the signal power is reflected and returns along the cable path. Hence this portion is also called backward-travelling wave. The amplitude ratio between the go- and the return wave can be indicated by the return loss. For the components used in a CATV network a frequency-dependent quantity of 15–20 dB is required. When travelling through the cable path the reflected signal undergoes a second attenuation. Given that, as a rule, even the output of the first splitter is not totally identical with the characteristic impedance, a portion of the reflected signal power is reflected in its turn. The doubly reflected signal portion has a cumulative effect on the forward-travelling wave since it moves in the same direction as the main signal. It is delayed by the amount of time it requires for travelling twice through the cable path. Of course a portion of its power is again reflected at the connection point between the cable and the second splitter. The attenuation that an echo undergoes at both connection points and while travelling along the cable is generally so high that reflections of a higher order, i.e. signal portions which have travelled forward and backward more than once, can be ignored. The result is a transfer function of the following form:

$$H(f) = \frac{1 + \varrho \cdot e^{-j 2\pi f_c \tau} \cdot e^{-j 2\pi \Theta}}{\sqrt{1 + \varrho^2}} \qquad (10.8)$$

where

ϱ amplitude of the echo,
f_c carrier frequency,
τ delay between echo and main signal,
Θ additional phase shift of the echo signal.

The return loss of at least 15 dB of the components used, on the one hand, and an assumed low cable attenuation of only 2 dB, on the other, results in an amplitude of the echo of –34 dB, relative to the main-signal amplitude. Investigations [DIGISMATV] have shown that with a 64-QAM technique reflections can be neglected if the relative amplitude is smaller than –30 dB.

10.2 User Requirements of the Cable Standard

When the CATV network in Germany was built up in the 1970s the basic supply of television signals to the public was by means of terrestrial transmission. The design of the network parameters for CATV was therefore based upon the parameters required for terrestrial systems. The chronological development of the various standards within the DVB Project began with the drawing up of the satellite specification [ETS 421]. The development of the cable specification [ETS 429] did not begin until after the first results of the satellite transmission emerged. There was a catalogue of user requirements which had to be taken into account:

(1) The CATV network, in its conventional configuration, must be usable for digital signals.
(2) The feeding of the DVB signals into the cable networks must not noticeably impair the quality of the usual analogue services.
(3) As many data as possible should be transmitted on one cable channel so as to ensure an adequate useful data rate which is compatible with all satellite channels.
(4) Cost is a crucial parameter in the introduction of a new system. This applies not only to content providers but also to network operators. Most important, of course, are the costs to be borne by the consumer. For this reason a high priority in the development of systems was the reasonable pricing of decoder hardware in the integrated receiver/decoder (IRD). The cable IRD, just as the IRD for terrestrial reception (see chapter 11), should be as similar as possible to the satellite IRD.
(5) It should be possible for the cable IRDs to be introduced on the market at the same time as the satellite IRDs.

The conditions for the future transmission of DVB signals were established by requirements 1 and 2 above. The adaptation of the signal to the existing cable channel is effected by a QAM technique which yields the best results in terms of systems theory (see chapter 7).

A study of the available network structures concerning master antenna television systems revealed the coexistence of various network configurations and quality classes. As the programmes distributed in these systems are mainly received from a satellite or satellites, the systems are called satellite master antenna television (SMATV). In accordance with the requirement expressed in point 5 it was decided to develop an additional standard [ETS 473] which extended the CATV standard so as to make it applicable to SMATV systems.

The modelling of the cable standard on the satellite standard, as required under points 3 and 4, is achieved by the input signal processing. From baseband interface via energy dispersal and use of the outer error protection to convolution interleaver, the signal processing is identical in both systems. For the cable standard the application of an internal error protection is not required. There are two main reasons for this. Firstly, each additional process

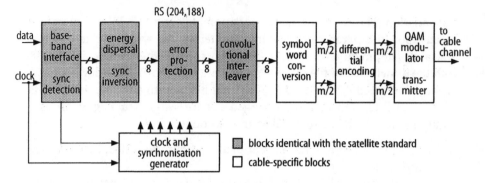

Fig. 10.3. Block diagram of a DVB cable encoder

that inserts redundancies into the data stream reduces the useful data rate, which is contrary to requirement 3. Secondly, the cable networks are of a high quality (see section 10.1), so that the DVB signal is sufficiently protected against transmission errors by the use of the outer error protection.

10.3 Signal Processing at the Encoder

This section introduces the DVB cable specification as defined in the ETSI standard [ETS 429]. As only the signal processing in the encoder is described in this standard, the present section will be restricted to the discussion of the processing blocks at the encoder. Moreover, the signal technique, from baseband interface to convolution interleaver, will not be reiterated here, as these were already discussed in the chapter dedicated to the satellite standard. The DVB cable encoder is shown as a block diagram in figure 10.3. The shaded processing blocks are identical with those in the satellite standard (see section 9.3).

The cable networks, contrary to the situation in the case of satellites, ensure a relatively constant and high transmission quality. For this reason the cable standard provides for several bits to be combined to form a symbol and to be transmitted by a higher-level quadrature amplitude modulation. There is a choice between five variants. The modulation type with the lowest efficiency provided in the standard, i.e. an efficiency of 4 bits per symbol, is 16-QAM, followed by 32-QAM which combines 5 bits per symbol. The modulation type with the highest efficiency in the first version of the standard was the 64-QAM, which enabled a parallel transmission of 6 bits per symbol. The first revision of the standard provided for a transmission efficiency of 7 and 8 bits per symbol with a 128-QAM and a 256-QAM respectively.

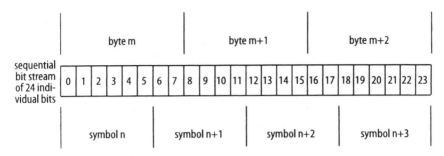

Fig. 10.4. Conversion of the bytes to symbol words, taking the 64-QAM as an example

10.3.1 Conversion of Bytes to Symbol Words

After the 8-bit data words have passed through the channel coding they are fed into a symbol-word converter (referred to as "byte-to-m-tuple converter" in the standard). This has the task of combining the m bits which are to be transmitted per symbol into symbol words of a corresponding width. During transmission with 16-QAM the individual bytes are divided into two 4-bit symbol words. The MPEG transport packets, which are augmented from 188 bytes to 204 bytes in the Reed-Solomon encoder by appending 16 redundancy bytes (see section 6.4.3), have a length of 408 symbol words after conversion of the symbol words. For the conversion of the 8-bit data words to 6-bit symbol words, as required when using 64-QAM, the individual bits must be reordered beyond the byte boundaries. Four symbol words result from 3 data words, so that 24 bits form one framework. The packet length following the reordering increases to 272 symbol words. Figure 10.4 shows the conversion of a 24-bit framework, using the example of a 64-QAM.

In the case of the 32-QAM the fixed frame length is augmented to 40 bits. After the reordering of the 5-bit symbol words the 204 bytes of an extended transport packet result in a frame length of 326.4. This means that a 204-byte packet cannot be assigned to a whole-number frame length.

10.3.2 Differential Coding of MSBs

Following the conversion of the 8-bit data words to symbol words, the latter are differentially pre-processed as described below. The reason for this measure is to be found in the loss of absolute phase information by the suppression of the carrier signal at the transmitter. The receiver can only recover the reference phase up to an uncertainty of $n \cdot 90°$, due to the known signal statistics of the QAM technique (see figure 7.22). In order to achieve this, a frequency-synchronous demodulator separates the received signals into two components which bear an orthogonal relation to each other. Following this, the demodulator interprets the positions of the symbols received and readjusts the amplification and the signal phase until the mean values of the individual symbol states coincide as closely as possible with the desired values. This leads to a constellation as described in section 7.5, figure 7.24. The values of the two subsignals correspond to the values of the in-phase and quadrature components of the transmitted symbol. The information relating to the algebraic sign of both components is, however, lost by the suppression of the carrier. For this reason the information is specially pre-processed in the transmitter, firstly, by separating the two MSBs from the rest of the actual symbol word and, secondly, by differential coding of only these MSBs, which are given in (10.9) by A_K and B_K:

$$I_K = (\overline{(A_K \oplus B_K)} \cdot (A_K \oplus I_{K-1})) + ((A_K \oplus B_K) \cdot (A_K \oplus Q_{K-1}))$$

$$Q_K = ((\overline{A_K \oplus B_K}) \cdot (B_K \oplus Q_{K-1})) + ((A_K \oplus B_K) \cdot (B_K \oplus I_{K-1})).$$

$$(10.9)$$

I_K and Q_K are referred to as the results of the Boolean operation, i.e. the differentially encoded MSBs, and their temporal predecessors are I_{K-1} and Q_{K-1}. As according to (10.9) the outputs of the processing block are fed back into the inputs, this is a recursive algorithm. The truth table 10.2 illustrates the dependence of the output variables on the input variables.

If, for example, the values $A_K = 0$ and $B_K = 0$ are valid at the input of the differential encoding unit, at the output there result the same values I_K and Q_K as the ones of the symbol previously transmitted. As both values have remained unchanged, the actual symbol is transmitted in the same quadrant as the one temporally preceding.

According to table 10.2, if the input occupancy of $A_K = 0$ and $B_K = 1$, I_K assumes the value which was allocated to the inverted Q-component in the previous symbol, while Q_K assumes the value which the I-component previously had. This results in a 90° phase shift in a mathematically positive direction. The constellation diagram known from the chapter on satellite transmission (figure 9.9) shall serve as an example, as its allocation requirements concerning bits and quadrants also apply to the cable standard. If I_{K-1} is defined as having had the value of 0, and Q_{K-1} the value of 1, then the previously

Table 10.2. Truth table for differential encoding

Inputs		Outputs		Rotation
A_K	B_K	I_K	Q_K	
0	0	I_{K-1}	Q_{K-1}	0°
0	1	$\overline{Q_{K-1}}$	I_{K-1}	+90°
1	0	Q_{K-1}	$\overline{I_{K-1}}$	−90°
1	1	$\overline{I_{K-1}}$	$\overline{Q_{K-1}}$	180°

transmitted symbol bad just been in the fourth quadrant. In the case under consideration, i.e. $A_K = 0$ and $B_K 1$, both I_K and Q_K have the value 0 (see table 10.2). The actual symbol can be found in the first quadrant in accordance with figure 9.9. The phase shift from the centre of the fourth quadrant to that of the first quadrant is identical with the $+90°$ rotation as indicated.

The advantage resulting for the receiver from the differential pre-processing is that, for recovering the MSBs, the receiver only needs to evaluate the changes in those quadrants in which a transmitted symbol is found. The receiver carries out a subtraction, in which process the absolute phase information is dropped. Hence the phase uncertainty of $n \cdot 90°$ as described above suffices for the recovering of the MSBs. For the sake of completeness it should be mentioned that the LSBs, which have not yet been discussed, possess the information about the state of the symbol within the quadrant selected by the MSBs. Figure 7.22 shows the constellation of a 64-QAM. In the first quadrant the 16 possibilities are drawn as vectors resulting from the four remaining LSBs.

10.3.3 Modulation

The feeding of the symbol words with the differentially encoded MSBs into the quadrature amplitude modulator is the final step in the processing chain of the encoder. The multiplication with a complex carrier signal results in a band-pass signal, at the output of the modulator, whose centre frequency is equal to that of the carrier signal. This frequency is referred to as intermediate frequency (IF). Concerning the modulation, it is particularly important that the unequivocal allocation of the symbol words to the complex amplitudes of the symbols to be transmitted be guaranteed. Depending on the number m of the bits which are to be transmitted per symbol, a constellation results in which a definite amplitude and phase position are modulated onto the carrier by every possible bit combination (see chapter 7). Figure 10.5 shows the allocation for the 16-QAM, the 32-QAM and the 64-QAM as defined in the standard. The 16 amplitude phase states of the 16-QAM have been plotted as a proper subset directly in the constellation diagram of the 64-QAM. In the 6-bit symbol words of the 64-QAM it is necessary to delete only the two middle bits b_4 and b_3 in order to achieve an allocation of the 4-bit words to the amplitude phase conditions of the 16-QAM. Here again it can be clearly seen that the MSBs of the individual bit combinations never change within one quadrant and therefore are of no relevance to any decision taken within a quadrant.

The bit combinations of the symbol words for the 16-QAM and the 64-QAM have been chosen such that if a symbol state is altered in a way that causes it to take on a neighbouring value only one single bit is changed. This,

32-QAM $(b_4 b_3 b_2 b_1 b_0)$ 64-QAM $(b_5 b_4 b_3 b_2 b_1 b_0)$ 16-QAM $(b_3 b_2 \, x \, x \, b_1 b_0)$

Fig. 10.5. Constellation of the 16, 32 and 64-QAM

however, only holds for those cases in which the invalidation causes no crossing of quadrant borders. Within one and the same quadrant the symbol words are thus coded in accordance with Gray [LÜKE 1]. If the transmitted symbol has been invalidated so as to assume a state which lies in a neighbouring quadrant, there can occur up to 5 bit errors per symbol in the DVB signal. Compared to a system that is entirely Gray-coded this results in a disadvantage, which, however, cannot be avoided because of the introduction of the differential coding. The decoding algorithm requires a rotationally symmetrical arrangement of the individual quadrants; however, within the quadrants it enables the implementation of a Gray code.

The revised version of the standard includes a 128-QAM and a 256-QAM, the constellation diagram of the latter being shown in figure 10.6.

The layout of the matched filter at the receiver is described in section 7.1. The roll-off factor defined in the cable standard has the value:

$$\alpha = 0.15 \ . \tag{10.10}$$

The transfer function shown in figure 7.6a refers to the total transmission path. Hence, in the pre-processing at the transmitting end there must be a subfilter with a square-root raised-cosine characteristic.

There is a template in the annexe to the ETSI standard which gives the required frequency response of the amplitude of all filters used in the transmitter (see figure 10.7). Apart from the matched filter, the effect of the D/A conversion and the transfer functions of all further filters need to be taken into account. The resulting in-band ripple in the pass band may not exceed 0.4 dB, while an out-of-band rejection greater than 43 dB is required in the stop band.

Q

10110000 10110010 10111010 10111000 10011000 10011010 10010010 10010000 | 00100000 00100001 00100101 00100100 00110100 00110101 00110001 00110000

10110001 10110011 10111011 10111001 10011001 10011011 10010011 10010001 | 00100010 00100011 00100111 00100110 00110110 00110111 00110011 00110010

10110101 10110111 10111111 10111101 10011101 10011111 10010111 10010101 | 00101010 00101011 00101111 00101110 00111110 00111111 00111011 00111010

10110100 10110110 10111110 10111100 10011100 10011110 10010110 10010100 | 00101000 00101001 00101101 00101100 00111100 00111101 00111001 00111000

10100100 10100110 10101110 10101100 10001100 10001110 10000110 10000100 | 00001000 00001001 00001101 00001100 00011100 00011101 00011001 00011000

10100101 10100111 10101111 10101101 10001101 10001111 10000111 10000101 | 00001010 00001011 00001111 00001110 00011110 00011111 00011011 00011010

10100001 10100011 10101011 10101001 10001001 10001011 10000011 10000001 | 00000010 00000011 00000111 00000110 00010110 00010111 00010011 00010010

10100000 10100010 10101010 10101000 10001000 10001010 10000010 10000000 | 00000000 00000001 00000101 00000100 00010100 00010101 00010001 00010000

— I

11010000 11010001 11010101 11010100 11000100 11000101 11000001 11000000 | 01000000 01000010 01001010 01001000 01101000 01101010 01100010 01100000

11010010 11010011 11010111 11010110 11000110 11000111 11000011 11000010 | 01000001 01000011 01001011 01001001 01101001 01101011 01100011 01100001

11011010 11011011 11011111 11011110 11001110 11001111 11001011 11001010 | 01000101 01000111 01001111 01001101 01101101 01101111 01100111 01100101

11011000 11011001 11011101 11011100 11001100 11001101 11001001 11001000 | 01000100 01000110 01001110 01001100 01101100 01101110 01100110 01100100

11111000 11111001 11111101 11111100 11101100 11101101 11101001 11101000 | 01010100 01010110 01011110 01011100 01111100 01111110 01110110 01110100

11111010 11111011 11111111 11111110 11101110 11101111 11101011 11101010 | 01010101 01010111 01011111 01011101 01111101 01111111 01110111 01110101

11110010 11110011 11110111 11110110 11100110 11100111 11100011 11100010 | 01010001 01010011 01011011 01011001 01111001 01111011 01110011 01110001

11110000 11110001 11110101 11110100 11100100 11100101 11100001 11100000 | 01010000 01010010 01011010 01011000 01111000 01111010 01110010 01110000

256-QAM $(b_7 b_6 b_5 b_4 b_3 b_2 b_1 b_0)$

Fig. 10.6. Constellation of the 256-QAM

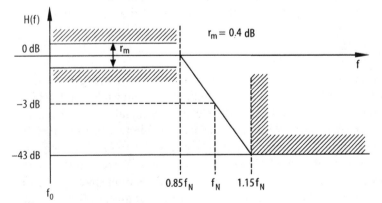

Fig. 10.7. Required overall frequency response of the signal
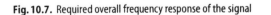

The maximum variation of the group delay in the pass band is 10% referred to the transmitted symbol duration T_S. In a channel of an 8-MHz bandwidth, utilising the given roll-off factor from (10.10), the maximum transmittable symbol rate will be:

$$R_S = 2f_N = \frac{B}{1+\alpha} = 6.96 \text{ Mbaud} \qquad (10.11)$$

which results in a symbol duration T_S of

$$T_S = \frac{1}{2f_N} = \frac{1+\alpha}{B} = 143.75 \text{ ns}. \qquad (10.12)$$

The variation of the group delay shall therefore not exceed 14.4 ns. The in-band ripple of the amplitude and the group-delay variation must be observed up to the Nyquist frequency f_N.

10.4 Decoding Technique

As the ETSI standard ETS 300 429 is limited to the description of the processing algorithm in the encoder there are various possibilities for the realisation of the individual system units in the decoder. These sometimes differ fundamentally from one another, as the required signal processing can be carried out in the digital as well as in the analogue domain. The input interface of the cable tuner and the cable tuner itself are basically similar to the structural components used in conventional receivers for analogue television signals. The input signal is converted from its RF range to a constant intermediate frequency in accordance with the superheterodyne principle. At this point the channels are separated with the aid of a steep-edged intermediate filter. For the subsequent quadrature amplitude demodulation the suppressed carrier signal must be recovered. Furthermore, the sampling of the transmitted signal requires a recovering of the clock signal used at the transmitter.

10.4.1 Cable Tuner

The cable tuner converts the signals received, after a pre-filtering by means of a tunable selection filter, to a fixed intermediate frequency (IF) range. Here it is of no importance whether an analogue television signal or a DVB signal is concerned. For cost-saving reasons it seems sensible to use the same frequency converter for the conversion of both kinds of signals. However, it has to be borne in mind that the DVB signal cannot only be transmitted with a bandwidth which differs from the commonly used channel spacing of 7 or 8 MHz, but in principle also with a bandwidth of, for instance, 2 MHz (see table 10.5).

In order to achieve what is generally known in analogue television as the IF characteristic [SCHÖNFD 2] a VSB-modulated signal from the RF transmission channel must be down-converted with a sinusoidal signal of a local oscillator whose frequency is 38.9 MHz above the image-carrier frequency. Given that the upper as well as the lower sideband of the DVB signal need to be transmitted, the suppressed carrier in this case is placed at the centre of the transmission channel. Hence a difference of 2.75 MHz in frequency results between the two carrier positions. If for both the analogue and the digital signal the same oscillator is to be used for the conversion to the IF range, then the difference in frequency between the image carrier of the analogue TV signal and the imaginary DVB carrier remains unchanged, so that, under the given conditions, the following DVB IF ensues:

$$f_{IF} = 38.9\,\text{MHz} - 2.75\,\text{MHz} = 36.15\,\text{MHz}. \tag{10.13}$$

The cable tuners used in conventional receivers are usually only adjustable to frequencies whose value is an integral multiple of 62.5 MHz. That is why the DVB IF will usually be 36 MHz.

Concerning this method it has to be borne in mind that the signal spectrum is inverted after the frequency conversion. This phenomenon is already known from analogue television. All portions of a signal which in the HF range had frequencies above the carrier frequency, lie below the carrier frequency in the IF range and vice versa. Mathematically these changes can be described as the inversion of the quadrature component. Thus the constellation of the complex-modulated QAM symbols (see figures 10.5 and 10.6) is mirrored on its I-axis. In order to achieve an error-free decoding the necessity of this new correspondence between the complex amplitudes of the received symbols and their respective symbol words has to be taken into account. However, if the transmitter also carries out a spectrum inversion, as is common in analogue television, then the signal spectrum will be inverted twice and will therefore be back in its original state.

10.4.2 IF Interface

The actual channel selection is carried out in the IF range, as already mentioned. The conventional filters used in this range are constructed as surface acoustic-wave (SAW) filters. Owing to the required VSB Nyquist edge and the suppression of the modulated audio signals, the IF filter for the analogue TV signal differs from the IF filter for the DVB signal, and therefore it is indispensable, at this point in the process, to separate the two signal paths. The IF filter for the DVB receiver can have half the Nyquist frequency response. Since it may be implemented as a band-pass filter its transfer function needs to run symmetrically to the carrier frequency position. Due to the fixed distance between the two points of 50% amplitude response, only a signal whose symbol rate is exactly twice the Nyquist bandwidth can, as shown in (10.11), be optimally filtered. The same is true if after demodulation an analogue Nyquist filtering is carried out in the baseband. It is only through the use of a digital filter that a flexible adaptation to various symbol rates is possible. In this case the IF filter only serves to separate adjacent channels and should have a steep-edged transfer function.

10.4.3 Recovery of the Carrier Signal

A synchronous demodulation process requires a coherent carrier. This must be regenerated by a local oscillator in the receiver and synchronised with the signal received. A possible situation would be a frequency- and phase-selective control loop, on the principle of the phase-locked loop (PLL), working on a predetermined IF level. The carrier signals which are orthogonal to each other in phase are generated by means of an oscillator. According to whether the PLL is to be used for analogue or digital signals it would either be a voltage-controlled oscillator (VCO) or a digitally-controlled oscillator (DCO). The information about the phase difference between the carrier which was used in the transmitter for the modulation and the carrier which was generated in the receiver is obtained by a phase discriminator. This can be realised as an analogue or a digital unit, exactly like the loop filter.

An example of an analogue technique is shown in figure 10.8a. Non-linear processing must ensure that a discrete line is originated in the spectrum of the QAM signal at the carrier frequency or an integral multiple of the carrier frequency. However, a quadratic characteristic curve will not suffice, as

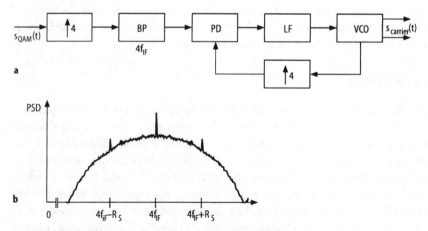

Fig. 10.8. Example of a technique for carrier recovery by quadrupling the IF range.
a Block diagram, **b** Power spectral density in the range of the quadruple IF

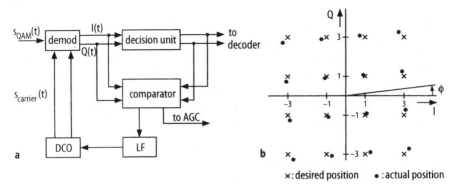

Fig. 10.9. Example of a technique for carrier recovery by decision feedback of the symbol error.
a Block diagram, **b** Effects of phase error in the constellation of a 16-QAM

shown by the PSD in figure 10.10b. For this reason the signal is affected by a non-linear characteristic curve of the fourth order which may consist of a series connection of two squarers. Besides the various sum and difference frequencies this process yields the desired line at the quadruple carrier frequency. This range of the PSD is shown in the diagram of figure 10.8b. The signal is then preselected by a band-pass filter and serves the subsequent PLL as a reference variable input. With its quadruple frequency it has a phase uncertainty of $n \cdot 90°$, which, however, is of no importance because of the differential coding of the DVB signal (see section 10.3.2). The controlled variable is generated by a quadrupling of the frequency of the output signal of the oscillator. After a phase comparison in the phase discriminator (PD) an error signal results at the output. Subsequent to passing through the loop filter this error signal is used to lock the oscillator.

The recovery of the phase difference between the carrier signal suppressed in the transmitter and the carrier signal generated in the receiver can alternatively be performed directly in the baseband by means of the evaluation of the complex signal amplitudes (see figure 10.9). For this the QAM signal is demodulated with two carrier signals which are orthogonal in phase. At first these have a random reference phase position. After sampling with double the Nyquist frequency each symbol is compared with the M known reference values in the decision unit and assigned to the reference value which has the minimal Euclidean distance d. These reference values are then made available to the decoder for further processing. Within the loop a comparator follows the decision unit, which computes the sum and phase of the minimal Euclidean distance d. Besides the value for the phase error a value for the erroneous amplitude results. This value can be fed to the automatic gain control (AGC) as a reference signal. The phase error signal passes through a loop filter and then controls the DCO. Using the example of the 16-QAM, figure 10.9b shows how a phase error causes the received symbols – indicated here by dots as actual positions – to undergo a constant rotation in the complex constellation. Each desired position is shown by an x. For reasons of symmetry this technique has a phase uncertainty of $n \cdot 90°$. If the received QAM signal has not been converted correctly to the desired frequency, the phase deviation accumulates from sample to sample and the constellation of the received signal rotates around its own centre. This error can be compensated by the PLL as long as the frequency offset is within the lock-in range.

10.4.4 Generating the Clock Signal

Apart from the carrier signal, the clock signal with which the digital symbols in the transmitter are processed must also be generated in the receiver. The information about the clock is present, for example, in the envelope of the modulated QAM signal and can be recovered by non-linear signal processing. Therefore similar laws apply as for the recovering of the reference carrier (see section 10.4.3). The basic block diagram of a technique which relies on purely analogue signal processing is shown in figure 10.10a. A simple squarer in front of the phase discriminator PD is sufficient as a non-linear pre-processing block, as compared with that required for carrier recovery. The PSD of the signal at the output of the phase discriminator is depicted in diagram b of figure 10.10. The spectral ranges of the sum and difference frequencies which were generated by the multiplication of the IF signal with itself are easy to identify. The envelope of the QAM signal additionally generates a discrete line in the spectrum at the frequency which corresponds with the symbol rate transmitted. This line is identical with the clock frequency used at the trans-

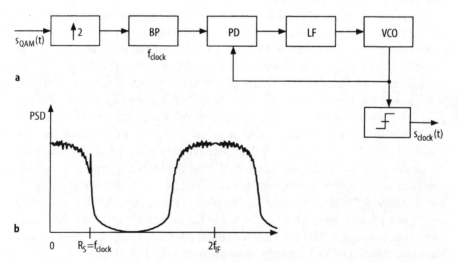

Fig. 10.10. Example of a technique for clock recovery. **a** Block diagram, **b** Power spectral density

mitting end. The synchronisation of the clock frequency generated at the receiver is effected by means of a band-pass filter and a subsequent PLL, where the centre frequency of the oscillator corresponds with the clock frequency, which in turn corresponds with the symbol rate. After an amplitude limitation the clock signal is ready for the digital signal processing.

There are further techniques described in the literature [LEE, PROAKIS, KAMMEYER, MÄUSL] which generate the error signal for the oscillator within the analogue as well as within the digital range.

10.4.5 Demodulation of the QAM Signal

The demodulator has the task of converting the QAM signal to the baseband and simultaneously splitting it into its in-phase and quadrature components. After the temporal multiplication with two carrier signals which are orthogonal to each other in phase, the resulting sum-frequency signals of the two components must be suppressed by one low-pass filter each. If the multiplication and the filtering are achieved by means of analogue signal processing, then both components will subsequently be available for A/D conversion in the baseband.

By using conventional analogue technology a reasonably priced demodulator can be manufactured. The disadvantage of this solution is to be found in the two parallel paths through which the in-phase and quadrature components have to pass before sampling as well as in the parallel sampling itself.

The parallel signal paths can be transferred to the digital level if the modulated signal is fed into a single A/D converter in its entirety. In this case the demodulation is carried out by the multiplication with two digital carrier signals. The required sampling rate for the sampling of the QAM signal at the IF level, in accordance with the sampling theorem, will have to be at least twice that of the highest frequency to be found in the signal. For transmission in an 8-MHz channel the resulting minimal sampling frequency would be:

$$f_{min} = 2\left(f_{IF} + \frac{B}{2}\right) = 2\left(36.15\,\text{MHz} + \frac{8\,\text{MHz}}{2}\right) = 80.3\,\text{MHz}. \tag{10.14}$$

The high clock rate has various disadvantages when used in consumer receivers (e.g. high cost, high power consumption), so that a totally digital solution must be ruled out at this moment.

A demodulation alternative arises if the QAM signal from the IF range at 36.15 MHz is converted to a second IF range. The centre frequency of the signal should be chosen so as to correspond to the transmitted symbol rate. This technique offers the advantage that the converted QAM signal as a whole can be oversampled by a clock signal of a frequency which is exactly four times the value of the transmitted symbol rate. In the case of a symbol

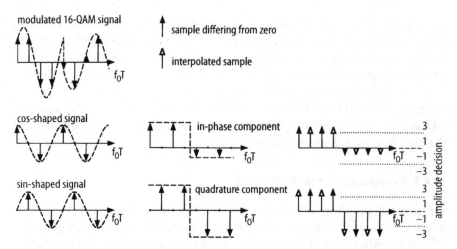

Fig. 10.11. Time graph of signals during processing in a system with a lockage between the suppressed carrier signal and the transmitted symbol rate with subsequent quadruple oversampling

rate of 6.875 Mbaud, as mentioned in section 10.5.1, this results in a clock frequency of:

$$f_O = \frac{1}{T_O} = 4 R_S = 4 \cdot 6.875 \text{ MHz} = 27.5 \text{ MHz}. \tag{10.15}$$

Hence there are exactly four samples for each modulated symbol. Figure 10.11 shows idealised signal forms. The first graph shows the received QAM signal after oversampling, the vertical arrows indicating the oversampled amplitude values. For better understanding it also shows the shape of the equivalent analogue signal. Two symbols of a 16-QAM signal are used as an example. The digital carriers required for the demodulation can be seen in the second and third graphs. These also have four samples per period: 1, 0, -1, 0 for the cosine-wave oscillation, and 0, 1, 0, -1 for the sine-wave oscillation, respectively. Here, too, the cosine waves and sine waves of the equivalent analogue signals have been plotted. The in-phase and quadrature components are generated by multiplying the sampled QAM signal with one of the two digital carriers. These can be seen in the middle diagram of figure 10.11. The same result is achieved by replacing the multipliers with a simple inverter which inverts the QAM signal for the second half of each symbol period. A subsequent de-multiplexer divides the signal into two partial data streams in which every second sample equals zero.

It follows that an individual symbol, viewed in isolation from the data stream, has a mean value not equal to zero and an AC component whose direct wave has a period duration of half the length of the symbol duration. The fundamental frequency is therefore identical with the Nyquist frequency. The

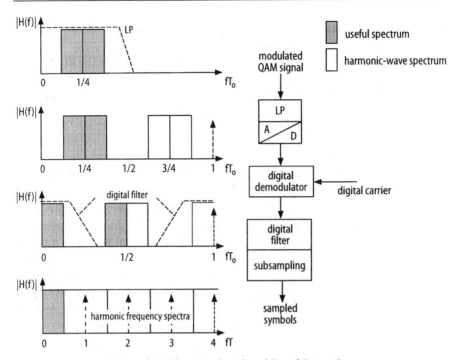

Fig. 10.12. Frequency spectra of signals processed in a demodulator if the sampling rate has the quadruple value of the symbol rate

digital representation of the signals also results in the corresponding harmonic-wave spectra.

The signal processing described is shown again in the frequency domain in figure 10.12. The signal spectra here are rectangular templates. The signal spectrum shown in the second diagram is generated subsequent to sampling with the oversampling clock. The centre frequency of the QAM signal in the useful spectrum is $fT_O = \frac{1}{4}$. Apart from this the image frequencies of the first harmonic-wave spectrum are plotted at $fT_O = \frac{3}{4}$. Following the multiplication with the digital carrier signals the useful spectrum shifts to the baseband. At the same time a second signal portion is formed at the frequency $fT_O = \frac{1}{2}$ (see third diagram). This portion of the spectrum, which consists of the sum of the carrier frequency and the frequencies of the QAM signal, causes every second sample in the wave-form of the in-phase and the quadrature component to be equal to zero. Each second sample must therefore be suppressed by a subsequent digital low-pass filter. In the time domain, the filtering results in an interpolation of the missing samples. Utilisation of one of the four samples suffices for the subsequent amplitude decision. Therefore the signal is subsampled by a factor of four. Since the oversampling was higher by a factor of four than that required according to Nyquist, this subsampling just satisfies the first Nyquist criterion (see section 7.1).

In reality no rectangular-shaped wave-forms occur, so that the instants of the subsampling – or rather, Nyquist-sampling – have to be carefully chosen. Finally, the conversion to the transmitted bit combination, the demapping of the symbol, is carried out with the help of a table.

10.4.6 Differential Decoding of MSBs

The transmitted m-bit symbol words must be differentially decoded in the receiver, i.e. the differential encoding of the symbol words (see section 10.3.2) performed in the transmitter must be revoked in the receiver. For this, as a first step, the two most significant bits of each m-bit symbol word transmitted are separated from the rest of the symbol word. The actual values I_K and Q_K have resulted in the encoder by the combination of their temporal predecessors I_{K-1} and Q_{K-1} with the actual MSBs of the uncoded symbol words A_K and B_K. These must now be recovered in the decoder from the transmitted I_K and Q_K. The rule for computing this process is found by solving the system of equations specified in the standard (see (10.9)) for A_K and B_K:

$$A_K = (\overline{(Q_K \oplus I_{K-1})} \cdot (I_K \oplus Q_{K-1})) + ((I_K \oplus I_{K-1}) \cdot (Q_K \oplus Q_{K-1}))$$

$$B_K = (\overline{(I_K \oplus Q_{K-1})} \cdot (Q_K \oplus I_{K-1})) + ((I_K \oplus I_{K-1}) \cdot (Q_K \oplus Q_{K-1})).$$

(10.16)

In accordance with (10.16) the values received for I_K and Q_K are to be stored for one clock period and are thus denoted as I_{K-1} and Q_{K-1}. By a comparison with those values of I_K and Q_K which are received with a one-symbol delay, information can be gathered as to whether the successively transmitted symbols are located in the same quadrant or in different quadrants. The actual values for A_K and B_K result from the rotation of the symbols.

10.4.7 Conversion of Symbol Words to Bytes

Before the information received can be fed to the error-correction unit it has to be reconverted to its original form of 8-bit data words. Care must be taken that the same data-packet structure is generated as existed in the transmitter before the symbol-word conversion (see section 10.3.1). In the case of the 64-QAM, for example, there are three different possibilities of performing the required allocation (see section 10.4). Information is required by which the original byte boundaries are recognised. This information can be gained from the MPEG sync bytes. Therefore each IRD requires a sync-byte detector which searches the data stream for the required bit combination -47_{HEX} for the sync byte, or $B8_{HEX}$ for the inverted sync-byte – and transfers the information to the symbol-word converter at the receiving end.

10.4.8 Detection of MPEG Sync Bytes

The detection of the MPEG sync bytes is carried out before the error correction, hence in the uncorrected data stream. At this point a typical maximum bit-error rate (BER) would be $2 \cdot 10^{-4}$. A sync-byte detector will search the incoming data stream for data words with the values 47_{HEX} or $B8_{HEX}$. In principle, two kinds of false decisions can occur. Firstly, there is a certain probability that the transmitted sync words will be invalidated by the superposition of unwanted signals in the channel. Assuming that, on average, the disturbances affect all bits to the same extent, then, by approximation, the following sync-byte error rate holds:

$$P_{e,Sync} = 1-(1-BER)^8 \xrightarrow[BER=2 \cdot 10^{-4}]{} 1.6 \cdot 10^{-3} . \tag{10.17}$$

Secondly, there are data words within the transport packets which have the above-mentioned sync-byte values as useful information. If all data words have a pseudo-random character and if all values occur with the same regularity, then an occurrence probability for a data byte of the value 47_{HEX} or $B8_{HEX}$ can be established as follows:

$$P_{47} = P_{B8} = 0.5^8 = \frac{1}{256} = 3.91 \cdot 10^{-3} . \tag{10.18}$$

While the first type of error prevents the detection of a transmitted packet, the second type wrongly indicates to the detector the beginning of a packet.

The length of the packets before passing through the error correction is exactly 204 bytes. Apart from the sync byte it comprises 187 useful bytes and 16 redundant bytes. Assuming that each one of the 203 useful and redundant bytes can take on one of the two sync-word values independently of each other, this will result in an occurrence probability of at least one value of 47_{HEX} or $B8_{HEX}$ within one packet:

$$P_{e,Packet} = 1-(1-P_{47})^{203} \cdot (1-P_{B8})^{203} \cong 0.8 . \tag{10.19}$$

This implies that, through the second type of error alone, approx. 8 out of 10 packets can be erroneously detected.

By exploiting the temporal redundancy in the periodic packet structure of the MPEG data this unacceptable number can be considerably reduced. The detector output is then only activated if a random combination of the two values 47_{HEX} or $B8_{HEX}$ occurs exactly n times during a packet length. The probability of all 203 combinations occurring (as a function of n) is computed as:

$$P_{e,Packet} = 1-(1-P_{47}^n)^{203} \cdot (1-P_{B8}^n)^{203} . \tag{10.20}$$

The probabilities for $n = 2, 3$ and 4 are given in table 10.3.

Table 10.3. Occurrence probability of data words of the value 47_{HEX} or $B8_{HEX}$ as a function of the number of MPEG packets required for the detection

n	2	3	4
$P_{e,Packet}$	$6.18 \cdot 10^{-3}$	$2.42 \cdot 10^{-5}$	$9.32 \cdot 10^{-8}$

The first type of error, however, leads to less favourable effects. The probability that with a non-ideal channel at least one erroneous sync byte will occur among the n consecutive ones increases with increasing n.

$$P_{e,Sync} = 1-(1-BER)^{8n} \cong n \cdot 1.6 \cdot 10^{-03} . \tag{10.21}$$

When dimensioning a detector, which might be realised as a correlator, both error probabilities must be taken into account as a function of the number n. The whole of the following decoder processing relies on safe sync information, which has already been described in sections 9.4.5 and 9.4.6 using the example of the satellite decoder.

10.5 Performance Details of the Standard

When using the DVB standard for the cable-based transmission of DVB signals there are two aspects of particular importance. The first is to know how much useful information can be transmitted, practically error-free, over a transmission channel with a given bandwidth, and the second is to know the amount of the required carrier-to-noise ratio.

10.5.1 Determination of Useful Data Rates

In accordance with (10.11) the maximum symbol rate which can be transmitted in an 8-MHz channel is limited to a value of 6.96 Mbaud by the roll-off factor $\alpha = 0.15$ which is specified in the standard. The gross bit rate results from multiplying the number of bits m transmitted per symbol. The gross bit rates for various QAM techniques are given in table 10.4. To obtain the useful data rate, the portion of the data rate provided for the error correction must first be deducted. The DVB standard uses a Reed-Solomon error protection of code rate $R = {}^{188}\!/_{204}$. Since the data rate can only be given in combination with a channel bandwidth, it is usual to divide the useful data rate by the channel bandwidth. The resulting value is referred to as spectral efficiency. Its unit is bit/s per Hz. The spectral efficiency, which is only theoretically possible, is identical with the number of bits transmitted per symbol. Through the introduction of the error correction the values decrease by the

Table 10.4. Performance details of various QAM techniques

QAM technique	Number of bits per symbol m	Gross bit rate in one 8-MHz channel [Mbit/s]	Useful bit rate in one 8-MHz channel [Mbit/s]	Bandwidth efficiency [bit/s per Hz]
256-QAM	8	55.65	51.28	6.4
64-QAM	6	41.73	38.45	4.8
32-QAM	5	34.78	32.05	4
16-QAM	4	27.82	25.63	3.2

factor R, and through the utilisation of a matched filter they decrease by the factor $1/(1 + \alpha)$.

The original plan of Deutsche Telekom was for the signals to be fed into the CATV network at a symbol rate of 6.875 Mbaud [SCHAAF]. A transmission with 64-QAM would have resulted in a gross data rate of 41.25 Mbit/s and, after deducting the error protection, a useful bit rate of 38.01 Mbit/s. This is exactly the bit rate transmitted in each transponder in DVB satellite transmissions (see section 9.5.1). Signals which are contributed to a head-end as described above can generally be fed into the cable network in two ways. One option is to feed a complete data stream in its entirety into a cable channel after its having come from a transponder, subsequent to a QPSK demodulation, a Viterbi decoding, and a QAM remodulation. In this case it is not necessary to carry out de-interleaving, RS-decoding, energy dispersal, and the respective processing steps at the encoder. The other option is to combine new MPEG transport streams from various partial data streams, which in their turn were transmitted in different transponders, and to subsequently feed them into various cable channels. In this second case a complete decoding is necessary. The simultaneous utilisation of at least two transponders must be possible. The signals transmitted in the transponders are not generally in synchronism with each other. The exact values of their actual data rates are subject to the usual tolerances. As the multiplex units operate on a synchronous time base, the individual MPEG transport streams have to be adapted to the highest transmitted data rate. This can be achieved by inserting stuffing packets into the individual data streams until the data rates of the various signals are identical. For this reason Deutsche Telekom has increased the symbol rate for a DVB transmission by cable from 6.875 Mbaud to 6.9 Mbaud. This results in a sufficient useful data rate of approximately 38.15 Mbit/s. The corresponding gross data rate is 41.4 Mbit/s.

Table 10.5 is taken from appendix B of the standard and gives further examples of a system configuration. It illustrates the flexibility of the DVB cable standard. Because there is a possibility of combining the channel bandwidth

Table 10.5. Examples of system configurations

Useful bit rate [Mbit/s]	Gross bit rate [Mbit/s]	Symbol rate [Mbaud]	Bandwidth used [MHz]	Modulation scheme
38.1	41.34	6.89	7.92	64 QAM
31.9	34.61	6.92	7.96	32 QAM
25.2	27.34	6.84	7.86	16 QAM
31.672 PDH	34.367	6.87	7.90	32 QAM
18.9	20.52	3.42	3.93	64 QAM
16.0	17.40	3.48	4.00	32 QAM
12.8	13.92	3.48	4.00	16 QAM
9.6	10.44	1.74	2.00	64 QAM
8.0	8.70	1.74	2.00	32 QAM
6.4	6.96	1.74	2.00	16 QAM

and the number of bits transmitted per symbol, the useful data rate can be increased from approx. 6 Mbit/s to more than 38 Mbit/s. With the DVB cable standard it is possible to achieve a data rate of, for instance, 31.672 Mbit/s by using at least a 32-QAM. Therefore the standard is also compatible with the terrestrial PDH (plesiochronous digital hierarchy) networks.

10.5.2 Carrier-to-noise Ratio Required in the Transmission Channel

The spectral efficiencies of the modulation types specified in the standard can be seen in figure 10.13 (except for the 128-QAM), each as a function of the required carrier-to-noise ratio. A bit-error rate of $2 \cdot 10^{-4}$ was chosen for this example. In this case the power of the subsequent error correction is just sufficient for a practically error-free signal to be interfaced at the decoder output (see section 6.2.6). The theoretical values for a redundancy-free transmission in accordance with section 7.5 are marked in figure 10.13 as crosses. The conversion of E_b/N_0 to C/N was carried out in accordance with (10.5). The positions of the DVB signals are represented by dots, for which purpose (10.7) was used to compute the C/N values. Each transfer from the theoretical signal to the DVB signal improves the carrier-to-noise ratio by approx. 1 dB. These improvements are mainly due to the introduction of the matched filter in the receiver, which results in a decrease of the actually effective noise power (see Section 10.1). This gain, however, is paid for by a reduction in the useful data rate, which manifests itself in a lower bandwidth efficiency.

Furthermore, figure 10.13 shows the upper limit of possible C/N values, which follows from (10.4) if the DVB signals are fed into the cable networks with a power reduction of 13 dB.

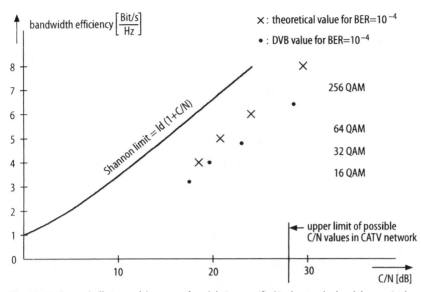

Fig. 10.13. Spectral efficiency of the types of modulation specified in the standard and the required signal-to-noise ratios at a bit error rate of $2 \cdot 10^{-4}$

10.6 DVB Utilisation in Master Antenna Television Networks

In the RACE Project "DIGISMATV" extensive work was carried out on the transmission of DVB signals in SMATV cable networks. On the basis of the obtained results the standard for the transmission of DVB signals in CATV networks [ETS 429] and that for the transmission of DVB signals in SMATV systems [ETS 473] were both adapted by the DVB Project. The results introduced here are based essentially on the findings by [DIGISMATV].

SMATV networks make available to their customers DVB signals which can be decoded with commercial IRDs. These must be transparent from the SMATV head-end right up to the user outlet, without the signal being converted to the baseband. Two different systems were defined to perform this task:

(1) System A includes the conversion of the DVB satellite signal (in accordance with [ETS 421]) into a DVB cable signal (in accordance with [ETS 429]). This conversion consists of a demodulation of the received QPSK signal and its Viterbi decoding as well as a subsequent remodulation by means of a quadrature amplitude modulator. The use of a de-interleaver, an RS-decoder and a unit for the energy dispersal as well as the corresponding processing steps at the encoder are not absolutely necessary. In [ETS 473] a differentiation is made between 'full implementation' and

'simplified implementation'. In both cases the customer requires a cable IRD for the decoding of the signal provided.

(2) System B enables a direct distribution of the DVB satellite signals in the cable network. Here, too, there are two variants, which depend on the intended frequency ranges. For the SMATV-IF system the signals in the satellite IF range (i.e. above 950 MHz) are fed into the distribution network. In this case the customer requires a satellite IRD in order to decode the signals delivered. In the SMATV-S system the satellite IF signal is downconverted to a range within the frequency band between 230 MHz and 470 MHz for distribution over cable. Before it can be decoded by a conventional satellite IRD decoder, it must be reconverted to the regular satellite intermediate frequency.

An essential factor in the choice of a SMATV system is the cost-performance balance. System B is the more economical solution. However, the application of the QPSK technique and the high transmission bandwidth that this technique entails exact a certain toll. The high frequency range of the satellite IF used in the SMATV-IF system for transmission in cable networks has the disadvantage of considerable cable attenuation, a fact which limits the system to short cable lengths. This disadvantage can be partially avoided by the SMATV-S system. However, in this case there is an additional requirement for frequency converters, which drives up the costs. It is only systems with a large number of subscribers that justify the expenditure entailed by the transmodulation, as recommended under (1). Numerous systems, however, do not attain the limit values for CATV networks discussed in section 10.1 because of the costs involved. Of particular importance are the differences in the domain of echo suppression. Investigations have shown that in SMATV systems there can occur echoes whose amplitudes differ considerably from those in CATV networks. Under adverse conditions they can have an amplitude which is only 10 dB below that of the main signal. Due to the cumulative effect of two time-shifted subsignals the frequency response assumes a ripple characteristic. The ripple r_m is expressed in logarithmic form as the ratio between maximum and minimum amplitudes:

$$r_m = 20 \log\left(\frac{1+\varrho}{1-\varrho}\right). \tag{10.22}$$

In the above case the ripple is approx. 6 dB.

The difference in delay time between the main signal and the echo results from the ratio of the echo path, which is twice the cable length, to the velocity of propagation of the wave in the cable. Depending on the type of cable used, the velocity of propagation is approx. 70% that of the speed of light, so that with a cable-connection length of 10 m the difference in delay time would be approx. 100 ns.

The distortions caused by the effect of the echo on the main signal can be cancelled by an equaliser at the receiver. Additional information is not required for the equalisation [BENVEN].

Simulations have shown that the transmission of a 64-QAM signal makes the use of an equaliser indispensable if the ripple exceeds 1 dB and if more than two ripples are to be found within the transmission channel. The deterioration of the signal as opposed to a transmission in an undistorted channel can be equalised up to 1 dB. Equalisers which can simultaneously process nine subsequent symbols will be quite adequate for this purpose. The equalisation of the channel impairment is mainly performed in the baseband so that the corresponding equalisers have a complex structure. The forms of implementation can nevertheless vary. In the literature the various algorithms are discussed in detail (e.g. [PROAKIS; LEE; KAMMEYER; GERDEN]).

10.7 Local Terrestrial Transmission (MMDS)

The DVB cable standard is intended not only for the transmission of DVB signals over cable but also for local terrestrial transmission in the microwave domain below 10 GHz. The systems operating in this microwave domain are called MMDS (microwave multichannel/multipoint distribution system, see section 9.6) and differ fundamentally from the traditional terrestrial broadcasting system by the designated frequency bands in which they operate and the related propagation characteristics of the electromagnetic waves. Distances of more than 60 km between transmitter and receiver, which are quite usual in traditional terrestrial broadcasting, are not possible in the microwave domain. The future technique, which can operate in the UHF and VHF bands in conventional transmission networks, is described in chapter 11. For transmission in the microwave region there must not only be short transmission paths, but also a direct line-of-sight from the transmitter to the receiver.

The DVB specification with the designation DVB-MC (DVB Microwave cable-based), has been ratified as European standard EN 300 749 [EN 749]. An important reason for adopting the cable standard was to ensure that identical signal processing would be used in as many applications as possible. A bandwidth-efficient and robust technique for modulation and channel coding was achieved with the cable standard and this should also be suitable for microwave distribution.

Symbols used in Chapter 10

A_K	MSB before differential coding and after differential decoding
B	bandwidth
b_i	bit of value $2i$
B_K	MSB-1 before differential coding and after differential decoding
C/I_{Ix}	carrier-to-interference ratio (as defined)
C/N_{Ix}	carrier-to-noise ratio (as defined)
E_b	energy per bit
f	frequency in general
f_{Ix}	frequency (as defined)
f_N	Nyquist frequency
$H(f)$	transfer function
$I, I(t)$	in-phase component of a signal
i	running variable, integer
I_K	MSB of the I-component after differential coding and before differential decoding
I_{K-1}	I_K delayed by one clock
M	number of constellation points of m bits which are transmitted per symbol, $M = 2^m$
m	bits per symbol or running variable
N	noise power
n	running variable, integer
N_0	noise-power density
PSD_{Ix}	power spectral density (as defined)
P_{Ix}	probability (as defined)
$Q, Q(t)$	quadrature component of the signal
Q_K	MSB of the Q-component after differential coding and before differential decoding
Q_{K-1}	Q_K delayed by one clock
R	code rate
r_m	in-band ripple
R_S	symbol rate, baud rate
S_{Ix}	signal power (as defined)
T_{Ix}	clock period (as defined)
T_S	symbol duration
α	roll-off factor
θ	phase rotation of echo channel
ϱ	amount of transfer function in the echo path
τ	delay time of echo

11 The Standard for Terrestrial Transmission and Its Decoding Technique

Of the standards for digital broadcasting via the "classical" media (cable, satellite, terrestrial transmitters) the specification for terrestrial transmission (DVB-T) was developed last. In the early days of the DVB project the main priority for digital television was the transmission by satellite and cable, so it was only after completion of these two standards that a draft of the specification for terrestrial transmission was prepared, which was adopted in December 1995. Terrestrial transmission is far more complex than satellite or cable transmission, from the point of view of user requirements, with regard to the characteristics of the transmission path, and in view of the technical solutions that are called for. Furthermore, while the first two systems were being developed it was possible to draw upon practical experience acquired in the professional field, for instance with respect to modulation techniques for QAM and QPSK. In the first half of the 1990s, the terrestrial transmission of digitised broadcasting signals had only been tested in digital audio broadcasting (DAB), and here, too, without the experience of having worked with large numbers of receivers. Only in November 1998, the first DVB-T network became operational in the United Kingdom. During that same period a network was built up in the northern part of Germany, which by now includes some 20 operational transmitters. Since 1998 DVB-T has been introduced in European countries like Finland, Sweden and Spain. Over and above that, Australia, Singapore and Taiwan also decided on DVB-T, especially since practical tests had proven that mobile reception of DVB-T is possible, and launched DVB-T services. In recent years, the proven ability of DVB-T to address mobile receivers has been an important criterion in deciding on an ideal digital terrestrial system in several countries. Even though mobile reception was not required when the terrestrial standard was developed, the robustness of the standard has attracted considerable interest. Fostered by the rapid progress in the development of receivers, DVB-T in the meanwhile offers the possibility to address receivers in all kinds of environments, like stationary, portable and mobile reception.

The preparation of the specification presented here required the close cooperation of many dedicated and expert partners from various European countries. A prominent role was played not only by individual industrial companies, broadcasting corporations and network providers, but also by the research consortia already mentioned in section 1.2.2.

11.1 Basics of Terrestrial Television Transmission

The distribution of television programmes by way of terrestrial transmitters is the "classical" technology of broadcasting. In contemporary analogue television in Europe (PAL, PALplus, SECAM) a typical content provider assumes the responsibility to provide a single television programme for the greatest possible section of the population. For example, in Germany this means that a typical viewer in a metropolitan area requires up to three different rooftop antennas in order to receive - at a good quality level - six or seven television programmes from up to three transmitting stations at different locations. "Quality" should be interpreted here as the overall effect, the sum of various influences including the many possible disturbances created by noise and interference. The concept of the provision of services to viewers and the extent of permissible disturbances are defined in the "guidelines for the assessment of the provision of TV services from the first and second German public programmes and the German postal services (now Deutsche Telekom AG)" (Richtlinie für die Beurteilung der Fernsehversorgung bei ARD/ZDF und DBP) [IRT].

A typical home receiver operates with a rooftop antenna, which should have a gain of approx. 7 dB (VHF) or 10 dB (UHF). The considerable directivity of the rooftop antenna, apart from the inherent increase in the signal power obtained by the gain, can partially reduce echo impairments caused by reflection from hills, buildings or suchlike. It is also possible to partially reduce the harmful effects of other transmissions in the same channel or in adjacent channels.

A good rooftop antenna guarantees television viewers in the service area of a transmitter a satisfactory video and audio quality, although some identifiable disturbances in the form of slight echoes, some noise and sporadic sputter can occur.

The reception conditions within the service area can, ideally, be described by the so-called Gaussian channel model, which is based on a direct signal path from transmitter to receiver, overlaid with additive white Gaussian noise (AWGN) which is mainly produced in the receiver itself.

In order to include the impairment caused by echoes it is necessary to enlarge the channel model. The transmission path known as "Ricean channel", takes into account the effect of multipath signals in addition to noise and to the dominating direct signal path between transmitter and receiver. The statistics of these multipath signals are approximated by a Ricean distribution [LÜKE 1]. As opposed to analogue television, in which echoes are more or less visible in the form of double contours on the screen, echoes in the case of DVB signal reception - when the delay time has exceeded a certain value - cause an increase in intersymbol interference (ISI), which ultimately results in an increase in the bit-error rate. This increase must then be corrected, for example, by increasing the transmission power. A simulation of the power requirements for the terrestrial DVB standard applied the following mathematical model to describe the Ricean channel, where 20 echoes were taken into consideration [ETS 744]. The output signal $y(t)$ of the channel model is described as a function of the input signal $x(t)$

$$y(t) = \frac{\rho_0 \cdot x(t) + \sum_{i=1}^{N_e} \rho_i e^{-j2\pi\Theta_i} x(t - \tau_i)}{\sqrt{\sum_{i=0}^{N_e} \rho_i^2}} \qquad (11.1)$$

where:

ρ_0	=	attenuation in the direct signal path
N_e	=	number of echoes (here 20)
ρ_i	=	attenuation in echo path i
θ_i	=	phase rotation in echo path i
τ_i	=	relative delay time in echo path i

The Ricean factor $K = \rho_0^2 / \sum_{i=1}^{N_e} \rho_i^2$ denotes the ratio of the signal in the direct path to the sum of the signals in all echo paths. $K = 10$ dB was used for the simulation.

In table 11.1 the values given for the simulation of the performance data for terrestrial DVB standards are for attenuation, relative delay time, and phase rotation. Figure 11.1 shows the attenuation of the 20 echo paths as a function of the echo delay time.

A comparison of the results of simulations, in terms of carrier-to-noise ratio (C/N) required for quasi error-free (QEF) reception of a DVB signal in the AWGN channel, with the respective values in the Ricean channel shows, as expected, that the Ricean channel has higher requirements. The additional requirement (see section 11.6) is actually in the range of 0.3 to 1.1 dB, according to the choice of system parameters. This additional requirement is valid for a system with non-hierarchical modulation (see section 11.5). An increase in the transmission power by a maximum of 30% is therefore required to compensate the effect of echoes.

Reception with a rooftop antenna can be viewed as stationary reception and the directivity of the antenna can be used either for the selection of the

Table 11.1. Data for simulation of the terrestrial transmission channel. The values relate to the parameters in (11.1) and (11.2)

i	ϱ_i	τ_i [μs]	Θ_i [rad]
1	0.057662	1.003019	4.855121
2	0.176809	5.422091	3.419109
3	0.407163	0.518650	5.864470
4	0.303585	2.751772	2.215894
5	0.258782	0.602895	3.758058
6	0.061831	1.016585	5.430202
7	0.150340	0.143556	3.952093
8	0.051543	0.153832	1.093586
9	0.185074	3.324866	5.775198
10	0.400967	1.935570	0.154459
11	0.295723	0.429948	5.928383
12	0.350825	3.228872	3.053023
13	0.262909	0.848831	0.628578
14	0.225894	0.073883	2.128544
15	0.170996	0.203952	1.099463
16	0.149723	0.194207	3.462951
17	0.240140	0.924450	3.664773
18	0.116587	1.381320	2.833799
19	0.221155	0.640512	3.334290
20	0.259730	1.368671	0.393889

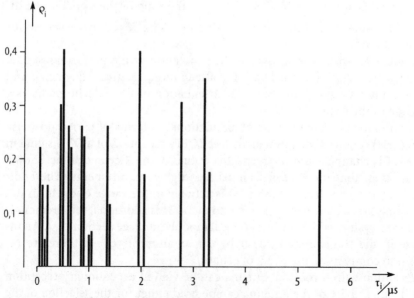

Fig. 11.1. Example of an echo profile used for the representation of the terrestrial transmission channel

direct signal or at least for choosing a dominant echo signal as the main reception signal. The situation changes in two main ways with the transition to a portable receiver. Firstly, the movement of the receiver during reception can cause changes to the reception condition, particularly inside buildings. Secondly, rod antennas on portable receivers bring no noticeable antenna gain or directivity.

The first version of the commercial and user requirements for terrestrial digital television in Europe defines the "stationary reception with a portable receiver" as one of its objectives. Mobile reception was thus deliberately not included as an objective. In comparison to the reception with a rooftop antenna, however, this limitation means it can no longer be taken for granted that a direct signal path would dominate. Consequently, for the simulation of the channel model, which is representative of this case, the term which describes the direct signal path was eliminated in (11.1). This leads to (11.2).

$$y(t) = \frac{\sum_{i=1}^{N_e} \rho_i e^{-j2\pi\Theta_i} x(t - \tau_i)}{\sqrt{\sum_{i=0}^{N_e} \rho_i^2}} \tag{11.2}$$

The transmission channel modelled in this way, with echoes of more or less equal priority, is called "Rayleigh channel" [LÜKE 1] since the distribution of echoes corresponds to a Rayleigh distribution. Like the Ricean channel, the Rayleigh channel – as compared to the AWGN channel – requires a higher carrier-to-noise ratio. Section 11.6 presents simulation results which show that for an error-free reception - according to the choice of parameters in the standard - carrier-to-noise ratio is required which is up to 9 dB higher. This ratio, too, is valid in the case of non-hierarchical modulation (see section 11.5). An increase in the carrier-to-noise ratio by 9 dB would require an eight times higher transmission power. Naturally, an increase of that magnitude hardly seems realistic.

The portable receiver therefore is affected by the lack of gain in the receiving antenna (for example, approx. 10 dB in the UHF band, which is the band most relevant to DVB), and by the signal deterioration in the Rayleigh channel. Additionally, further losses occur due to the fact that the reception antenna is often inside a building and often close to ground level, while an antenna height of 10 m from ground level is assumed in the calculation of the coverage according to [IRT].

In 1997 the "Chester 1997 Multilateral Co-ordination Agreement" was developed to describe technical criteria, co-ordination principles and procedures for the introduction of DVB-T in Europe and certain regions around it [CHESTER]. It contains, among other things, statements about the signal levels required for DVB-T and field strength values. This agreement is an amendment of the Stockholm agreement of 1961 and constitutes an interim solution to define planning parameters for the protection ratios in the course of the introduction of DVB-T. Therefore, it will be necessary to revise the Stockholm agreement in a follow-up conference, which is planned in 2005. This conference needs to define coverage objectives and frequency requirements for DVB-T; existing determinations will be adapted to the experiences gained in real networks and to recent technical developments.

Three different receiving conditions were looked at in this new agreement: fixed antenna reception, portable outdoor reception and portable indoor reception at ground floor. In the case of a portable receiver outside a building the change of the antenna height from 10 m to 1.5 m results in a reduction in field strength of 12 dB. Taking this value as a basis, the result is a requirement for an overall increase by approx. 30 dB in field strength for stationary reception with a portable receiver outside buildings (7-8 dB due to the Rayleigh channel, 10 dB loss in antenna gain, and 12 dB due to a decrease in height above ground level). With reception inside buildings the field-strength loss is additionally increased owing to the attenuation caused by the walls of the buildings. [CHESTER] recommends using a building penetration loss of 7 dB with a standard deviation of 6 dB. In theory, the resulting overall effect is an increase in field strength by up to 36 dB (approx. 7-8 dB + 10 dB + 19 dB), if stationary reception using a portable receiver on the ground floor of a building is to be guaranteed. However, [CHESTER] does not take into consideration the local environment but instead uses generalised assumptions meaning that in practice the theoretical increase in field strength is not required. Besides, practical values highly depend on reception technology used. For instance, within the framework of [CHESTER] values were based on a receiver noise figure of 7 dB. In the meantime, noise figures, in the range of about 2 dB can be achieved using adapted antenna pre-amplifiers (see section 11.4). Further improvements can be expected in future.

The reception of DVB-T by portable receivers inside buildings was not a planning criterion in the United Kingdom. In countries like Germany, however, where the percentage of TV households equipped with cable or satellite receivers exceeds 90 % "portable indoor" is a must. Therefore the networks will have to be planned accordingly. The latest investigations which are based on topographical data have shown that in the area considered, indoor portable reception will be possible [REIMERS 6]. The investigations were carried out for Schleswig-Holstein, a rather flat part of Germany based on the existing transmitter locations. A reduction of the total transmitter power currently in use for the broadcasting of one analogue TV programme by a factor of 3.6 and the use of single frequency networks were assumed.

In Berlin DVB-T services were officially launched in 2002. Other German countries will take part in 2004. The coverage area for these services will be defined in such a way that in the metropolitan areas portable indoor receivers can be used whereas portable outdoor reception will be possible throughout the rest of the country.

The hierarchical modulation described in section 11.5 will offer an option for the terrestrial transmission of DVB signals by which the sudden disruption of the reception, at least for individual programmes within a data container, can be mitigated.

11.2 User Requirements for a System for Terrestrial Transmission of DVB Signals

For the first time in the history of television engineering, terrestrial transmissions were not introduced first, the initial services being delivered by satellite or cable transmissions instead. Regarding the DVB Project, the systems for transmission by satellite and for subsequent distribution over cable have established many of the conditions for the standard of terrestrial television. The original user requirements of the system for terrestrial transmission of DVB signals have to be seen against this background. Only the most important criteria are listed below from the catalogue of user requirements.

(1)   The system for terrestrial transmission should have as much similarity as possible with the systems for cable and satellite. This will ensure that the home receiver technology will have as much similarity as possible with that of cable and satellite.

(2)   DVB programmes should be transmitted in data containers and their capacities should be as large as possible. The channel bandwidth in Europe should be chosen so that a channel spacing of 8 MHz can be supported. Channels with 7 MHz spacing need not be supported, which implies that the introduction of DVB is not intended for bands I and III (VHF range).

(3)   The system should have optimum area coverage for stationary reception with a rooftop antenna. The support of stationary reception with portable receivers is desirable; mobile reception is not a development objective.

(4)   It should be possible for DVB signals to be transmitted in terrestrial single-frequency networks. Single-frequency networks consist of transmitters which transmit exactly identical data streams in synchronism with each other, using the same transmission frequency. Neighbouring transmitters support each other in their coverage task. The topography of the network is essentially characterised by the maximum permissible distance to the next transmitter.

(5) The standard should allow the commencement of terrestrial transmission of DVB signals in 1997. The technology of the home receiver must be designed to enable the production of reasonably priced equipment by 1997.

(6) Finally, "hierarchical modulation" should be included as an option.

For the technical layout of the system for the terrestrial transmission of DVB, requirements 1 and 4 are of paramount importance. In order to achieve the greatest possible similarity between the standards for terrestrial and satellite transmission the very same forward error correction was chosen which had been specified for satellite transmission. This comprises inner and outer error protection as well as interleaving (see section 6.4 and 9.3). The requirement for the support of single-frequency networks leads automatically to the choice of an orthogonal frequency division multiplex (OFDM) as a modulation technique (see section 7.6). The combination of this modulation technique with the known methods of error protection has been termed "coded orthogonal frequency division multiplex" (COFDM).

Changes which were later incorporated in the commercial and user requirements were the result of non-European countries beginning to ask for solutions to 6-MHz and 7-MHz channel spacing and of the growing demand for mobile reception. As a consequence, DVB-T was eventually made available in versions appropriate for 6-MHz, 7-MHz and 8-MHz channel spacing and can even be received by fast-moving receivers.

For the design of a standard for the terrestrial transmission of DVB signals on the basis of COFDM it is the required maximum distance between mutually supporting transmitters in a single-frequency network which primarily decides what the minimum length of the required guard interval should be (see section 7.6). In simplified terms it can be said, that the required duration of this guard interval can be computed if, for each receiver location at which it becomes impossible to decode the signal of one of two transmitters in the presence of signals from the other without the existence of the guard interval, the time difference between the two signals has been determined, which is computed by taking into account the distance from transmitter to receiver and the velocity of propagation. The guard interval should then be longer than the greatest time difference resulting from this computation. In the United Kingdom, in particular, investigations into the relationship between the chosen length of the guard interval and the percentage of the population that can be provided with DVB services were carried out [LAFLIN]. According to these investigations, 97 % of the population can be supplied with a national single-frequency network if the length of the guard interval is 500 µs, 91% at 250 µs, 76% at 125 µs, and 63% at 62,5 µs. This analysis provides a possible approach to the determination of the length of the guard interval. When national single frequency networks need to be supported, the guard interval should be at least 200 µs. This represents a maximum difference of approx. 60 km in distances between neighbouring transmitters and the reception point.

In the process of generating the specification for the terrestrial transmission of DVB, the determination of the guard interval led to a particularly complex co-ordination problem between the participating nations. Probably none of the technical parameters with respect to DVB caused such prolific correspondence, even between the ministries responsible for broadcasting in the various countries, as did the length of the guard interval. It will be shown in section 11.3.2., that the choice of a guard interval length of 200 µs represents a decision in favour of a rather complex variant of COFDM (8K COFDM). The adoption of this variant, particularly in the opinion of the representatives from the United Kingdom, meant that the complexity of the terrestrial receiver would be so great that the estimated costs for such equipment would be too high and the introduction schedule (1997) could never be met. Against the background of a White Paper published by the government in London in 1995 (see chapter 1) the relevance, in the United Kingdom, of a national single-frequency network seemed to have taken second place to the importance of the adherence to the implementation schedule. The representatives of other nations, such as Spain, strongly insisted on the guard interval with a length of 200 µs because of the belief that only with the introduction of national single-frequency networks terrestrial DVB would be possible in their countries. The Spanish partners of the DVB Project, as well as several others, attached less importance to the possible introduction of the system in 1997. As will be shown, it was possible to satisfy both positions by finally arriving at a specification which allowed for different complexities and at the same time provided a choice of guard intervals with different duration.

11.3 Encoder Signal Processing

Numerous parts pertaining to the system for the terrestrial transmission of DVB signals are identical with those of the other systems previously described. Figure 11.2 depicts the block diagram of the encoder, in which the shaded components are exactly the same as those used in the satellite standard and therefore require no further explanation.

1.3.1   Inner Interleaver and Symbol Mapping

A first new processing element is an inner interleaver, which follows the inner error protection. Since OFDM is a multicarrier technique, which can,

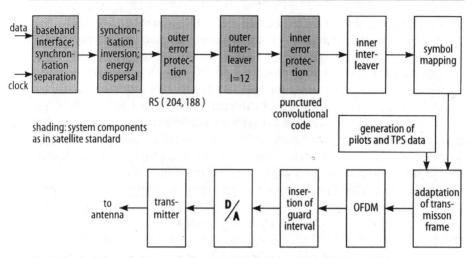

Fig. 11.2. Block diagram of an encoder for terrestrial DVB with non-hierarchical modulation

among other things, be used to minimise the effects of frequency-selective impairments as a result of either multipath-reception interference or external sources of interference, the distribution of successive data over the large number of available carriers is suggested. The distribution pattern should help disperse the effects of disturbances – even of a longer duration – affecting individual neighbouring carriers or a whole group of these, such that a correction will hopefully be achieved by the bitwise-functioning inner error protection.

The interleaving takes place in two steps. As a first step, 126 successive bits are combined into a block. Within this block the bits are then interleaved (bit interleaver). Subsequently a large number of blocks, generated from interleaved data in this way, is defined as a new block, within which whole groups of bits (symbols) are then interleaved (symbol interleaver).

It will be shown in sections 11.3.2 and 11.3.3 that in the two variants of the COFDM technique either 1512 (in 2K COFDM) or 6048 (in 8K COFDM) single carriers each are simultaneously modulated with useful data, and combined in an OFDM symbol. In addition to these 1512 or 6048 useful carriers another 193 or 769, respectively, are contained in the same OFDM symbol for synchronisation etc. The entire OFDM symbol therefore comprises either 1705 or 6817 single carriers. The process of interleaving refers to this structure of the OFDM symbol in various ways.

According to the choice of method for the modulation of the individual useful carriers with useful data, each of the carriers requires several useful bits. Alternative modulation procedures are QPSK, 16-QAM and 64-QAM. For QPSK, 2 bits are required for the modulation of each carrier (1 bit for the real and 1 bit for the imaginary axis in accordance with 4 possible constellations), for 16-QAM, 4 bits are required (2 bits for the real and 2 bits for the imaginary axis in accordance with 16 possible constellations) and for 64-QAM, 6 bits are required (3 bits for the real and 3 bits for the imaginary axis in accordance with 64 possible constellations). Consequently, for the creation of QPSK-modulated OFDM carriers two of the bitwise-functioning interleavers will also be connected in parallel. For the 16-QAM there are four, and for the 64-QAM there are six such interleavers. The interleaver structure for 16-QAM is shown as an example in figure 11.3. The choice of blocks of 126 bits for the bitwise interleaving makes it possible for a whole number of such blocks to be used in both COFDM variants ($1512 = 12 \cdot 126$, $6048 = 48 \cdot 126$).

The second level of the interleaver works on the basis of the symbols required for the modulation of the useful carriers. In order to make room for the already- mentioned 193 or 769 carriers which are required for synchronisation, the symbol interleaver generates no continuous stream of useful symbols at its output, but outputs an intermittent stream, which has intervals at points at which the additional carriers can be inserted.

In view of the great number of possible variations of the inner interleaving, which depend not only on the modulation technique but also on the chosen COFDM variant, the possible types of interleavers cannot be dealt with at this point. For further details the reader is referred to [ETS 744].

Each of the individual useful carriers of the OFDM signal is therefore separately modulated. A choice can be made between the modulation techniques QPSK, 16-QAM and 64-QAM. Furthermore, for the hierarchical-modulation technique (to be discussed in section 11.5) a so-called multiresolution QAM of type MR 16-QAM or type MR 64-QAM may also be selected.

The allocation of sequential bits to the symbols of the chosen modulation technique is carried out in accordance with a method named after Gray [LÜKE 1]. The reason for the selection of this technique is that it produces only one bit error in the case of the most probable transmission errors, i.e. when just one of the decision thresholds between two neighbouring constel-

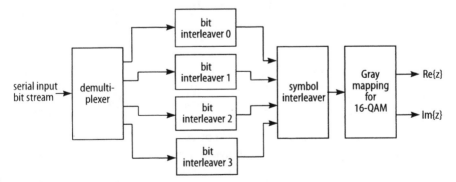

Fig. 11.3. Structure of an inner interleaver for the generation of 16-QAM symbols

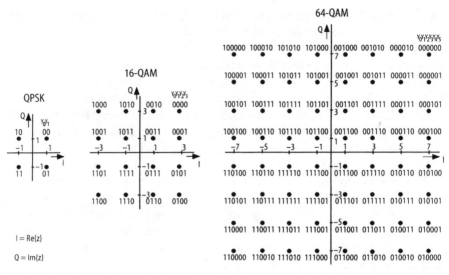

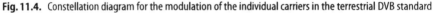

Fig. 11.4. Constellation diagram for the modulation of the individual carriers in the terrestrial DVB standard

lation points is crossed. Consequently, the ordering of the symbols in the I/Q plane must be effected in such a way that the symbols running parallel to the I-axis and the symbols running parallel to the Q-axis may only differ from each other in one place. The result of the "Gray mapping" is shown in figure 11.4.

11.3.2 Choosing the OFDM Parameters

As already explained in section 7.6, various parameters in OFDM can be chosen separately to determine the performance of the whole system. One of the crucial parameters in the use of OFDM for DVB is the length of the guard interval. If single-frequency networks are to be realised in which the maximum permissible difference in distance between neighbouring transmitters and the corresponding receiver locations (further referred to, rather imprecisely, as transmitter spacing) is, for example, approx. 60 km, then the length of the guard interval T_g can no longer be freely chosen. This must then be at least 200 μs (200 μs · 300,000 km/s = 60 km). Now, the duration of the guard interval implies a reduction in the time available for the actual data transfer, which reduces the available channel capacity. Therefore the length of the guard interval is always kept relatively small as compared to the symbol duration T_S. On the other hand, long symbol durations automatically result in long durations of the useful part of the symbol T_U and therefore in reduced spacing s between the individual carriers. Close carrier spacing, however, leads to complex circuitry in the receiver, which, after all, needs to be in a position to recover the information of the single carrier (see section 11.4). T_S might actually be fixed, for example, at $5 \cdot 200$ µs = 1 ms, and T_U would then be computed as $4 \cdot 200$ µs. Consequentially the spacing between the single carriers amounts to 1/800 µs = 1.25 kHz. On the basis of the chosen assumptions, 6000 neighbouring single carriers are transmitted in a UHF channel with a bandwidth of 8 MHz, of which, after deduction of protection zones at the band limits, approximately 7.5 MHz are effectively usable. With the assumption that each of these carriers is modulated with a 64-QAM, i.e. that 6 bits per carrier can be transmitted, the gross transmission capacity can be computed as (6000 · 6 bit per 1 ms =) 36 Mbit/s. From this gross capacity, also containing the required redundancy for the forward error correction, some portions must be reserved for synchronisation requirements, etc..

In many applications the assumed guard interval of 200 µs is not appropriate. For example, if there is no necessity to create national single-frequency networks with complete coverage of the whole nation, but if, instead, only the coverage of a metropolitan area is required, it would be desirable to be able to work with shorter guard intervals in order to save channel capacity. Even when a transmission network is to be constructed which is based on conventional planning (an independent transmission frequency per transmitter site) the utilisation of a long guard interval is not desirable. When the transmitter spacing only needs to be in the range of 15 km for instance, a guard interval of 50 µs is sufficient such as in the case of the coverage of a metropolitan area by a single transmitter supplemented by low-power transmitters in the „canyons of houses" or on the outskirts of a coverage area. On the analogy of the previous analyses this example would lead to a symbol duration $T_S = 5 \cdot 50$ µs = 250 µs. T_U is then computed as $4 \cdot 50$ µs. Therefore the spacing between the single carriers results in 1/(200 µs) = 5 kHz. Within one UHF channel approx. 1500 single carriers can be transmitted. When using 64-QAM for the modulation, the gross transmission capacity for each carrier again computes to (1500 · 6 bit per 250 µs =) 36 Mbit/s. Of this gross capacity, which, of course includes the redundancy required for forward error correction, appreciable amounts must again be reserved for synchronisation and other purposes.

The study of the required length of the guard interval has shown that for terrestrial DVB, large numbers of carriers are required, which in the case of national single-frequency networks would be in the range of 6000 and in the case of regional single-frequency networks, in the range of 1500. Since OFDM signals (see section 7.6) are generated by means of an IDFT, a solution would be to select, at the modulator, one-chip implementations of an IFDT which can carry out such a Fourier transform. These components, however, are so designed that the number of the samples used is equivalent to some power of two. The next power of 2 greater than 6000 is $2^{13} = 8192$ (8K). The next power of 2 greater than 1500 is $2^{11} = 2048$ (2K). The specification for terrestrial DVB had to be prepared for both types of SNFs due to the insurmountable differences of national interests explained in section 11.2. Therefore it not only includes a variant based on an 8K-IDFT but also one based on a 2K-IDFT. The working title chosen for this combination is "Common 2K/8K Specification".

Based on the two variants six possible values for T_g were agreed: 224 μs, 112 μs, 56 μs, 28 μs, 14 μs, and 7 μs. The four longer guard intervals belong to the 8K variants of the specification and the four shorter ones belong to the 2K variants. The reason for selecting precisely these values will be explained in the following. In any case, the order of magnitude shows, that national SFNs can be realised with the longest guard intervals (transmitter spacing < 67 km) that the middle range of values is conceived for regional networks (transmitter spacing 17 km or 33 km) and that the short guard intervals (transmitter spacing 2 km, 4 km or 8 km) are aimed at local coverage, where a master transmitter would, for instance, be supported by one or more gap-filling transmitters.

There now remains the sampling frequency to be determined for performing the IDFT. This was set at 64/7 MHz = 9.143 MHz. With this task completed, the time T_U is now clearly defined.

For the 8K variant, $T_U = 8192 \cdot (1/[64/7]$ MHz$) = 896$ μs. The spacing between the individual OFDM carriers can be computed as $(1/896$ μs$) = 1.116$ kHz. Following from this number not all 8192 carriers can really be transmitted. The actual bandwidth used has to be within the 8-MHz channel spacing and must remain limited to approx. 7.5 MHz. 6817 carriers were actually chosen within the limit of $k_{min} = 0$ and $k_{max} = 6816$. The OFDM signal therefore occupies a total bandwidth of $6817 \cdot (1/896$ μs$) = 7.609$ MHz.

For the 2K variant, $T_U = 2048 \cdot (1/[64/7]$ MHz$) = 224$ μs. The spacing between the individual OFDM carriers can be calculated as $(1/224$ μs$) = 4.4643$ kHz. It is logical that not all 2048 carriers can be transmitted, as the actual bandwidth used for this variant also has to be within the 8-MHz channel spacing and must remain limited to approximately 7.5 MHz. 1705 carriers were chosen within the limit of $k_{min} = 0$ and $k_{max.} = 1704$. The OFDM signal therefore occupies a total bandwidth of $1705 \cdot (1/224$ μs$) = 7.612$ MHz.

The aim pursued in making the seemingly arbitrary choice of a sampling frequency of 64/7 MHz can now be explained. For if this value is reduced to (64/8 MHz) while retaining an 8K-IDFT, then this results in the parameters $T_U = 1.024$ ms, carrier spacing 977 Hz, and bandwidth 6.66 MHz. This means that after a simple change of the sampling frequency it is possible to use a 7-MHz channel if required. It goes without saying that another change to the appropriate value allows the use of DVB-T in 6-MHz channels ($64/7 \cdot 6/8$ MHz, $T_U = 1.19$ ms, carrier spacing = 837 Hz, bandwidth = 5.71 MHz).

The three values for the length of the guard interval can now be deduced from the time T_U (again only taking into account the specification for 8-MHz channels). For the variants 8K and 2K, these values are each at a ratio of $\Delta = (T_g / T_U) = 1/4$ or $1/8$ or $1/16$ or $1/32$.

The total symbol duration T_S then originates from the sum of the duration of the usable symbol and the duration of the guard interval, which in the case of the 8K variant is 1120 µs or 1008 µs or 952 µs or 924 µs. In the case of the 2K variant this results in 280 µs, 252 µs, 238 µs or 231 µs. With the above assumption, that each single carrier is modulated with a 64-QAM, thus carrying 6 bits per carrier, the gross transmission capacity can be calculated, for example in the 8K variant and in the case $\Delta = 1/4$, as ($6817 \cdot 6$ bit per 1.120 ms =) 36.52 Mbit/s.

11.3.3 Arrangement of the Transmission Frame

The transmission of a complex signal such as the OFDM signal described above, comprising either 6817 or 1705 carriers modulated individually with QPSK or QAM with a spacing of about 1.1 kHz or 4.4 kHz requires the addition of a considerable amount of synchronisation and signalling data. It is only due to this additional information that the receiver is enabled to utilise algorithms for certain control functions in order to realise an error-free demodulation of the signal. To keep the data rate sacrificed for the additional data within limits, it is sensible to combine a certain amount of the OFDM symbols and additional data to form a transmission frame. All the additional data relevant to that frame can then be transmitted only once. In the following, the term "symbol" denotes all 6817 or 1705 carriers, which are transmitted simultaneously for the symbol duration T_S.

In the case of the specification for the terrestrial transmission of OFDM signals there are actually 68 symbols combined to form a transmission frame. Four such frames are defined as a superframe. Figure 11.5 depicts a section of the transmission-frame structure. The 6817 or 1705 subcarriers of the OFDM signal are shown next to each other in the horizontal direction. The abscissa therefore represents the frequency axis. The succession of the symbols is shown in the vertical direction.

Apart from the carriers, which are modulated with the actual useful information, three further types of carriers required for the synchronisation or the transmission of additional information are used. Each of the carriers used for the synchronisation is individually modulated. The rules for choosing a modulation word for each one of these carriers are known to the receiver.

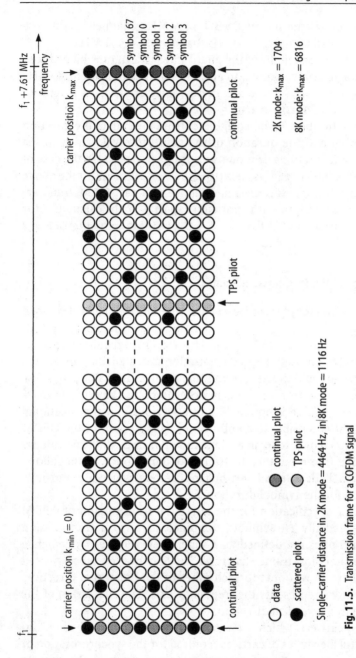

Single-carrier distance in 2K mode = 4464 Hz, in 8K mode = 1116 Hz

Fig. 11.5. Transmission frame for a COFDM signal

The "continual pilots" are present in each symbol at the same carrier position (in the case of 2K, for instance, at the 45 positions 0, 48, 54, 87 ... 1323, 1377, 1491, 1683, 1704; in the case of 8K, for instance, at the 177 positions 0, 48, 54, 87 ... 6435, 6489, 6603, 6795, 6816). The distribution along the frequency axis is chosen, so that within one channel no periodicities can occur. Therefore the continual pilots can be used for the coarse adjustment of the frequency of the local oscillator in the receiver. In order to make the synchronising information as robust as possible against transmission errors, the continual pilots have an amplitude which, in comparison to the carriers containing useful information, is increased by the factor 4/3. That is why they are also called boosted pilots.

The "scattered pilots" are scattered over the transmission frame according to a definite plan. This plan can be easily identified in figure 11.5. The aim of the scattering is, firstly, to make a large number of pilots within each symbol for the fine adjustment of the receiver available. Secondly, the scattered pilots can contribute to the temporal synchronisation. Thirdly, the receiver will be able to perform a nearly continuous analysis of the frequency- and the temporal plane, which can be used for the evaluation of the current channel condition ("channel estimation"). The scattered pilots are also transmitted with increased amplitude.

The "transmission-parameter signalling pilots" (TPS pilots) permit a transmission of additional information with which the temporal synchronisation, i.e. the recognition of the beginning of the transmission frame, can be ensured. Moreover, they provide the information about the transmission parameters used. The details of this will be discussed in the following. The TPS pilots, just as the continual pilots, can be found in precisely fixed carrier positions (in the case of 2K, for instance at the 17 positions 34, 50, 209, 346, 413 ... 1262, 1286, 1469, 1594, 1687; in the case of 8K, at the 68 positions 34, 50, 209, 346, 413 ... 6374, 6398, 6581, 6706, 6799). TPS pilots are transmitted with an amplitude, which represents the average amplitude of all carriers transmitting useful information.

The numbers of the various types of pilots named per symbol cannot just be added to determine the channel capacities lost for the transmission of use-ful information, as the possibility that a continual pilot may be at a position intended for a scattered pilot cannot be ignored (see figure 11.5). On closer examination it may be seen that each of the continual pilots is co-located in every fourth symbol with a scattered pilot. This results, on average, in 1512 carriers with useful information per symbol in the 2K case, and in 6048 carriers with useful information in the 8K case, which means that approximately 13% of all carriers have to be utilised for synchronisation, etc.

Attention has already been drawn to the fact that the continual pilots and the scattered pilots are individually modulated so that from an analysis of this modulation the receiver can, among other things, derive information for the tuning. The modulation is carried out in accordance with the structure of a maximum-length sequence, which is known to the transmitter and to the receiver.

Maximum-length sequences (m-sequences) are often referred to as pseudo-random sequences. They have some prominent features. One of these is the ideal, pulse-shaped, periodically recurring auto-correlation function, which is of particular importance in this connection [LÜKE 2].

A maximum-length sequence can be generated by a shift register, in which the positions indicated by the descriptive polynomial are linked by means of EX-OR circuits (see also section 9.3). The polynomial used here has the form

$$g(X) = X^{11} + X^2 + X^0 .$$

X^{11} represents the input of the first cell; X^0 represents the output of the shift register. All sequences commence with the initialisation condition $X^5 = 1$, in such a way that the first bit at the output of the shift register coincides with the first carrier of a symbol. Further shifting within the register takes place such that there is a shift step to each carrier, independently of whether it is a pilot or not. When the temporally coincidental carrier happens to be a continual pilot or a scattered pilot, a modulation symbol is allocated to the pilot by the bit at the output of the shift register. When this bit is 0, then the pilot is modulated with the positive real part +4/3, when the bit is 1, then the pilot is modulated with the negative real part -4/3. The imaginary part of the modulation symbol is always zero. If the adjacent carriers in figure 11.5 which form part of a symbol are examined from left to right, i.e. along the frequency axis, a subsampled maximum sequence will be detected at the 176 or 701 positions per symbol at which continual or scattered pilots are to be found. This maximum sequence can now be used for fine-tuning in the receiver.

In every transmission frame the data denoted as TPS are transmitted within each symbol. Basically, they constitute a special data channel, which can carry important information for the duration of every transmission frame. The receiver requires this information for the demodulation of the subsequent transmission frame. The purpose of the TPS data is therefore similar to that of the so-called fast-information channel of the DAB system [ETS 401]. Within each transmission frame there are 68 successive symbols. Each of these symbols contains one bit from a 68-bit TPS word. For the 2K variant 17 carriers per symbol are made available for TPS, and for the 8K variant 68 carriers. In each symbol these TPS carriers all transmit exactly the same individual bits in parallel.

The information prepared by the TPS refers to the modulation procedure (QPSK, 16-QAM, 64-QAM), the constellation model for the case of hierarchical modulation with the possible values $\alpha = 1$ or $\alpha = 2$ or $\alpha = 4$ (see section 11.5), the coderate of the inner error protection (1/2, 2/3, 3/4, 5/6, 7/8), the duration of the guard interval ($\Delta = 1/32$, $\Delta = 1/16$, $\Delta = 1/8$, $\Delta = 1/4$), the OFDM variations (2K and 8K) and the position of the transmission framework within the framework (1,2,3,4). In addition, a cell identifier is also transmitted to identify the network cell from which the received signal is being transmitted [LADEBUSCH].

In view of the importance of an error-free reception of the TPS data, these are transmitted in parallel many times over and are also separately error protected. A shortened BCH code (67, 53) is used. The individual bits are modulated for the TPS carriers with differential 2-PSK (see section 7.2).

A maximum of 68 bits can be used for TPS. 31 bits are required for the word recognition and error protection. 31 bits are used at the moment. The remaining 6 bits are reserved for further use.

11.4 Decoding Technique

The DVB-T standard EN 300 744 only describes the transmitted signal and thus does not specify the receiver side. So various possibilities for the realisation of the units in the decoder are conceivable. The implementation strongly depends on the reception environment. Basically, the processing steps which were carried out at the encoder have to be reversed. In figure 11.6 the fundamental construction of the corresponding receiver is shown

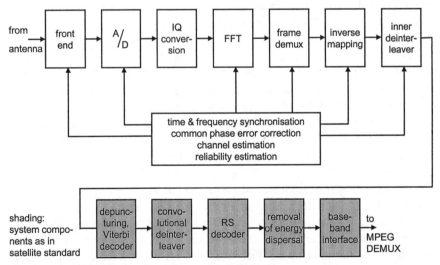

Fig. 11.6. Block diagram of signal processing at the receiving end

[MØLLER]. During the introduction of the first regular DVB-T digital TV service in the UK in November 1998, the first integrated digital receiver chip set was based on 3 chips and only supported the 2K mode. In the meantime single-chip solutions combining A/D conversion, COFDM processing and channel decoding are available and fully standard compliant.

A DVB-T receiver needs to cope with several impairments that occur during terrestrial transmissions (see section 11.1). In addition, the receiver front end will add noise and phase noise to the signal [KLINKEN]. Receiver implementations may differ in the way they handle time and frequency synchronisation and channel estimation. The pilots, which are spread in time and frequency within the OFDM signal, have a known amplitude and phase. Therefore, the measuring of the channel frequency response for attenuation and phase can be carried out on the basis of the continual pilots and on the scattered pilots. Knowing the amplitude and phase of all the individual pilot signals, the receiver is able to equalise each subcarrier. This ideally reverses the effects of the transmission channel. Various so-called channel estimation techniques have an important effect on the overall performance of the receiver [STRUBENH].

The stability of the oscillators required in the receiver is of great importance for the overall performance. On the one hand, the single carriers, which are approx. 1.1 kHz or 4.4 kHz apart, can only be separated free of crosstalk if it can be guaranteed that the sampling is carried out exactly at the zero crossings of all $\sin(x)/x$ functions not belonging to the desired carrier (see section 7.6). On the other hand, it must remain possible for the phase information, which is integrated into the QPSK- or QAM-modulated individual carriers, to be recovered. With the aid of the pilots described before, it is possible to accurately set the oscillator frequency and also the static phase of the local oscillator in the receiver. However, these static quantities are superimposed by phase noise. This results, first of all, in a phase error which affects all the individual carriers within a symbol in the same way, but which shows little correlation with the phase errors within the subsequent symbols. Secondly, the phase noise can lead to phase errors, which affect different carriers within the same symbol in different ways. This is why both the level and the spectral distribution of the phase noise in the local oscillator are of interest. Nowadays local oscillators of consumer type equipment are able to master the problem of phase noise even using conventional technology [MUSCHALL].

In the following the implementation aspects of the network interface part of DVB-T, i.e. the part from the "air interface", the antenna, to the MPEG-2 transport stream generation are described. An overview of several types of receivers is presented and their main requirements are outlined in the first paragraph. For the development of reception techniques for DVB-T signals extensive experiences of the analogue TV reception can be used. The required components and their specific requirements for DVB-T signals are explained in the next section, followed by some new technologies which evolved with the development of digital television.

11.4.1 Receiver Classes

With the focus on implementation aspects it is most important to first determine in which reception environment, namely stationary, portable, or mobile, the receiver device will be operated. While the option "stationary" is mostly associated with reception by a rooftop outdoor antenna and a device which is not very often moved from one place to another, "portable" means that the device can easily be carried or taken from one point to another. It will contain a built-in omnidirectional antenna and will mostly be operated in a nomadic mode; i.e. it will not be operated while moving fast. "Mobile" means reception while moving at high speeds in automobiles, buses, trains, and so on.

In table 11.2 various DVB-T receivers and their main operational environment are outlined. For each of these devices a different reception concept is required. Some characteristic key points related to these devices are:

- A high-end IDTV set will need a high-end reception system which gives optimum reception under difficult (but stationary, sometimes indoor) receiving conditions. It may contain one antenna within the housing of the whole IDTV but can also use an external indoor antenna as well as an outdoor rooftop antenna.
- Set-top boxes are quite similar, maybe with less performance. They will possibly contain an integrated antenna.

Table 11.2. General classification of DVB-T receivers

	stationary	portable	mobile
Integrated Digital TV set (IDTV)	X		
Set-top Box (STB)	X		
Portable TV		X	
Personal Digital Assistant (PDA)		X	X
Home-PC extension	X		
Notebook-PC extension		X	X
Car receiver			X

- The emphasis of handhelds and PDAs is on a lowest possible power consumption combined with smallest packaging. Nevertheless they will be used under quasi-mobile receiving conditions. Also these receivers often have to be "world receivers", allowing reception in all possible environments (frequency range, channel arrangement, bandwidth) and they should be able to handle all possible DVB-T parameter combinations. These devices have to work with simple rod antennas or antennas integrated into the housing. Sensitivity and dynamic range of their tuners will become a classification and quality mark.
- PC-Extensions (like PCI-cards) or USB-adapters will suffer from the highest price pressure. Often manufacturers of these devices have a limited know-how of RF and IF design techniques and technologies, so they have to buy ready developed modules (so-called NIM, see below). One design aspect will be the interference of the PC environment emitting into the DVB-T receiver device.
- Car receivers have to cope with the most difficult receiving conditions, described by the mobile channel. They will use advanced technologies, for instance antenna diversity. Technically they undoubtedly will be the most sophisticated of all types of receivers. The advantage of a car receiver is that it can be adapted to the (most of the time) window-integrated antenna and pre-amplifier system to get the best performance out of DVB-T. This receiver also has to be a "world receiver" (see above), sometimes a receiver which also features other reception standards, like analogue TV, ATSC or ISDB-T. Due to the requirement to integrate the receiver into the infrastructure of an automobile, most of these devices will be OEM devices. Aftermarket devices will probably only play a small role.

Going deeper into technical aspects of the different devices a further classification is possible. Important factors may be dynamic range, input sensitivity, channel adaptation speed, power requirements and EMC immunity (see table 11.3).

11.4.2 Straight Forward Technology – the Classical Approach

For a first rough overview of a DVB-T reception system one can start with a very "classical" approach of a receiver design. It contains a tuner for down-conversion of the incoming RF signal to the first IF, some IF processing like filtering of the downconverted spectrum, maybe a further down-conversion to a second IF, and finally the DVB-T decoder for demodulating and decoding the baseband signal to MPEG-2 data. Figure 11.7 summarises this approach. It is generally valid for most receivers.

Table 11.3 Classification of receivers by important technical design goals

	Dynamic range	Input sensitivity	Channel adaptation speed	Importance of low power consumption	EMC immunity
Integrated Digital TV set (IDTV)	Low	High	Slow	Low	Low
Set-top Box (STB)	Low	Moderate	Slow	Moderate	Low
Portable TV	Low	Moderate	Slow	Moderate	Low
Personal Digital Assistant (PDA)	High	High	Moderate	High	Moderate
Home-PC extension	Low	Low	Slow	Low	High
Notebook-PC extension	Moderate	Moderate	Moderate	High	High
Car receiver	High	High	Fast	Moderate	High

Fig. 11.7. Top-level block diagram of signal processing for DVB-T reception

In the following section, the main functional modules and components needed for reception of DVB-T signals according to this top-level block diagram are described in more detail.

11.4.2.1 Antenna

While for IDTVs, STBs and Home-PC-Extensions the use of a directional rooftop outdoor antenna with high gain in the direction of its main lobe may be possible, a common case may also be a small antenna mounted in the device or nearby. This antenna should have the following characteristics:

– non-directional, i.e. no main lobe and no notches in the radiation diagram,
– gain of approx. $G \approx 0$ dB

- dual polarised, i.e. ability to receive horizontally and vertically polarised signals,
- broadband [1]

These are requirements that characterise the ideal antenna, which will not exist. Small rod antennas, which can be found on handheld-, PDA- and USB-card-receivers, have the disadvantage of a very restricted performance with horizontal polarisation. They are to a certain extent broadband but not broadband enough to cover both the UHF bands and the VHF band III. The big advantage of these antennas is that they are easy to produce and as a result cost-effective. Classic indoor antennas like the common "rabbit ears" may work well with a DVB-T receiver. However these are mostly active antennas and are traditionally optimised for analogue reception. The performance of the integrated amplifier may not be good enough for DVB-T. Active indoor antennas optimised for DVB-T are emerging in the market. They have a flat panel where an antenna structure is printed on a printed circuit board (PCB) and a low noise amplifier is connected to the antenna.

11.4.2.2 Tuner

The tuner amplifies the RF signal, tuning a local oscillator in such a way that the downconverted product falls onto the fixed first IF, and finally amplifies the IF signal for further signal processing. It is the only performance dominating key component, since performance losses in the first signal processing stage cannot be gained back at later stages. For DVB-T reception scenarios the performance of the tuner is even more important than that of the DVB-T decoder.

Tuners available in the market which were designed for analogue TV typically do not work sufficiently well with DVB-T signals due to mainly one reason: the tuner's local oscillator introduces phase noise to the received signal. Phase noise is more critical for narrowband signals than for wideband signals [MUSCHALL2]. Analogue TV signals are wideband signals (≥ 5 MHz) and are hardly effected by tuner phase noise. DVB-T signals are wideband signals as well, but due to the fact that the individual OFDM carriers need to be demodulated, which may have a bandwidth as low as 1 kHz in the 8K mode, phase noise is extremely relevant for DVB-T. Characterisation of tuner phase noise requires description of the noise spectrum adjacent to the carrier of the local oscillator. Good tuners have phase noise values of:

- > -70dBc at 100Hz
- > -80dBc at 1kHz
- > -85dBc at 10kHz

---

[1] The UHF band has a bandwidth factor of about 1:2. If the VHF band III is also to be included, the bandwidth factor grows to about 1:5

Another important tuner characteristic is the noise figure, i.e. the amount of *additive* noise introduced by the tuner's components. This noise source, since it is the first noise source in the whole signal processing chain, is mainly responsible for the overall carrier-to-noise ratio *C/N* of the received signal. Hence, in order to improve tuner sensitivity and therewith the overall receiver's sensitivity, the noise figure should be as low as possible. Low noise tuners are most important during the migration phase from analogue to digital TV since during this phase the digital TV transmitters are operated with very limited power. Top-performing tuners have noise figures of about 5dB over the whole UHF band.

Further tuner parameters that need a closer look are

- sideband suppression,
- image rejection,
- 1st IF suppression.

The manufacturers of tuners offer a wide range of different tuner concepts. Some of these are outlined below:

- Single band tuner (UHF band IV/V only) with low noise amplifier (LNA), phase locked loop (PLL) and mixer to 1st IF (standard 36.166 MHz). This may be called the "classical" tuner.
- Three band tuner (VHF bands I and III, UHF band IV/V).
- Tuner with integrated DVB-T compatible SAW-Filter at 1st IF and integrated AGC.
- Combi tuner with integrated SAW-Filter for the digital signal and with broadband output for analogue reception,
- Tuner with integrated SAW-Filter and down converter to 2nd IF (typically 4.5714 MHz or 7.225 MHz).
- NIM-Tuner with integrated DVB-T decoder (see section 11.4.3.3).

All tuners are available in many mechanical versions (horizontal/vertical mounting, different connectors, loop through input, phantom supply for external pre-amplifiers) and are controlled via the I^2C bus.

Figure 11.8 shows a standard product featuring a UHF-only tuner with integrated SAW filter and compensation amplifier, AGC, narrowband output for digital services and broadband output for analogue services.

11.4.2.3 IF Processing

First generation DVB-T decoder chips required a very sophisticated IF signal processing. They did not contain any input amplifier or internal AGC amplifier and needed a "perfect" spectrum shaping with a rather high input voltage. The internal sampling frequency used was the 2nd IF. Despite the fact that a classical tuner was required as the front end of the receiver the "bill of material" (BOM) for IF processing alone was rather long:

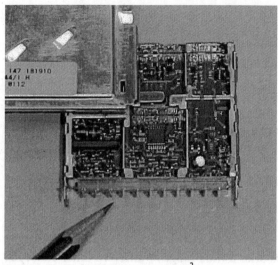

Fig. 11.8. Tuner module TD1344 from Philips [2]

- 1st SAW filter,
- one-stage amplifier to compensate insertion loss of 1st SAW filter,
- 2nd SAW filter for better sideband suppression (adjacent channel rejection) and spectrum forming,
- one-stage amplifier to compensate insertion loss of 2nd SAW filter,
- AGC detector and amplifier,
- down converter to 2nd IF,
- converter to symmetrical output format.

With newer DVB-T decoder chips the IF can "collapse" down to one straight connection between the tuner and the decoder chip. This is because the decoder can sample on the 1st IF, has an integrated input amplifier and a sophisticated sub-sampling concept accepting channel bandwidths of 6, 7, and 8 MHz without requiring re-timing or precise IF spectrum shaping. This simplest concept requires a tuner with integrated AGC and one 8 MHz SAW filter – irrespective of the channel bandwidth actually used.

More common is the use of a "classical" tuner (see above) followed by a SAW filter and, depending on the dynamic range needed, an external AGC amplifier. Some possible IF concepts are shown in figures 11.9 to 11.11.

---

[2] With kind permission of Philips Components, Business Unit Tuners

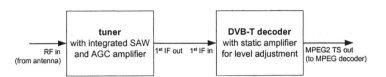

Fig. 11.9. Simplest configuration, using sophisticated tuner concept

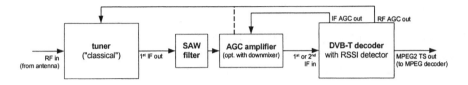

Fig. 11.10. Configuration using standard ("classical") tuner

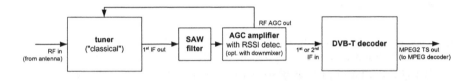

Fig. 11.11. Same configuration as figure 11.9, but with different AGC implementation

Since the sub-sampling technique on the first IF is still sub-optimum compared to the sampling on the second IF, some high performance receivers like measurement equipment may still use the sampling on the second IF. Also a second SAW filter is useful to increase the adjacent channel rejection. A very crucial point is the dimensioning of the AGC for difficult reception conditions. It has to be adapted to the expected receiving conditions such as the dynamic range acceptable for the input signal (stationary/mobile), the dynamic range of the A/D conversion in the decoder chip, the expected channel profile, and the sensitivity of the tuner. Therefore two possible AGC concepts are depicted in figures 11.10 and 11.11, respectively.

11.4.2.4 DVB-T Decoder Chip

The first decoder solutions employed up to three chips: one for the A/D conversion of the 2nd IF signal, one for the COFDM signal processing, incl. FFT (only the 2K mode was supported) and the third for FEC decoding and MPEG-2 transport stream output processing. Current solutions are much more powerful. Figure 11.12 shows a basic block diagram of a typical DVB-T decoder chip architecture.

Some key features found in the data sheets of products available in 2003 are:

- input of 1st IF (36.166 MHz), with direct sampling
- automatic identification of signal bandwidth (6, 7, or 8 MHz), processing requires only one external crystal by internally adjusting clock frequency
- automatic identification of all possible DVB-T modes, including all possible 8k modes
- enhanced and adaptive filtering for co-channel interference (CCI) and adjacent channel interference (ACI) rejection
- packages as small as 40 pin LQFP (Low Profile Quad Flat Pack) with 0.13 micron technology
- power dissipation of less than 300 mW compared to more than 1000 mW of first generation solutions.
- prices starting at 6 US$, compared to more than 40 US$ for first generation solutions.

In 2003 third generation decoder chips from approx. 10 chip suppliers were available through distribution plus some estimated 2 or 3 chips were exclusively developed and produced for a closed group of CE-manufacturers. The choice of products increased rapidly at the beginning of 2002. It is expected that it will further grow. It is remarkable that many "young" companies are among these chip manufacturers.

11.4.3 Enhanced Technologies for DVB-T Reception

Recently further technical developments and experiences gained from analogue reception and adapted to digital receivers were used to increase the quality of the DVB-T receiver. Various enhanced techniques have allowed the optimisation of the reception quality not only for mobile applications of DVB-T. In this section it will become clear that the performance limits of DVB-T have not been reached. To the contrary, further optimisation can be expected.

11.4.3.1 Antenna Pre-Amplifiers for DVB-T

Reception of DVB-T signals requires a certain minimum C/N ratio in order to achieve QEF signals. This minimum value depends on the DVB-T transmission parameters chosen by the broadcaster and on the current transmission channel. There are several ways to increase the C/N ratio at a given receiver location and a given field strength at this location:

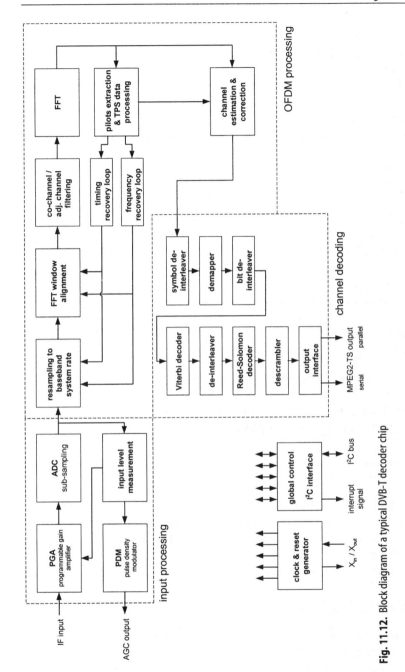

Fig. 11.12. Block diagram of a typical DVB-T decoder chip

(1) by optimising antenna position and antenna orientation: the power density C_o directly depends on the gain G of the receiving antenna in the direction towards the transmitter or towards the strongest signal,

(2) by using RF pre-amplifiers: with a high gain and a low noise figure the noise power can be minimised while maximising the signal input level at the decoder input,

(3) by selecting IF SAW filters which closely match the frequency shape of the received signal: in this way it can be assured that the noise bandwidth is the same (and not larger) as the signal bandwidth, thus minimising N.

The antenna problem has been an issue for a while now. Practically all known receivers use SAW filters adapted to the signal spectrum. The following section explains the advantages of installing an antenna preamplifier:

Let us assume a tuner noise figure of F_t and a pre-amplifier with noise figure $F_p < F_t$ and gain g_p. Then the overall combined noise figure of the pre-amplifier and tuner F_{t+p} becomes (in linear notation) [UNGER2].

$$F_{p+t} = F_p + \frac{F_t - 1}{g_p}. \tag{11.3}$$

Thus the effective noise figure is reduced, and so is the noise entering the decoder. Since the gain of the amplifier is the same for the noise and the signal, the advantage in C/N ratio achieved by introducing an antenna pre-amplifier can be calculated directly by (in linear notation):

$$\Delta C\!\!\!\Big/\!\!_N = \frac{F_t}{F_{p+t}} = \frac{F_t \cdot g_p}{F_p \cdot g_p + F_t - 1} \tag{11.4}$$

Assuming typical values of $F_t = 7$ dB, $F_p = 0.9$ dB and $g_p = 19$ dB, $\Delta C/N$ is 5.9 dB.

Antenna pre-amplifiers are being built into active indoor antennas and are used in conjunction with car antennas – especially those consisting of elements integrated in the window panes.

11.4.3.2 One-Chip Silicon Tuner

With further advances in silicon technology first solutions for one-chip tuners have become available. These no longer require the integration of a "classical" tuner module into the DVB-T receiver, but consist of a single chip which accepts the RF signal on one input, and outputs the first IF signal. Only a minimum of external passive components are required (see figure 11.13).

Fig. 11.13 Microtune single-chip tuner integrated onto the receiver's main PCB [3]

The advantages of this concept are the following: Tuners are small, light-weight, are integrated in the receiver board and are cost-effective in production (no manual placement and soldering of the tuner module). On the other hand the disadvantages in comparison to the more classical tuners are the following: the power dissipation is higher, the performance is slightly lower and the cost of the components is higher – but this list is from a snapshot from the year 2002, and time will tell.

11.4.3.3 Network Interface Module (NIM) Technology

One interesting technology for reception of DVB signals in general – be it DVB-S, DVB-C, or DVB-T- is the implementation of a so-called Network Interface Module (NIM). A NIM integrates in one module all signal processing from RF input to MPEG-2 transport stream output, containing the whole tuner functionality, the IF processing and the DVB-T decoder chip. The size of such a module is roughly double that of a standard tuner module. A NIM shows a number of advantages for the design of a DVB-T receiver:

- It has all the technology and the related know-how "on board", in one module. The receiver designer is therefore not forced to bother with RF problems like crosstalk etc.
- The RF and IF processing and also the PCB layout are optimally adapted to the integrated DVB-T decoder chip.
- NIMs from various manufactures can be interchanged. The control software needs to be adapted, which usually is rather simply done by re-

---

[3] With kind permission of Microtune (TEMIC-Tuners)

designing the appropriate I²C bus routines, and a rather small PCB layout adaptation is typically required.
- Since NIMs are available for the different DVB transmission systems, namely satellite, cable, and terrestrial, they allow for the design of exactly the same receiver for all DVB systems.

Figure 11.14 shows a typical Network Interface Module (NIM).

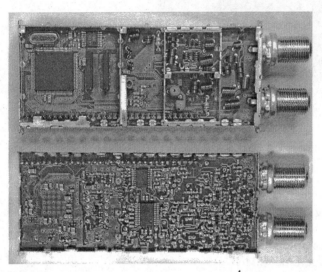

Fig. 11.14 Network Interface Module (NIM) from Philips [4] (top and bottom view)

Integrating a NIM into a receiver design is a good choice for PC-extension cards and for set-top boxes. If the performance is good enough they may be used for IDTVs. But for other receiver classes, NIMs have two main disadvantages:

- The size of a NIM is still quite large, so it may not be the perfect choice for small receivers like in PDAs or USB card extensions.
- It is not possible to control the RF- and IF-processing parameters, especially of the AGC from outside the NIM. These parameters, therefore, cannot be adapted or optimised for difficult reception conditions like the mobile channel. The NIM is therefore less suited for automotive receivers and PDAs
- The designer has to take the module "as it is". Exchanging parts, e.g. the DVB-T decoder or the SAW filter, for a part from a different supplier is not possible.

---

[4] With kind permission of Philips Components, Business Unit Tuners.

11.4.3.4 Antenna Diversity

The basic idea underlying antenna diversity reception is that if two or more independent antennas are installed, the signals delivered by the individual antennas will fade in an uncorrelated manner. It follows that a signal composed of a suitable combination or selection of segments derived from the various individual antennas will encounter much less fading than any individual antenna signal. In this way, antenna diversity can cope with fading problems in severe channel conditions. When employing antenna diversity for DVB-T receivers, the processing of the received signals is conceivable on different levels, for instance a choice of signal elements can be made on the transport stream level or on the OFDM subcarrier level [LISS].

For high-end IDTVs an antenna diversity solution can be conceived, where one or two antennas are integrated into the TV set and another connector is implemented for an external antenna. For smaller devices like PDAs or handheld receivers the antenna diversity approach is not as advantageous, because a certain physical distance between the antennas which is normally in the order of one wavelength of the RF signal is needed to achieve uncorrelated fading.

The antenna diversity strategies make most sense in the automotive mobile receiver where severe channel fading conditions are pre-dominant. It may even be inevitable for difficult reception environments while moving fast. In such implementations, up to four antennas can be connected via a switch matrix to the tuner input(s). Another advantage of multiple antennas is that they help to cope with the non-omnidirectional single antenna patterns of automobile integrated antenna systems creating an omnidirectional pattern with all single antenna patterns combined via antenna diversity. All diversity concepts require that not only one reception path consisting of tuner, IF processing, and decoder is implemented but at least two. If a car travels through a DVB-T network it will be necessary to scan the frequency spectrum for possible alternative RF channels which carry a transport stream including the program which is currently decoded by the receiver. In SFNs a change of RF channel is only required at the borders of the network. In MFNs such a change is required more frequently. DVB has developed a set of technologies, which assist in such changes [LADEBUSCH]. Mobile receivers may require a third reception part only devoted to scanning the frequency spectrum.

Although diversity decoder ICs are available, diversity reception is still a very complex and costly architecture, which is only useful for automotive mobile reception.

11.5 Hierarchical Modulation

It is a fact that all DVB transmission systems exhibit an abrupt failure characteristic. A variation in the carrier-to-noise ratio of only a fraction of a dB
can cause a complete breakdown of the transmission path. This effect is, of
course, of less importance in cable systems. A signal of sufficient quality can
always be guaranteed in cable networks when the failure criteria have been
taken into account during the planning of the network. A fluctuation of the
reception conditions is not to be expected. In satellite transmission, fluctuations in the carrier-to-noise ratio can be caused by changes in the meteorological conditions. The use of sufficiently-sized reception dishes (diameter $\geq$
60 cm) directed exactly at the satellite, should ensure that there are very few
moments of breakdown in the course of a year (see chapter 9). Weather conditions can affect a wide area causing widespread failure. A storm front
traversing a region, for instance, can cause a breakdown in many places at
the same time. During good weather conditions and with a suitable receiving
installation, every viewer can receive the offered programmes, free of transmission errors. This possibility only exists, of course, if the antenna is installed in line-of-sight with the satellite.

Similar conditions to those described for satellite reception will prevail for
the reception of terrestrially transmitted DVB signals with a rooftop antenna. Those viewers, who are able to receive sufficiently high-powered signals, either directly or as echoes, will receive their choice of programmes free
of transmission errors. Fluctuations in the carrier-to-noise ratio can also occur in terrestrial transmissions, for instance due to seasonal changes in the
echo impairments or due to aeroplanes flying past. However, these are far
less drastic than the meteorologically induced disturbances in satellite systems. The real difference between terrestrial and satellite reception is characterised, on the one hand, by the objective for terrestrial transmitters to also
facilitate, within certain limits, the coverage of stationary portable receivers
with a rod antenna and, on the other hand, by the expectations of the viewers. The viewer of analogue TV programmes is used to, and will therefore
expect, a flawless reception of at least the main television programmes, provided that certain rules are observed concerning the type and height of the
rooftop antenna required at the place of residence. Reception on portable
receivers equipped with set-top aerials is regarded as a bonus. These viewer
expectations can only be met with considerable expenditure on the part of
the broadcasters, even allowing for the "graceful degradation" of analogue
signals. DVB signals with their abrupt failure behaviour complicate the
situation considerably. For example, it is conceivable that while large areas
of a city might be well served with terrestrial DVB signals, one district might
be situated on lower lying ground where no reception whatsoever is possible.

To supply this district, a general increase in the transmission power would of course be possible, and/or instead of each single carrier being modulated with 64 QAM, a QPSK modulation could be applied and/or a lower code rate for the inner error protection could be utilised. The reception could also be improved by using fill-in transmitters. All these procedures have one distinct disadvantage; they would result in a less efficient use of the available channel capacity while at the same time only few viewers might profit from the above measures.

The hierarchical transmission technique is based on the division of the television channels used for the terrestrial transmission of DVB signals into two parts. The first part enables relatively low data rates (a few Mbit/s) to be transmitted in such a way that they can still be received in the case of relatively poor carrier-to-noise ratios. Parallel to this, the second part enables the transmission of considerably higher data rates, however with higher carrier-to-noise ratio requirements. The concept of the "data container" can be maintained if a "hierarchical transmission technique" is understood as dividing the data container into a robust and a fragile portion. Both "subcontainers" are nevertheless transmitted in the same channel. Within the robust portion of the data container, for example, some basic (public) programmes could be transmitted (although with a somewhat more modest audio and video quality) and the less robust portion could accommodate new, additional programmes. Alternatively, the basic programmes could be transmitted once again, but with a considerably improved audio and video quality. Due to the commercial and user requirements described in section 11.2 the standard for terrestrial transmission of DVB signals does not include any mechanisms for the so-called hierarchical source coding. Had that been implemented, the robust portion of the data container could be used to transmit a programme in modest quality, while the less robust portion of the container could transmit additional data for the same programme. This could in turn exploit all the data of the entire container in order to enhance that programme to, for example, HDTV quality.

A concept for the hierarchical transmission is easily defined. Whilst retaining the normal transmission frame it should be possible to transmit two data streams at the same time; one modulated with QPSK and with a lower code rate for the inner error protection, the other modulated with either 16-QAM or 64-QAM but with a higher code rate for the inner error protection. The necessary interleaving of the two data streams could either be designed as a temporal multiplex or a frequency multiplex or even as a modulation multiplex. The hierarchical modulation in the intended form utilises the modulation-multiplex method. This means that each single carrier in the transmission frame transmits two data words simultaneously, of which one comprises a high-priority data stream, which needs to be specially protected, and the other a lower-priority data stream with less need for protection.

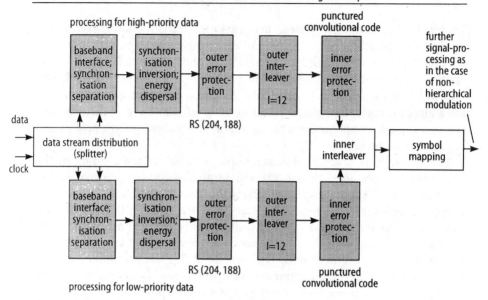

Fig. 11.15 Block diagram of an encoder for terrestrial DVB with hierarchical modulation

Figure 11.15 shows a part of the block diagram of an encoder for terrestrial DVB with hierarchical modulation. A splitter divides the input data stream. Both partial streams subsequently pass through identical processing steps; however, the inner error protection is chosen differently in each case. Subsequent to the inner interleaver the two partial streams are reunited in the inner interleaver and then further processed as shown in figure 11.2.

The reuniting of the partial streams is outlined in figure 11.16. The case of the generation of a 64-QAM symbol with six bits per symbol is shown as an example. Two of the six bits come from the high-priority data stream and the other four bits are from the low-priority data stream. As an alternative, it would also be possible to generate a 16-QAM symbol with four bits per symbol, with two of the four bits coming from the high-priority data stream and

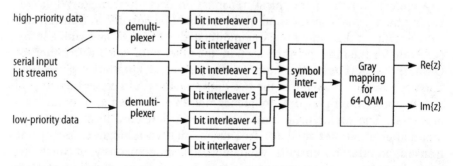

Fig. 11.16 Structure of an inner interleaver for the generation of 64-QAM symbols with embedded QPSK symbols

two from the low-priority data stream. In both cases the essence of the mapping lies in the fact that the two bits which come from the high-priority data stream (y_0, y_1) define the quadrant in which the symbols are found (figure 11.17). Hence it could be said that hierarchical modulation enables the embedding of a QPSK symbol, as defined by the high-priority data, in either a 16-QAM or a 64-QAM symbol. The term "multiresolution QAM" adequately describes this situation.

Figure 11.17 could be found puzzling in that the constellation points within each quadrant have a different spacing from the constellation points in neighbouring quadrants. The spacing measure is defined by the characteristic quantity α, the value of which can also be transmitted within the TPS (see section 11.3). The values 1, 2, and 4 are given for α. Figure 11.4 shows the condition for $\alpha = 1$. In figure 11.17 the constellation for $\alpha = 4$ is shown. It can be seen by a comparison of the two diagrams that α represents the ratio between the spacing of two adjacent constellation points of two neighbouring quadrants and the spacing of the constellation points within one quadrant.

For a given type of QAM, an increase in α leads, on the one hand, to an increase in the average transmission power, which follows from the extension of the length of the vector between the origin of the ordinates and the mean constellation point. For given transmission power, on the other hand, an increase in α leads to an increase in the required carrier-to-noise ratio for the demodulation of the QAM symbols, namely to an increase by some 4 dB at each doubling of α. Moreover, from figure 11.17 it is obvious that the permissible noise for an error-free demodulation of the QPSK component increases with α. Hence it can be said that an increase in α renders the robust

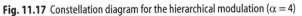

Fig. 11.17 Constellation diagram for the hierarchical modulation ($\alpha = 4$)

portion of the data container even more robust, while at the same time the less robust portion becomes more error-prone.

The implementation of a receiver which is able to process hierarchical modulation can be carried out in two ways. A simple receiver differs from a receiver for non-hierarchical transmission only in that the demodulator/demapper must be able to split off either a high-priority data stream or a low-priority data stream, in accordance with the information about the type of modulation (that is, the value of α and the inner code rate continuously transmitted in the TPS), and must then forward the data stream to the (single-channel) channel decoder. The disadvantage of using this simple receiver technology is that in the case of a deterioration of the reception conditions, the receiver needs to switch from the low-priority data stream over to the high-priority data stream, and this causes a short-term disruption in the continuity of the audio and video. This disruption is caused by the fact that the receiver needs to adapt to the new reception parameters. Alternatively, a more expensive receiver is conceivable, which is designed with two channels after the demodulator/demapper and which enables an immediate switching of at least the channel decoder. The effects of the switching of the source decoder, however, can only be avoided if two such components are operated in parallel.

11.6 Features of the Standard

The performance of the system for the terrestrial transmission of DVB signals can be analysed according to two criteria: (1) the available useful data rate and (2) the carrier-to-noise ratio in the transmission channel required for QEF reception.

However, the analysis of the performance data can lead to very intricate results. This is due to the vast number of parameters that can be chosen. It should be remembered that all of the following parameters can be selected independently of one another: the 2K and the 8K variant, the code rate of the inner error protection, the type of modulation of each single carrier (with the further option of a choice of either hierarchical or non-hierarchical modulation), the length of the guard interval, and finally, in the case of hierarchical modulation, the value of α. The required carrier-to-noise ratio in the transmission channel has to be investigated for the three channel models (Gaussian, Ricean and Rayleigh). It is evident from the above that in this paragraph a limitation to just a few important combinations is unavoidable.

In recent years, mobile reception of DVB-T has attracted considerable interest despite the fact that DVB-T had originally been designed for fixed and portable reception. Mobile reception of DVB-T has often been demonstrated, for instance, within the DVB-T field trial in northern Germany (Modellversuch DVB-T Norddeutschland) and is even used for regular services in Singapore. Therefore various aspects on mobile reception of DVB-T also will be described in the following section.

11.6.1 Determination of Useful Data Rates

The basis for the determination of the useful data rates is the determination of the number of packets of the MPEG-2 transport stream that can be accommodated in a transmission frame. The fact must be taken into account that the outer error protection turns 188-byte packets into 204-byte packets by the addition of 16 bytes redundancy. The inner error protection then adds further redundancy according to the code rate. One should think that it is impossible to accommodate a whole number of transport packets in one transmission frame for all possible code rates and all possible types of modulation. Stuffing data seems to be unavoidable. But in fact it was possible to define the capacity remaining for useful data within a transmission frame such that for all conceivable combinations complete transport packets can be accommodated in a remaining superframe consisting of four transmission frames. The numerical background has already been considered in section 11.3. It was shown that for the 2K variant, 1512 carriers per symbol remain for the useful data, and for the 8K variant, 6048. In a superframe this would be $272 \cdot 1512 = 411,264$ carriers and $272 \cdot 6048 = 1,645,056$ carriers respectively. Since, depending on the chosen kind of modulation of the single carriers, 2 bits per carrier (QPSK), 4 bits per carrier (16-QAM), or 6 bits per carrier (64-QAM) are transmitted, at least $272 \cdot 1512 \cdot 2 = 822,528$ bits will fit into a superframe. If this "elementary quantity" is divided by $(8 \cdot 204 =) 1632$ (the number of bits of a transport packet with error protection), the result is 504, which is a multiple of 8, 6, 4, 3, 2, i.e. of the divisors of the puncturing rates of the inner error protection.

Table 11.4 Number of transport packets in one superframe

Code rate	QPSK "2k"	"8k"	16-QAM "2k"	"8k"	64-QAM "2k"	"8k"
1/2	252	1008	504	2016	756	3024
2/3	336	1344	672	2688	1008	4032
3/4	378	1512	756	3024	1134	4536
5/6	420	1680	840	3360	1260	5040
7/8	441	1764	882	3528	1323	5292

Table 11.4 documents the number of 204-byte packets in both variants as a function of the code rate of the inner error protection and the chosen modulation.

It can easily be seen that, for 16-QAM, the number of transport packets per transmission frame is always double that for QPSK, and that for 64-QAM it is three times that for QPSK. In the case of QPSK a gross total of 822,528 bits, including the bits for the outer error protection - as already computed using the number of carriers available for the transport of useful data in a superframe - can be conveyed per superframe with the 2K mode. This value is computed in the case of code rate 1/2, from the relation (204 bytes per transport packet) · (8 bits/byte) · (1/code rate) · 252 packets = 822,528 bits. With this method it is easy to compute the net data rates if the duration of the superframe is known. The net data rates should be interpreted as referring to the 188 net bytes in each transport packet. Of course, such a calculation includes the respective symbol duration T_S and therefore the length of the chosen guard interval. Let the data rate be calculated by way of an example (2K, QPSK, code rate 1/2, $\Delta = 1/4$ corresponding to $T_S = 280$ µs). The calculation leads to 822,528 bit · (1/2) · (188/204) · [1/(4 · 68 · 280 µs)] = 4.98 Mbit/s. If this is changed to the 8K mode (8K, QPSK, code rate 1/2, $\Delta = 1/4$ corresponding to $T_S = 1120$ µs), the calculation of the data rate leads to (204 byte per transport packet) · (8 bit/byte) · (1/code rate) · 1008 packets · (1/2) · (188/204) · (1/(4 · 68 · 1120 µs)) = 4.98 Mbit/s. In other words, the data rate is identical to that in the 2K mode. This result, which is somewhat surprising at first glance, is explained by the fact that for the 8K mode four times the number of transport packets are transmitted as for the 2K mode (see also table 11.4), whereas at the same time the symbol duration is also exactly four times as long with the same relative guard interval (Δ). All possible data rates are given in table 11.5.

Using table 11.5 it is also possible to determine all the various useful data rates for the different variants of the hierarchical modulation. For example, to determine which useful data rate results from hierarchical modulation in the case of "QPSK embedded in 16-QAM", the data rate for the high-priority data stream can be found in the corresponding line of the block of figures relative to QPSK, in accordance with the chosen error protection and the Δ used. The data rate for the low-priority data stream can also be found in the corresponding line of the block of figures relative to QPSK, in accordance with the chosen error protection and the Δ used. In the case of hierarchical modulation of the type "QPSK embedded in 64-QAM", the data rate for the high-priority data stream can be found in the QPSK block of figures for the relevant combination of code rate and Δ, and the data rate for the low-priority data stream, in the 16-QAM block of figures for the relevant combination of code rate and Δ. If in this last example, for instance, the high-

Table 11.5 Useful data rates for non-hierarchical modulation [Mbit/s]

Modulation	Code rate	Relative length of guard interval (Δ)			
		1/4	1/8	1/16	1/32
QPSK	1/2	4.98	5.53	5.85	6.03
	2/3	6.64	7.37	7.81	8.04
	3/4	7.46	8.29	8.78	9.05
	5/6	8.29	9.22	9.76	10.05
	7/8	8.71	9.68	10.25	10.56
16-QAM	1/2	9.95	11.06	11.71	12.06
	2/3	13.27	14.75	15.61	16.09
	3/4	14.93	16.59	17.56	18.10
	5/6	16.59	18.43	19.52	20.11
	7/8	17.42	19.35	20.49	21.11
64-QAM	1/2	14.93	16.59	17.56	18.10
	2/3	19.91	22.12	23.42	24.13
	3/4	22.39	24.88	26.35	27.14
	5/6	24.88	27.65	29.27	30.16
	7/8	26.13	29.03	30.74	31.67

priority data stream is protected by an inner error protection with the code rate 1/2, and the low-priority data stream is protected by an inner error protection with the code rate 5/6, then the data rate for the high-priority data stream is 4.98 Mbit/s. To this value, for the accompanying low-priority data stream, a data rate of 16.59 Mbit/s is added. The result is a total data rate of 21.57 Mbit/s.

It might be instructive to calculate the limits of the values possible for the bandwidth efficiency introduced in the standard. For this the respective data rate is divided by the channel bandwidth (8 MHz). This results in values in the form of bit/s per Hz. In table 11.6 these values are represented for $\Delta = 1/4$.

Furthermore it can be seen from table 11.5 that a decrease in the relative guard interval leads to a corresponding increase in the data rate. Considering the bandwidth efficiency, this means that, taking the example of a 64-QAM modulation and a code rate of 7/8, the bandwidth efficiency increases from 3.27 bit/s per Hz in the case of $\Delta = 1/4$ to 3.96 bit/s per Hz in the case of

Table 11.6 Bandwidth efficiency for non-hierarchical modulation in the case $\Delta = \frac{1}{4}$

Code rate	QPSK [bit/s per Hz]	16-QAM [bit/s per Hz]	64-QAM [bit/s per Hz]
1/2	0.62	1.24	1.87
2/3	0.83	1.66	2.49
3/4	0.93	1.87	2.80
5/6	1.04	2.07	3.11
7/8	1.09	2.18	3.27

$\Delta = 1/32$. Regarding these values for the bandwidth efficiency, it should be noted that the loss of efficiency caused by the unavoidable bandwidth limitation and by the insertion of reference signals, TPS data, error protection, etc. has been fully taken into account.

11.6.2 Required Carrier-to-noise Ratio in the Transmission Channel

The values given in the following are based on simulations of the system behaviour. They were computed on the assumption that a perfect correction of the channel frequency response has taken place in terms of attenuation and phase rotation. Phase noise, as a source of errors within the receiver, was considered to be non-existent in these simulations. As a result of these idealised assumptions the behaviour of the 2K variant no longer differs from that of the 8K variant. As with the investigations into the characteristics of the standards for transmission over satellite and cable, a carrier-to-noise ratio (C/N) was determined at which the bit-error rate – after decoding the inner error protection – is equal to or less than $2 \cdot 10^{-4}$. As explained in section 6.3.5, this condition leads to practically error-free signals (QEF) after the Reed-Solomon decoder. A compilation of the required C/N ratio (in dB) for all possible combinations of modulation procedures and code rates of the inner error protection is given in table 11.7. The data refer to non-hierarchical modulation.

A close analysis of the numerical values in table 11.7 leads to the following conclusions. A change from QPSK to 16-QAM with a constant code rate results in an increase in the required C/N by approx. 6 dB. The same applies to the change from 16-QAM to 64-QAM. The transition from the Gaussian channel to the Ricean channel, with the same type of modulation and constant code rate, results in a necessary increase in the C/N by a maximum of 1.1 dB. As explained in section 11.1, the simulation of the Ricean channel given by (11.1) represents an approximation of the actual conditions when receiving DVB signals via a rooftop antenna with high directivity. The transition from the Gaussian channel to the Rayleigh channel, with the same modulation type and the same code rate, results in a necessary increase in the required C/N by a maximum of 8.9 dB. The Rayleigh channel given by (11.2) is used to model the actual conditions when receiving DVB signals via stationary receivers, which have a rod antenna.

In the case of hierarchical modulation it is necessary to supply details of the required C/N not only for the QPSK portion, but also for the QAM portion. From the vast number of possible options table 11.8 shows only the data for the example "QPSK embedded in 64-QAM", with $\alpha = 2$.

Table 11.7 Minimum C/N ratio in the transmission channel in the case of non-hierarchical modulation required for QEF reception.

Type of modulation	Code rate	Gaussian channel [dB]	Ricean channel [dB]	Rayleigh channel [dB]
QPSK	1/2	3.1	3.6	5.4
	2/3	4.9	5.7	8.4
	3/4	5.9	6.8	10.7
	5/6	6.9	8.0	13.1
	7/8	7.7	8.7	16.3
16-QAM	1/2	8.8	9.6	11.2
	2/3	11.1	11.6	14.2
	3/4	12.5	13.0	16.7
	5/6	13.5	14.4	19.3
	7/8	13.9	15.0	22.8
64-QAM	1/2	14.4	14.7	16.0
	2/3	16.5	17.1	19.3
	3/4	18.0	18.6	21.7
	5/6	19.3	20.0	25.3
	7/8	20.1	21.0	27.9

When comparing the data in table 11.8 with the corresponding information in table 11.7, one notices that the C/N required for QEF reception of the high-priority data stream transmitted in the QPSK constellation points, must be 3.3 dB to 4.9 dB higher than in the case of non-hierarchical modulation. Hence, the hierarchical modulation reduces the robustness of the QPSK portion. This finding should not be surprising, as the QAM constellation points lead to an actual decrease in the spacing between the apparent QPSK constellation point (the cloud of 16 points in one quadrant) and the decision thresholds. The greater the value of α the less important the effect becomes; since an increase α results in a smaller relative size of the clouds forming the apparent QPSK constellation points.

For QEF reception of the low-priority data stream transmitted in the QAM constellation, a higher C/N is again required than for the non-hierarchical 64-QAM. The differences are between 1.7 dB and 3 dB. As already explained in section 11.5 these differences are mainly due to the increase in average transmission power described by α. In the case of $\alpha = 1$ there are, in fact, only minor differences between the required C/N for non-hierarchical and hierarchical modulation (maximum 1.2 dB).

In conclusion, the total effect of the introduction of the hierarchical modulation shall be discussed on the basis of a practical example ($\Delta = 1/4$). Let the hierarchical modulation be given in the form of "QPSK embedded in 64-QAM", with $\alpha = 2$. The total data stream should be receivable with a roof

Table 11.8 Minimum C/N in the transmission channel for hierarchical modulation in the form QPSK embedded in 64-QAM ($\alpha = 2$) required for QEF reception.

Type of modulation	Code rate	Gaussian channel [dB]	Ricean channel [dB]	Rayleigh channel [dB]
QPSK	1/2	6.5	7.1	8.7
	2/3	9.0	9.9	11.7
	3/4	10.8	11.5	14.5
in				
64-QAM	1/2	16.3	16.7	18.2
	2/3	18.9	19.5	21.7
	3/4	21.0	21.6	24.5
	5/6	21.9	22.7	27.3
	7/8	22.9	23.8	29.6

top antenna (Ricean channel), while the high-priority data stream should be receivable on a stationary receiver with a rod antenna (Rayleigh channel). For the high-priority data stream, let the code rate of the inner error protection be chosen at 2/3, while for the low- priority data stream the code rate of the inner error protection be assumed to be 5/6. The data rate of 6.64 Mbit/s for the high-priority data stream can be taken from table 11.6. The required C/N is 11.7 dB. The low-priority data stream transmits 16.59 Mbit/s. and for its reception a minimum C/N of 22.7 dB is required. For example, if sufficient C/N is provided in the transmission channel, a receiver with the rod antenna could receive one television programme, and a receiver with a rooftop antenna, three or four.

11.6.3 Features Relevant for Mobile Reception

Mobile reception suffers from all the impairments relevant for portable reception. In addition, Doppler shift is experienced and the properties of the transmission channel change over time. Doppler shift results in a frequency shift of the received OFDM carriers as a function of the speed and the direction of the movement. In the simplest case the receiver moves away from the transmitter at a speed v. The frequency shift, which is described by the Doppler frequency f_D can then be calculated as

$$f_D = \frac{v \cdot f_0}{c} \qquad f_0 : carrier\ frequency \qquad c : speed\ of\ light$$

Assuming appropriate signal processing in the receiver also the fast channel variations associated with mobile reception can be tracked. To achieve successful mobile DVB-T reception, a number of factors need to be observed:

- The receiver has to track channel variations in time and frequency. In addition, correct channel estimation needs to be provided. The density of the pilots in the DVB-T signal gives a theoretical limitation. It defines the theoretical maximum Doppler frequency $f_{Dmax} = 1/(8{\cdot}T_S)$ which is between 446 and 541 Hz for the 2K mode. At carrier frequency $f_o = 626$ MHz (UHF channel 40) this corresponds to a maximum speed of about 770 km/h - 930 km/h. For the 8K mode these values are exactly 4 times lower. This (theoretical) limit depends on FFT size and guard interval length but not on modulation and code rate. In practice, the performance is extremely dependent on the quality of the interpolation filters used. To achieve performance sufficient for mobile receivers, 2-dimensional interpolation filters need to be employed.
- The receiver has to handle noise-like distortions called FFT leakage which are caused by non-orthogonality of the DVB-T subcarriers due to a time varying channel.
- The receiver has to be correctly synchronised in time and frequency in a mobile channel. As explained in section 11.3, for synchronisation purposes the receiver may use the guard interval for coarse timing, the scattered pilots for fine timing and the continual pilots for frequency synchronisation.
- The received field strength and consequently the C/N ratio have to be sufficiently high at a sufficiently high number of locations to permit a reliable mobile service.

In recent years, several investigations, comparative tests of various DTV systems and field trials have looked into the performance of DVB-T for mobile reception. The DVB-T field trial in northern Germany demonstrates the efficiency of DVB-T especially for mobile users. Mobile reception of DVB-T proved a practical reality even in very densely built-up urban areas. Wherever the field strength surpassed a certain level, a perfect mobile reception was achieved. The measurements have shown that the required field strength for portable indoor reception is definitely higher than that needed for mobile reception [DVBTNORD].

The design of mobile receivers takes into account the fact that multiple antennas are used to receive radio and TV signals in cars. In consequence a RF pre-amplifier is used to combine these antennas and the effective noise figure of a mobile receiver is significantly lower than that typical for consumer type TV sets.

Significant gain in Quality of Service for mobile reception can be accomplished with the introduction of antenna diversity. Figure 11.18 shows the result of a simulation on the effect of antenna diversity (8K mode, 16-QAM, rate 2/3, $\Delta{=}1/4$). These simulations were carried out using a very critical channel, which is referred to as the Rural Area (RA) profile defined in [COST 207]. A receiver with only one antenna already requires a C/N of approx. 21 dB for perfect reception. The usage of diversity lowers the required C/N by approx. 12 dB for this reception scenario.

In figure 11.18, SC stands for the selection combining, a rather simple method, whereas MRC describes the more complex but more efficient maximum ratio-combining algorithm.

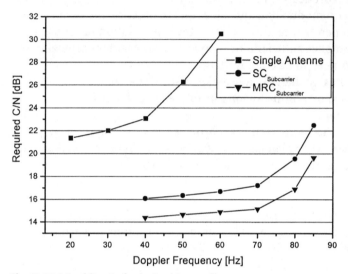

Fig. 11.18 Gain of diversity for the Rural Area profile

Symbols in Chapter 11

C_o spectral signal power density
C signal power
C/N signal to noise ratio
F_p pre-amplifier noise figure
F_t tuner noise figure
F_{t+p} combined noise figure of tuner and pre-amplifier
G antenna gain
g_p pre-amplifier gain factor
$g(X)$ generator polynomial defined in the time domain
I in-phase component
i running variable, integer
K Ricean factor
k_{max} maximum value of the OFDM carrier index
k_{min} minimum value of the OFDM carrier index
N Noise power

N_e	number of echoes in a channel model
Q	quadrature component
T_g	duration of guard interval
T_{Nutz}	duration of useful interval
T_S	duration of a symbol
t	time in general
X	polynomial argument defined in the time domain
$x(t)$	input signal of a channel model
$y(t)$	output signal of a channel model
z	complex-valued argument
α	distance measure for constellation points
Δ	relative length of the OFDM guard interval
θ_i	phase angle in echo path i
ρ_o	attenuation in the direct signal path
ρ_i	attenuation in echo path i
τ_i	relative time delay in echo path i

12 DVB Data Broadcasting

In analogue television we can find auxiliary information being broadcasted parallel to video and audio components. This information is in many cases connected to the actual programme and mostly transmitted via the teletext-mechanism. Teletext offers a preview of future events in the programme and other information like news or weather forecasts. This feature had to be retained by the digital TV system and possibly be enhanced to fit the increased need for data transmission. But over and above Teletext DVB now provides the means for a large variety of data services like software download, reception of Internet services or interactive TV.

12.1 Basics of Data Broadcasting

Usage of the MPEG-2 Transport Stream makes it possible to transmit not only video and audio data but also any other digitally coded information over the DVB distribution channels. To denote this, the picture of the Data Container (figure 12.1) is often used. The Data Container is conveyed inside the DVB multiplex or can even be seen as an alternative representation of the multiplex. Depending on the usable data rate of the broadcast channel, the size of the container varies and it can include data services in addition to PSI, SI, video and audio.

These data services can be either programme related or totally independent of any other service in the multiplex. Examples of programme related data services are teletext or other applications, which give further information about the presently received programme. In addition to this kind of services there are those services which are not linked to a specific programme and thus form an independent part of the multiplex. Possible areas of use are software download, MHP applications, information services or Electronic Programme Guides (EPG).

No matter, which kind of service the digital data represents, it can be inserted into the MPEG-2 Transport Stream (TS). But depending on the needs of the application (e.g. exact timing of presentations, synchronisation with other services), additional provisions have to be made in order to ensure a

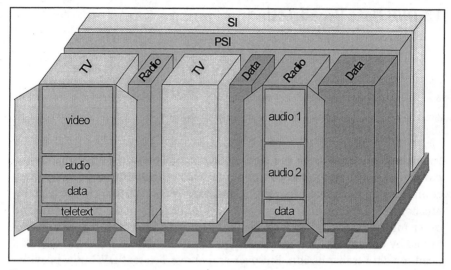

Fig. 12.1. Components inside the DVB data container

correct transmission and enable that the receiver understands the transmitted service. In consequence, four ways of transmission were defined in [EN 192]: Data Piping, Data Streaming, use of a Data/Object Carousel and Multiprotocol Encapsulation. The following four sections describe each of them:

12.2 Data Piping

Data Piping (figure 12.2), the simplest of the transport mechanisms of data in DVB, enables the user to send any non-synchronised data over the DVB channels.

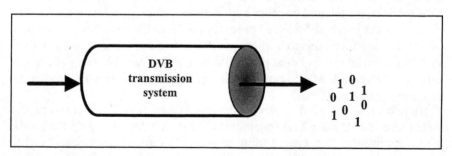

Fig. 12.2. Data Piping

The channel can be seen as a simple tube, where data is poured into one end and pours out at the other end. In technical terms this means that the data is directly inserted into the payload of the TS. In order to enable the receiver to understand the data pouring out at the end it is necessary that additional arrangements be made between sender and receiver prior to the transmission. I.e. the receiver has to be informed about how the data was distributed into the 184 Byte long payload packets and how it will need to re-assemble them. The aim of defining Data Piping is to provide a very simple mechanism suitable for any data service. Data Piping has the advantage that it is possible to include only the additional information absolutely necessary for the receiver to decode the transmitted data. An overhead caused by un-used but mandatory contents of a specific protocol does not occur because no fixed protocol is used. On the other side it is necessary that the receiver knows how to handle the data. Practically, Data Piping is not used a lot. It exists in certain broadcast environments, which are governed by proprietary technology and may be completely replaced by the more complex schemes described in 12.4 and 12.5

12.3 Data Streaming

Data Streaming (figure 12.2) can be classified into three areas of usage:

12.3.1 Asynchronous Data Streaming

Transmission can be made without the need of special synchronisation by using Asynchronous Data Streaming. This mechanism is similar to Data Pip-ing, but here the data is already packetised into segments 64 kBytes long. These segments are then inserted into PES packets which themselves are di-vided into units of 184 Bytes in order to fit into the payload of MPEG TS packets (s. Chapter 5). A possible usage of this streaming method could be the transmitting of data that is already existent in the form of large files.

12.3.2 Synchronous Data Streaming

In cases where a highly constant data rate is required, Synchronous Data Streaming can be used. A constant data rate is accomplished by transmitting the clock reference of the encoder, which is then used to synchronise the sys-tem clock of the receiver. In addition time stamps are provided within the data stream from which the data rate is directly connected to the system clock (s. Chapter 5.3). Different from the time stamps in the case of video

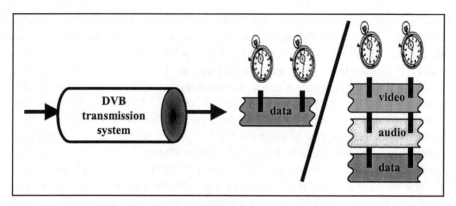

Fig. 12.3. Synchronous/Synchronised Data Streaming

transmission, these time stamps are extended by 9 additional bits ("PTS_extension") in order to reach a higher accuracy of the synchronisation. Also the presentation of the data rate ("ES_rate") is extended by an optional field "output_data_rate" which enhances the resolution of the data rate denotation from 400 bits/sec to 1 bit/sec. If applications require a synchronous output of data even after a loss of packets or when several decoders have to be synchronised, Synchronous Data Streaming is the right choice.

12.3.3 Synchronised Data Streaming

Some applications make it necessary to couple two elementary streams (PES) with each other. The classical example for this is the relationship between video and audio data where the presentation of both has to ensure synchronisation of the display of lip movements with the audio material. But also data other than audio may need accurate synchronisation to the video stream. Possible applications are for example subtitling, information for the hearing impaired and distribution of Karaoke song texts. Nevertheless, synchronisation to lip movement may not put such extremely strict demands on the accuracy of the data rate as in certain cases, the coupling of two data streams, (which is done by synchronous data streaming). Thus the field "output_data_rate" is not used and the field "PTS_extension" is only optional, because the highly precise time stamps offered in this field are not necessary in most cases.

12.4 Data/Object Carousel

Carousel mechanisms are well known to the users of teletext services accompanying analogue television programmes. The user selects a teletext page via a three-digit number, which one dials on their remote control and then waits for a certain period of time until the page is displayed. He or she therefore experiences a certain level of interactivity, which systematically would be called "local interactivity". The teletext pages are repeatedly played out of a data carousel, which the teletext provider operates and which all teletext pages are placed on. The operator can choose the number of pages on the carousel, the cycle period of the carousel and they can place certain pages on the carousel more than once – thereby reducing the maximum time the user will have to wait for these pages. The receiver does not have to store the teletext pages locally because they will be repeatedly made available. The user is not able to control the contents of the carousel but is able to locally select the teletext page displayed on his screen.

DVB data and object carousels offer comparable functionality – but instead of teletext pages, data files can be transmitted. In addition to the data a "table of contents" can be presented to the user. Two main groups of carousels are defined: the Data Carousel and the Object Carousel.

12.4.1 Data Carousel

The Data Carousel specified within DVB is based on the DSM-CC (Digital Storage Media – Command and Control) standard [ISO 13818-6]. The files belonging to one software program form one group and each file corresponds to one module. The identification of a module is unique for the whole carousel. It is not necessary that the numbering of the modules is continuous, the modules on the server can be numbered and then only the necessary modules can be sent. In figure 12.4 this results in a carousel including only the modules 2, 3 and 8. For transport purposes the modules are split up into blocks of equal size. Solely the last block of a module may be smaller than this block size. Several related modules (forming a software application) build up a group. For each group a "DownloadInfoIndication" (DII) is inserted, referencing and describing the modules that belong together. This structure constitutes a one-layer Data Carousel (figure 12.4).

The one-layer structure is often not sufficient. In case different versions of files or programs for different platforms are to be transmitted, each version needs to be placed into an own group and thus within an own data carousel, even though the transmitted data may differ only slightly. Another case where the one-layered structure reaches its limits is the description of groups with very large modules. Soon the description can reach a size where one single DII is not enough to contain it.

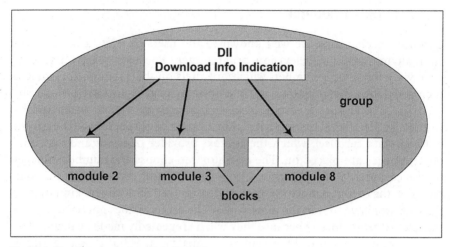

Fig. 12.4. Structure of a one-layer Data Carousel

These cases can be coped with by using a two-layer Data Carousel (figure 12.5) which combines several groups to one supergroup. The data field "DownloadServerInitiate" (DSI) includes a list of affiliated groups and their descriptions.

In case different groups contain similar files (= similar modules) these modules can be referenced even by different groups (one module can belong to different groups).

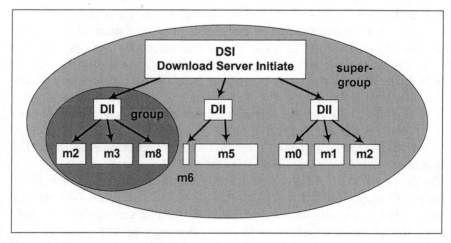

Fig. 12.5. Structure of a two-layer Data Carousel

The functional data structures used for Data Carousels are "DSM-CC sections" which are fragmented and inserted into the transport stream. These sections have the same structure as the sections described in chapter 5.4 (figure 5.12). The table_id of the section denotes whether it carries data ("Download Data Message", table_id 0x3B) or controls information ("Download Control Message", table_id 0x3C).

The only kind of "Download Data Message" used in a Data Carousel is the "DownloadDataBlock (DDB) Message" which carries exactly one block of a module. It also includes additional information about the block such as block number, version number, and identification number of the module to which the block belongs. A "downloadId" within the header makes it possible to assign the message to the respective carousel in the network (figure 12.6).

The "Download Control Messages" are the "DSI Message", the "DII Message" and the "DownloadCancel (DC) Message".

The "DSI Message", only used in Two-layer Data Carousels, lists the groups, which are included in the carousel. Each group is identified by a "groupId", and its size is given by the "groupSize". The list of descriptors

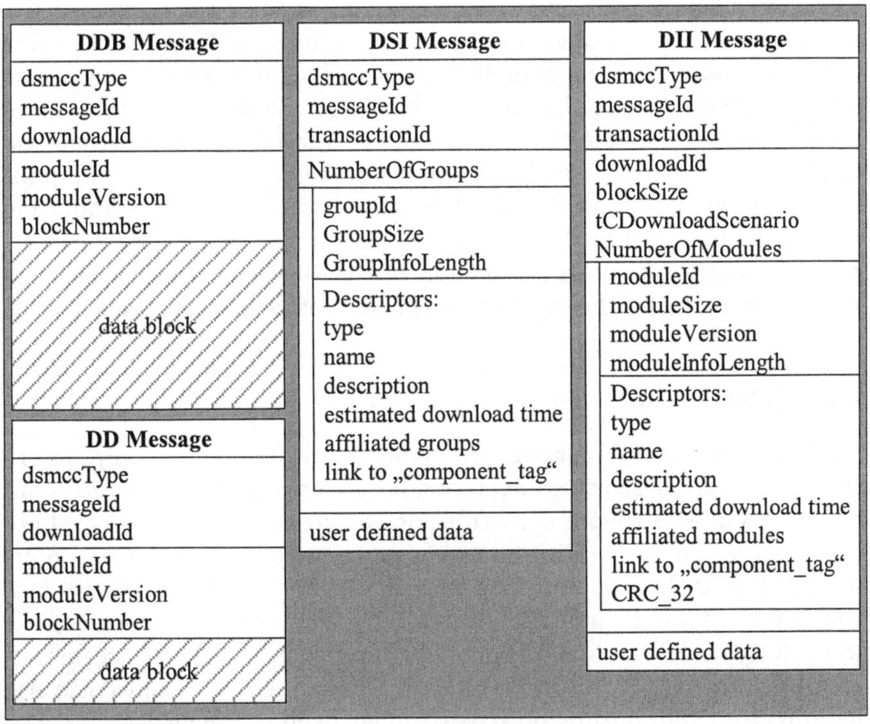

DDB Message	DSI Message	DII Message
dsmccType messageId downloadId	dsmccType messageId transactionId	dsmccType messageId transactionId
moduleId moduleVersion blockNumber	NumberOfGroups	downloadId blockSize tCDownloadScenario NumberOfModules
	groupId GroupSize GroupInfoLength	
data block	Descriptors: type name description estimated download time affiliated groups link to „component_tag"	moduleId moduleSize moduleVersion moduleInfoLength
DD Message		Descriptors: type name description estimated download time affiliated modules link to „component_tag" CRC_32
dsmccType messageId downloadId	user defined data	
moduleId moduleVersion blockNumber		user defined data
data block		

Fig. 12.6. DSM-CC Messages

includes the name, data type and further descriptions of the group. Furthermore, a "group_link_descriptor" provides the possibility to join groups together into units. If a group is to be sent in TS packets with a different PID, the "location_descriptor" can be found in the DSI. It provides a "component_tag" which can be resolved into the according PID by using the "stream_identifier_descriptor" in the corresponding PMT.

The "DII Message" lists the modules of each group. Again, for each module an identification ("moduleID") and size ("moduleSize") are indicated. In addition, the DII includes a version number. The descriptors in the DII have the same functionality as in the DSI, except for the "group_link_descriptor" which is here replaced by the "module_link_descriptor" for linking modules together into a unit. An additional "CRC_32"-word, calculated across the whole module enables error detection.

The "DC Message" only signals the cancellation of the download with no further structures to be presented here. The data carousel organises the control information and the data itself in the form of messages. No matter, in which order the data blocks are sent, the receiver is always able to reconstruct the right order by using the module numbers or block numbers. In consequence, only one cycle of the carousel rotation is necessary to collect all required data of one given module (all necessary data blocks) irrespective of the point in time at which the user requests the data. The time needed to collect the data is dependent on the speed of which the carousel rotates and on the position of the individual data blocks on the carousel. The usage of sections also enables the content provider to transmit more important sections more often, for example the DII or DSI messages in order to keep the receiver informed about the structure of the whole carousel without too much idle time. Thereby the access to certain groups and modules can be accelerated. (For each access to a module in a two-layer carousel the DSI and the corresponding DII have to be read out before acquiring the wanted module)

12.4.2 Object Carousel

While the data carousel supports the cyclic transmission of files only, the object carousel can also carry directories, file streams and other events, summarised as objects (figure 12.7). Directories provide a list of associated objects comparable to the directories known from computers. In case information has to be coupled to a specific stream in the broadcast (e.g. connecting additional information to a TV programme) a reference can be transmitted in the carousel (in figure 12.7 depicted by the block "stream (reference)"). Reference to a specific event in a stream (e.g. a TV show) can also be provided. Transporting objects is similar to that provided by the data carousel.

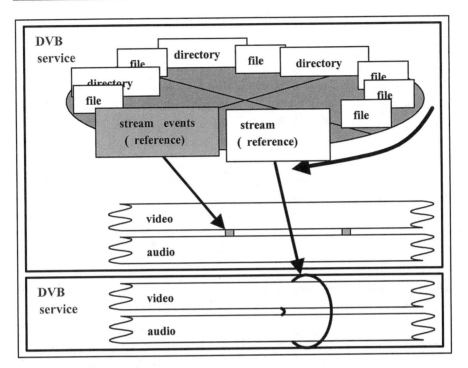

Fig. 12.7. Structure of an Object Carousel

The modules are separated into "Download Data Blocks" and inserted into "DSM-CC sections" (figure 12.8). The modules no longer include only the original data of the files but rather assemble themselves from "BIOP (Broadcast Inter ORB Protocol / ORB = Object Request Broker) messages".

The "BIOP messages" can have a variable length and include a header, an extended header and a payload. The header signals different characteristics of the message, such as the length. The extended header includes the "objectKey" i.e. the identification number of the object, and – depending on the object type – a list of attributes, which further describe the object.

The directory structure of the carousel is shown in figure 12.9. The root directory granting initial access to the carousel is called the "Service Gateway". Sub-directories are included in the service gateway by listing the objects of the lower layers. When the object is a file, the file data is inserted directly into the payload of the "BIOP message". "Stream" objects do not directly consist of data streams but include a link. This link can guide to elementary streams of the presently received service, of another service or to a service assembled from several elementary streams. Even services on different transport streams can be referenced (via original_network_id, transport_strema_id and service_id).

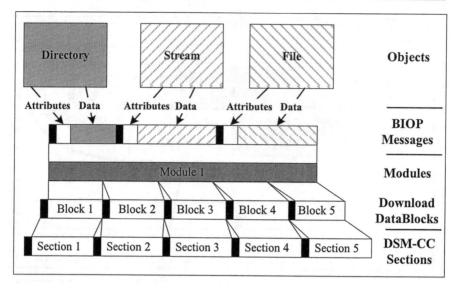

Fig. 12.8. Insertion of Objects into DSM-CC Sections

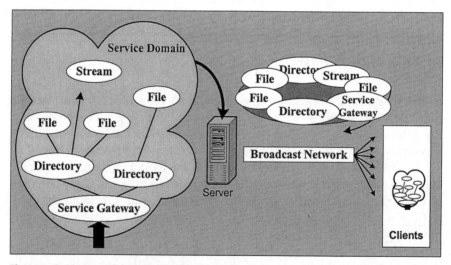

Fig. 12.9. Transmission of directories via the Object Carousel

In order to access the directories of the carousel, the receiver has to filter the DSI message, which points at a DII describing the group that contains the service gateway. Here the description of the module containing the service gateway can be found. After downloading the complete module, the "objectKey" within this module links to the BIOP message, which contains the service gateway.

The service gateway itself and any other directory include links to the objects belonging to this directory. Each link points at a DII message which itself denotes the affiliated module and the BIOP message embedded inside. To download the whole carousel it is thus necessary to go through the cycle "DII message – Module – BIOP message" repeatedly until all objects have been downloaded.

The Object Carousel as an extension of the carousel structures covers application areas that are broader than the ones of the Data Carousel. By adding the possibility of organising data in directories, data services beyond the possibilities of the Data Carousel can be implemented. One example in which data streams play an important role is the distribution of advertisements for large department store chains. From a central server, displays in all of the stores can be fed with information describing special offers, latest lists for pricing or even background music. Other examples are MHP applications transmitted by object carousels because of the need for complex directory structures and linking broadcast streams and programs.

Furthermore, the Object Carousel is an integral part of the specification for the network independent protocol stack for interactive services [ETS 802]. Using an interaction channel an even broader area for applications of the Object Carousel is created.

12.5 Multiprotocol Encapsulation

The transport mechanism termed Multiprotocol Encapsulation (MPE) opens the possibility to transport packets originating from a variety of communication protocols via DVB broadcast channels. MPE was optimised for the use of the Internet Protocol (IP), but the usage of the LLC/SNAP (Logical Link Control/SubNetwork Attachment Point) method [ISO 8802-2] makes it possible to include data from other and even future network protocols. MPE provides the means to address single receivers ("unicast"), groups of receivers ("multicast") and also all receivers ("broadcast") in the network.

The transport of the datagrams is based on DSM-CC sections [ISO 13818-6], which are compatible to the private sections defined in chapter 5. The DSM-CC standard includes a method for transmitting communication protocols based on MPEG-2 transport streams, but this method is not used within DVB, because it does not meet the DVB requirements, especially regarding security of the data.

Instead, DVB uses the semantics of a DSM-CC section with a DVB-specific interpretation of the header (s. table 12.1, showing the section structure in form of a table similar to the tables in chapter 5). The header of such a section (s. chapter 5.4) is denoted by the table_id 0x3E. In contrast to the sections used for transmission of the PSI/SI-tables, the section_syntax_indicator of a DSM_CC datagram section (following the table_id) indicates whether the common "CRC_32"-word or a checksum is **used for**

Table 12.1. Syntax of a datagram section

Syntax	Size (bits)	Unit
Datagram_section() {		
table_id	8	Uimsbf
section_syntax_indicator	1	Bslbf
private_indicator	1	Bslbf
reserved	2	Bslbf
section_length	12	uimsbf
MAC_address_6	8	uimsbf
MAC_address_5	8	uimsbf
reserved	2	Bslbf
payload_scrambling_control	2	Bslbf
address_scrambling_control	2	Bslbf
LLC_SNAP_flag	1	Bslbf
current_next_indicator	1	Bslbf
section_number	8	uimsbf
last_section_number	8	uimsbf
MAC_address_4	8	uimsbf
MAC_address_3	8	uimsbf
MAC_address_2	8	uimsbf
MAC_address_1	8	uimsbf
if (LLC_SNAP_flag == "1") {		
LLC_SNAP()		
} else {		
for (j=0;j<N1;j++) {		
IP_datagram_data_byte	8	Bslbf
}		
}		
if (section_number == last_section_number)		
for (j=0;j<N2;j++) {		
stuffing_byte	8	Bslbf
}		
}		
if (section_syntax_indicator =="0") {		
checksum	32	uimsbf
} else {		
CRC_32	32	rpchof
}		
}		

error detection. Addressing of the datagrams is done by a 48 bit long MAC-(Medium Access Control) address allocated to each receiver. The last two bytes of the MAC-address, which most likely distinguish between different receivers, are carried in the position of the header, which is normally used by MPEG for the "table_id_extension". The rest of the address follows behind the "last_section_number", i.e. outside of the header. By splitting up the address into two parts of which one is located in the header of a section, the demultiplexer is enabled to do the filtering of the more significant part of the MAC-addresses in hardware.

Transmission of protocols other than IP is signalled by the LLC_SNAP_flag which also indicates that the datagram starts after the LLC_SNAP-header (in table 12.1 denoted by "LLC_SNAP()"). The signalling within the PSI/SI-structures is done by the data_broadcast_descriptor (s. chapter 5.4.3.2). In the case of MPE it includes four additional fields: the MAC_address_range, indicating how many bytes of the MAC-address are significant for the filtering by the receiver, the MAC_IP_mapping_flag, indicating in which way the MAC-addresses are derived from the IP-addresses, the alignment_indicator, informing about an alignment of the data along 32-bit borders and a value giving the number of sections carrying the present datagram. This additional information is necessary for protocols other than IP to estimate the over-all length of the datagram prior to decoding in order to assign enough memory.

12.5.1 IP over DVB

One usage of the MPE mechanism is the transmission of IP packets over DVB broadcast networks. It is based on an extended form of Multiprotocol Encapsulation. The extension was required to solve some problems, which occurred in large networks where IP services are offered. Insertion of the Internet Protocol (IPv4 or IPv6) datagrams is done directly into the payload of the sections. In DVB the length of IP-datagrams is limited to 4086 byte, so that one datagram fits exactly into the payload of one section, without the need to fragment it over several sections. Accordingly, the additional information within the LLC_SNAP header, such as section_number, last_section_number and the over-all number of sections mentioned in conjunction with other protocols can be eliminated.

The DVB transport is used to convey all types of IP-based services from multicast file delivery, via multicast streaming, high speed Internet services to webcasting services. All these services are sometimes integrated into a single platform, running IP/DVB multiplexed services over DVB transports. The first version of the data broadcast specification was finalised in 1996. It did not provide a way to specify the location where IP/MAC addresses can be found on a DVB network. Therefore it was not easy for a receiver to find the traffic relevant to its IP/MAC address on a DVB network, providing an IP/DVB stream as part of a whole multiplex, carrying different kinds of services, and/or allocating more than one DVB transport stream for IP/DVB services. In order to solve this problem, a table was constructed which provides a possibility to link to a data stream within a DVB network. The revised data broadcasting specification [EN 192] introduces the IP/MAC platform concept. An IP/MAC platform represents a set of IP/MAC streams and/or receiver devices. Such a platform consists of a single IP network with unambiguous addresses in order to prevent address conflicts. It can include several transport streams within one or multiple DVB networks and thus allows operation of several data platforms on the same DVB network with overlapping address ranges in the respective platforms without clashes.

Table 12.2 represents the syntax of the INT.

Table 12.2. Syntax of the INT (IP/MAC Notification Table)

Syntax	Size (bits)	Unit
IP/MAC notification section() {		
table_id	8	uimsbf
section_syntax_indicator	1	bslbf
reserved_for_future_use	1	bslbf
Reserved	2	bslbf
section_length	12	uimsbf
action_type	8	uimsbf
platform_id_hash	8	uimsbf
Reserved	2	bslbf
version_number	5	uimsbf
current_next_indicator	1	bslbf
section_number	8	uimsbf
last_section_number	8	uimsbf
platform_id	24	uimsbf
processing_order	8	uimsbf
platform_descriptor_loop()		
for (i=0, i<N_1, i++) {		
target_descriptor_loop()		
operational_descriptor_loop()		
}		
CRC_32	32	rpchof

Most of the entries from table_id to last_section_number are already known from sections of other tables (s. chapter 5.4). Additionally, the action_type identifies the action to be performed. The only action specified in this context so far is the locating of IP/MAC streams in DVB networks (action_type 0x01). In order to assist receiver devices with limited filter capabilities in locating the appropriate INT, the platform_id_hash is transmitted. It is derived from the platform_id by a simple XOR-function forming a single byte value out of the three byte long platform_id. The platform_id itself labels a given IP/MAC platform in order to identify it.

The following part of the INT is divided into two independent loops. The first loop (platform_descriptor_loop) is used to describe the IP/MAC platform identified by the platform_id. The descriptors used here denote the name of the IP/MAC platform, the name of the platform provider, locate streams in the DVB TS which are valid for all devices, describe the access mode (e.g. dial-up) and convey the necessary telephone number for dial-ups.

The second loop itself is again composed of two loops, the target _descriptor_loop and the operational_descriptor_loop.

The target_descriptor_loop contains zero or more descriptors, which are used exclusively for targeting. The target_serial_number_descriptor in this loop for example is intended to target devices based on some manufacturing id, the target_smartcard_descriptor is able to target devices based upon their smartcard identifier. Descriptors are also defined for targeting MAC addresses, MAC address ranges and IP addresses in different notations (e.g. shortform slash format, IPv4/v6 notation).

The operational_descriptor_loop mostly contains descriptors relating to the localisation or assignment process, similar to the platform_descriptor _loop. Some of the descriptors are used in both, platform_descriptor_loop and operational_descriptor_loop. The IP/MAC stream_location_descriptor for example locates the IP/MAC stream in DVB transmissions and can be used within both loops. But in the operational_descriptor_loop it can override the descriptors of the platform_descriptor_loop and can give special information for the directly targeted devices.

12.6 System Software Update

For various reasons, manufacturers need to be able to update the receiver's software, of products which are already on the market. The specification for System Software Update (SSU) provides the tools to do this.

The specification includes two profiles. The "simple profile" describes the signalling of either a proprietary data transfer format or a standardised DVB data carousel. The second profile defines a new table, the Update Notification Table (UNT), which provides a standard mechanism for carrying additional data, e.g. update scheduling information, extensive selection and targeting information or filtering descriptors. Both the simple profile and the UNT based profile are described in one single document [TS 006].

The UNT has a similar structure as the INT. The OUI_hash in the UNT then replaces the platform_id_hash known from the INT. The OUI is the Organisation Unique Identifier as defined by the "Institute of Electrical and Electronics Engineers (IEEE)", identifying which organisation is providing the update. By this, a platform similar to the one in the INT case is defined, this time distinguishing the targeted devices and the provider, which runs the download. The hash value again (as with the INT) facilitates the handling of the OUI for receivers with limited filter capabilities.

Table 12.3. Syntax of the UNT (Update Notification Table)

Name	Size (bits)	Unit
Update_Notification_Table() {		
table_id	8	Uimsbf
section_syntax_indicator	1	Bslbf
reserved for future use	1	bslbf
Reserved	2	bslbf
section_length	12	uimsbf
action_type	8	uimsbf
OUI_hash	8	uimsbf
Reserved	2	bslbf
version_number	5	uimsbf
current_next_indicator	1	bslbf
section_number	8	uimsbf
last_section_number	8	uimsbf
OUI	24	uimsbf
processing_order	8	uimsbf
common_descr_loop()		
for (I=0, I<N1, I++) {		
CompatibilityDe-		
platform_loop_length	16	uimsbf
for (i=0, i<N2, i++) {		
target_desc_loop()		
operational_desc_loop()		
}		
}		
}		
CRC_32	32	rpchof
}		

The UNT shows 5 hierarchical loops (table 12.3). The first loop, the common_descriptor_loop contains a list of descriptors which, unless overridden in the operational_descriptor_loop, apply to all SSUs in this section. Descriptors here might for example be the update_descriptor, denoting if the update has to be done at once or upon the next restart of the receiver, or the SSU_location_descriptor, locating the software update service within the DVB transmission.

The second loop, the N1 loop, provides a mechanism by which many receiver devices may be addressed by a single sub-table section. It targets the devices by describing the receiver's system hardware and software rather than their affiliation to single update services.

The N2 or platform_loop associates an operational_descriptor_loop with a target_descriptor_loop, allowing multiple targeted and untargeted SSUs to be associated with a platform.

As in the INT case, the target_descriptor_loop contains zero or more descriptors used exclusively for targeting. The same descriptors as in the INT can be used, with slightly different functionalities.

The usage of the operational_descriptor_loop is also similar to the INT case, with descriptors like SSU_location_descriptor or scheduling _descriptor, overriding the same descriptors within the common _descriptor_loop for single targeted devices.

Thus the UNT enables receiver manufacturers to download new software with different versions to different groups of receivers by defining targeted groups of end devices.

13 DVB Solutions for Interactive Services

In digital television, video and audio signals are acquired, transmitted and presented as a sequence of numerical values. The techniques in processing the stream of digital data in the various components of the transmission chain have been described in the previous chapters. However, only the distribution of digital signals from the broadcasting service provider to the user, i.e. in one direction has been considered so far. With the increasing availability of digital transmission systems a desire emerged to augment classical broadcasting services with services providing a certain level of interactivity. The user would be able to react to information presented by the service provider or even to contribute to the information exchange. To achieve interactivity, a return channel needs to be added to the transmission system enabling data to travel from the user back to the service provider.

The solution for interactive services proposed by the DVB Project is presented in this chapter. It consists of a whole set of specifications describing protocols, and interfaces for all kinds of transmission media and network scenarios. After a short description of the service environment, protocols common to all types of interaction channels are explained in detail. In the following sections the interaction channels for individual types of networks are analysed. Two groups of networks have been identified that are able to provide interaction channel services. The first section is concerned with telecommunication networks used to enhance broadcasting networks with an interaction channel. In the second section, interaction channels specific to the various types of broadcasting networks are described.

13.1 Interactive Services

The term 'interactive services' may describe a whole range of different types of service offerings that require a varying level of interaction between the user and the service provider or the network operator. The most basic form of interactivity called 'local interactivity' can be achieved within the user terminal. For local interactivity, data belonging to certain interactive services is transmitted and stored in the terminal. That terminal can react to the inputs of the user without requiring further exchange of data across the network. This basic form of interactivity is not covered in this chapter.

The requirement of providing an interaction channel across the transmission network was established by the desire to enable the user to respond in some way to the interactive service and by the necessity of the service provider or network operator to listen and possibly react to that response. The user's response may take the form of some simple commands, like in 'voting' in favour of a particular participant in a game show or for 'purchasing' goods advertised in a shopping programme. In that case, it may be sufficient for the interaction channel to consist of a one-way, narrow-band path from the user to the service provider.

With a higher level of interactivity the user, who has made a response to an interactive service, will expect an acknowledgement. This may be the case if for example the 'purchase' involves a credit card transaction that must be accepted by the service provider. The consumer will require to receive notice of the result of the transaction. To transmit the acceptance note it will be necessary for the interaction channel to include a narrow-band path in the forward direction from the service provider to the individual user. The broadcast channel will not be sufficient since a single user is not individually addressable.

If the information expected or requested by the user of the interactive service is more complex or requires high transmission capacity, a further level of interactivity is reached. The shopping programme may for instance offer to provide upon request an additional presentation of a particular product. In this case, the forward interaction channel will need to be broadband. Applications are conceivable where even the user contributes to the interactive service with content that requires a broadband reverse path. At this point, the interactive service resembles two-way communication with similar requirements for transmission capacity and transmission quality in the forward as well as in the return channel.

Adding interactivity to the DVB infrastructure requires the system set-up to be extended by components providing communication means between the end user and the provider of the interactive service. The interactive service provider can be related to the broadcast service provider or even be the same organisation. In any case, it can make use of the high bit-rate DVB broadcast channels in delivering information to the user of the interactive service at typical rates of up to 20 Mbit/s per channel in terrestrial broadcast networks, and up to 38 Mbit/s per channel in broadcast networks via satellite or cable. The transmission capacity of the interaction channel depends largely on the type of network that is used. It may range from a few kbit/s if a simple telephone modem is used (DVB-RCP) to 12 Mbit/s via the CATV interaction channel (DVB-RCC).

Table 13.1. Set of specifications for interaction channels in DVB

	DVB acronym	Standard	Implementation guidelines
Network-independent protocols for interactive services	DVB-NIP	ETS 300 802	TR 101 194
Interaction channels			
• PSTN/ISDN	DVB-RCP	ETS 300 801	
• DECT	DVB-RCD	EN 301 193	
• GSM	DVB-RCG	EN 301 195	
• CATV	DVB-RCC	ETS 300 800, ES 200 800	TR 101 196
• LMDS	DVB-RCL	EN 301 199	TR 101 205
• Satellite	DVB-RCS	EN 301 790	TR 101 790
• SMATV	DVB-RCCS		TR 101 201
• Terrestrial	DVB-RCT	EN 301 958	

The current set of DVB specifications for interactive services describes solutions for a variety of possible network configurations covering the specified DVB broadcast options as well as interaction networks that are capable to provide interaction channel services to DVB systems. Table 13.1 summarises the technical areas that are specified and the appropriate references to ETSI deliverables. The tools enabling interactivity in DVB consist in general of a common part (DVB-NIP [ETS 802]) that is independent of the underlying transport network and a network-specific part.

13.2 Network-Independent Protocols for DVB Interactive Services

Network-specific solutions for the provision of interactive services across transmission channels based on DVB are described in sections 13.3 to 13.4. To enable the use of identical interactive applications over different physical return channels the higher layer protocols are network-independent. The concept of distinguishing between network-dependent and -independent protocols is explained in this section. The reference model and the definition of higher layer protocols specified in [ETS 802] are the common basis for bidirectional communications.

13.2.1 Protocol Stack

DVB uses a protocol stack which consists of the physical layer, the medium access control layer, the higher medium layers and the application layer.

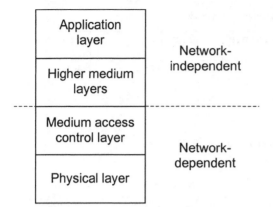

Fig. 13.1. Protocol stack for DVB interactive services

The physical layer contains the specification of the modulation scheme, channel coding, frequency range, filtering, equalisation and transmit power. The medium access control (MAC) protocol, which runs on the medium access control layer provides the interface to higher layer protocols in order to transmit and receive data transparently and independently of the physical layer. Both the physical and the MAC layer are network-dependent and, thus, differ for each technical solution. Protocols managing layers residing on top of the MAC layer are the network-independent protocols which can be used in any interactive system defined by the DVB set of specifications. The goal of this approach is to be able to develop interactive applications, which can be used commonly with different interactivity-enabled network solutions.

13.2.2 System Model

The system reference model describes the concept of interactive services for DVB. Two channels are established between the user and the service provider. The network adapter provides connectivity between the service provider and the network while the interface unit connects the network to the end-user. The broadcast service provider distributes the MPEG-2 transport stream over the unidirectional broadcast channel to the set-top box of the end-user. The interactive service provider offers an interaction channel for bi-directional communication which, in turn, is divided into a forward interaction path for the downstream direction and a return interaction path for the upstream direction. To offer high-speed services towards the end-user the interactive service provider can choose the broadcast link to embed data into the MPEG-2 transport stream. In this case, the broadcast channel may contain application control or communication data in order to connect

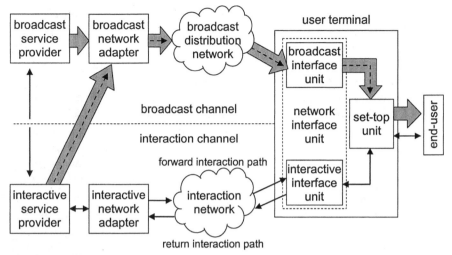

Fig. 13.2. Generic system reference model for interactive services

to the distribution network. The user may for example utilise a cable modem instead of a set-top box as a terminal device. A bi-directional application control and communication channel will also be required between the different service providers for synchronisation purposes.

13.2.3 Higher Layer Protocols

With the higher layer protocols a unified interface for interactive services is provided which is independent of the underlying network configuration. To limit the number of network-independent protocols, DVB focused on the definition of a certain class of protocol stacks. These protocol stacks send data from the broadcast or interactive service provider to the end-user's set-top box via the broadcast channel employing the DVB/MPEG-2 data transmission system or send data from the interactive service provider to the customer's terminal via the interaction channel and vice versa. The data packets processed by the higher layer protocols can contain audio, video or general computer data content as well as application control or application communication data. They may furthermore optionally be used to control data download and interactive sessions, and to implement a remote diagnostic service and network management.

The basis for the network independent definition of higher layer protocols is formed by the adoption of DSM-CC (Digital Storage Media - Command and Control) [ISO 13818-6]. It is complemented by the Internet Protocol (IP) with appropriate transport protocols in the interaction channel and by the MPEG-2 Data Carousel [EN 192] in the broadcast channel. Details of the application of DSM-CC in the DVB system for data communication are provided in section 12.5 of this book. Two sets of network-independent protocols have been specified.

The first set applies to the transmission of audio, video and data content. The media may be delivered using the broadcast or the interaction channel. On the broadcast channel it can be transmitted in its native format as specified in the DVB transmission system. Alternatively, media packets may be encapsulated in IP using MPEG-2 private sections (DSM-CC sections) as defined in [EN 192]. When TCP/IP is carried over the broadcast channel, an interaction channel shall be established to accommodate the flow of return acknowledgements. In the interaction channel, IP is used with the point-to-point protocol (PPP).

The second set of protocols describes the download of data across the broadcast or the interaction channel from the interactive service provider to the set-top box. It is also concerned with user-to-user interaction. The latter may take place likewise using the broadcast or the interaction channel. In the case of a broadcast channel, the data is organised in DSM-CC data and object carousels that are in turn transmitted as MPEG-2 private sections. In the interaction channel, DSM-CC is used over IP.

Session control signalling is usually not necessary in set-top boxes. However, to accommodate DVB interactive services requiring support for sessions (e.g. when traversing multiple networks), an optional protocol stack has been defined. This logical channel provides a bi-directional flow used for the exchange of session information between a set-top unit or service provider and a session control entity in a network. A core subset of the MPEG-2 Digital Storage Media-Command and Control User-to-Network Interface (DSM-CC U-N) is utilised.

A uni-directional flow is used to implement optional network management capabilities via the interaction channel. With this feature, set-top unit remote diagnostics are possible. The Simple Network Management Protocol (SNMP) was chosen as a means to exchange management information encapsulated in IP across the interaction channel.

13.3 Network-Dependent Solutions for PSTN, ISDN, DECT, GSM

A straightforward way to add interactivity to the communication system defined by DVB is to make use of the two-way capabilities of traditional telecommunication networks like the Public Switched Telephone Network (PSTN). By specifying an interface between the DVB broadcast system and a telecommunication network a narrow-band, bi-directional interaction channel becomes immediately available. Since telecommunication networks exhibit a widely available infrastructure, their application is sufficient in providing a basic level of interactivity without requiring additional investments in upgrading a broadcast network with a specific return channel.

Interaction channels using telecommunication networks are explained in the following sub-sections. The implicit advantage of the use of telecommunication networks for interactive services is that they are independent from the broadcast network (e.g. cable, satellite, terrestrial networks). The disadvantage is the fact that in a private home the telephone socket may be far away from where the broadcast network ends. The forward direction of the interaction channel might be embedded in the broadcast channel and delivered across the broadcast network or might be delivered through the telecommunication network. The return channel, i.e. the reverse direction of the interaction channel, is provided by the telecommunication network.

13.3.1 Interaction Channel through PSTN/ISDN

The interaction channel for both, Public Switched Telephone Network (PSTN) and Integrated Services Digital Network (ISDN) is specified in [ETS 801]. These systems use the existing telephone line. Depending on whether the local telephone connection is an analogue or a digital line, the interaction channel is established via the PSTN or the ISDN.

In order to allow access to the PSTN, the user terminal needs to be equipped with a modem, which can be internal to the device or external. An external modem is connected to the DVB user terminal with commonly used cables like serial, USB or Ethernet connectors. The network side of the modem is connected to the existing telephone line. Therefore, it shares the line with other telecommunication equipment that is already present at the customer's premises, e.g. telephones, fax machines, computer modems. The set-up of a bi-directional interaction channel via the PSTN is always initiated by the modem in order to avoid unsolicited calls by service providers. The interface between the modem and the PSTN network must be compliant to the national requirements for the attachment of terminal equipment to an analogue subscriber interface of the PSTN as described in [EN 001]. Calling procedures applied by the modem, including dialling, call set-up and call maintenance are described in [EN 001]. A detailed explanation would go well beyond the scope of this chapter. The modem uses DTMF tones for dialling. Optionally, a pulse-dialling mode may be implemented. The line status ('on-hook' or 'off-hook') is recognised by the modem in order to be able to perform a call repetition process if desired. The same applies for multiple call attempts if the called line of the interactive service provider is busy. Multiple calls to the same service provider need to observe a certain distribution in time as specified in [EN 001]. In some European countries, it is required to give absolute priority to emergency calls. For this reason, the modem has to force the disconnection of the line and the teardown of the interaction channel if the user picks up any other communication equipment sharing the same line with the modem.

The ISDN infrastructure can as well support the implementation of an interaction channel for the DVB system. The Basic Rate Access interface as specified in [EN 012] provides a bi-directional communication path with two payload (B-) channels at 64 kbit/s each and a signalling (D-) channel at 16 kbit/s. The calling procedures are specified in [ETS 402] and [EN 403]. Similar rules as for the PSTN interaction channel apply with regard to emergency services. However, as a digital signalling system, features can be implemented using upper layer protocols.

Obviously, the role of the DVB Project in specifying the use of PSTN and ISDN as the interaction channel was limited. The specification [ETS 801], therefore, is mainly a collection of references to the relevant telecommunication standards. The added value lies in the selection of profiles and/or components of these standards to meet the requirements of interactive services in the DVB environment.

13.3.2 Interaction Channel through DECT

The interaction channel for Digital Enhanced Cordless Telecommunications (DECT) is specified in [EN 193]. DECT provides a local, wireless, bi-directional access technology to a further network infrastructure connecting the user to the service provider. Figure 13.3 shows an adaptation of the general reference architecture for DVB interactive services as described in section 13.2. as it would be used, when the interaction channel is implemented through DECT. The DECT portable part (PP) constitutes the interactive interface unit of the network interface unit. Likewise it connects the user terminal to the interaction channel. It may be internal or external to the user terminal. With DECT covering the local access rather than the complete network, the DECT fixed part (FP) is part of the functional components establishing the interaction channel. The DECT FP is connected by the DECT fixed terminal (FT) to the local access portion of the network and provides an interface to the infrastructure extending the bi-directional connection to the service provider via the DECT interworking unit (IWU).

The connection of DECT to different networks at the FP, the operation of the DECT air interface and the signalling protocols for specific services and applications are all described in the DECT profiles. The default implementation for using DECT in the DVB interaction channel that is described here is based on the Radio in the Local Loop Access Profile (RAP) [EN 765] and the Data Services Profile (DSP) PPP interworking [EN 240]. However, since the

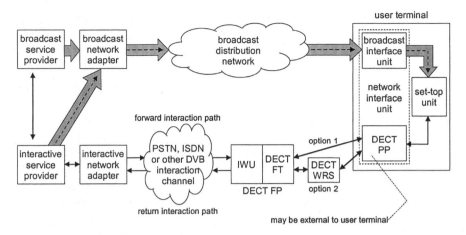

Fig. 13.3. System reference model with an interaction channel through DECT

DECT standard offers various additional options, several other possibilities for the implementation of an interaction channel based on DECT exist. Nonetheless, the chosen profiles have been evaluated against the requirements of interactive services in the DVB system. Even though there is no restriction for using one or another implementation variant, DVB strongly recommends abiding to the described solution.

In the default implementation two scenarios can be distinguished. If the DECT FP is external to the home, protocols defined in the RAP are used to implement a DSP PPP interworking in the local loop. In this case, the DECT PP communicates across a RAP data service infrastructure comprised of DECT FPs and possibly wireless relay stations (WRS). The WRS relays DECT signals between DECT FPs and DECT PPs. DECT can also be used in the home as a wireless interface to the network. The DECT PP recognises that the home FP does not support RAP features and will not, therefore, use RAP procedures to establish a DSP PPP interworking data service. Similar to the case when PSTN is used as the interaction channel in DVB, national requirements on the prioritisation of emergency calls need to be observed.

As already pointed out in the previous section, [EN 193] similarly focuses on the collection of references to the relevant telecommunication standards and on the selection of appropriate profiles and/or components of these standards. As such, the DVB Project conducted an important process for the implementation of DECT as an interaction channel.

13.3.3 Interaction Channel through GSM

The interaction channel through the Global System for Mobile Communication (GSM) is specified in [EN 195]. GSM provides a wireless, bi-directional access technology connecting the user to the service provider. The GSM wireless network may form a complete DVB interaction channel or may be complemented by a further infrastructure such as the PSTN or the ISDN in order to provide a complete transmission path. Figure 13.4 shows an adaptation of the general reference architecture for DVB interactive services as described in section 13.2 as it would be used for an interaction channel implemented through GSM. A GSM mobile station (MS) constitutes the interactive interface unit of the network interface unit, which provides access to the GSM network for the user terminal. It may be internal or external to the user terminal. General requirements and signalling protocols for the interface between the MS and the GSM network as well as between the GSM network and a complementing network like the PSTN are described in the appropriate ETSI specifications. An overview is given in [ETS 500]. Depending on the network linked to the GSM system, the MS must be configured to support the correct bearer capabilities to provide the complete interaction channel. Whenever possible it is preferred to implement GSM-ISDN interworking which provides an end-to-end digital link between the user terminal and the interactive service provider.

Concerning the role of the DVB Project in specifying the use of GSM as interaction channel, the situation is similar to what has been described in the previous sections. DVB's main contribution is the selection of appropriate profiles and/or components from the variety of available options. New data services are becoming available with the evolution of GSM towards the third generation of mobile communications. Particularly, the General Packet Radio Service (GPRS) will provide a suitable interaction channel for the DVB service scenario with a similar reference model as shown in figure 13.4.

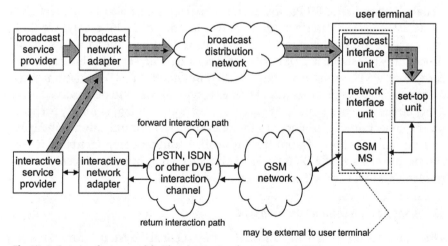

Fig. 13.4. System reference model with an interaction channel through GSM

13.4 Network-Dependent Solutions for DVB-C, DVB-S and DVB-T

In contrast to establishing a return channel using telecommunication networks, here the forward and reverse directions of the interaction channel are created inside the broadcasting network. In this case, the broadcasting network must be equipped with additional components to allow for the transmission in the backward direction. Such solutions are defined for cable (DVB-C), satellite (DVB-S) and terrestrial (DVB-T) broadcast networks.

13.4.1 Interaction Channel for Cable TV Distribution Systems

Cable TV networks are widely deployed in many countries for distribution of broadcast programs. The upgrade with a return channel enables additional interactive services on the existing infrastructure. Due to the fact that many customers are connected to the same coaxial cable, many users must share the available bandwidth. For the assignment of the resources a medium access control protocol is needed. [ES 800] specifies the medium access control layer and the physical parameters. Thus, the standard establishes a DVB Return Channel for Cable (DVB-RCC).

13.4.1.1 System Concept

The system concept follows from the reference model shown in figure 13.2. The interactive system consists of a forward interaction path in downstream direction and a return interaction path in upstream direction that are added to the existing broadcast system. The interactive network adapter (INA) at the service provider's side and the network interface unit (NIU) residing in the user terminal terminate the interaction channel. The underlying concept is to use the downstream transmission to provide synchronisation and information to all units. Two alternatives are defined to achieve that goal. They are called in-band (IB) and out-of-band (OOB) downstream signalling, respectively. IB signalling denotes the embedding of the control information into the MPEG-2 transport stream of a DVB-C channel, in contrast to transmitting control information in a forward interaction path separate from the broadcasting channel. The IB mode is most suitable for cable modem usage. It can offer an asymmetric data connection with a higher data rate in downstream than in upstream direction, which matches the traffic characteristics of many Internet applications. The OOB mode is more applicable to the operation of interactive set-top boxes. With the broadcast channel used for the reception of TV programs, the set-top box can establish additional connections to exploit interactive services. Both types of systems may exist on the same network under the condition that different frequency ranges are used for each system.

13.4.1.2 Physical Layer and Framing

Figure 13.5 indicates the possible spectrum allocation for broadcast and interactive channels. The frequency band for broadcast channels ranges from 60 to 862 MHz and is the same as for DVB-C. It is recommended to use parts of the frequency range from 70 to 130 MHz or from 300 to 862 MHz for the OOB forward interaction path. Frequencies from 5 to 65 MHz are assigned to the return interaction path. Due to ingress from external sources and in order to avoid interference with the intermediate frequency range used by set-top boxes or cable modems as well as analogue receivers in the same network, it may be necessary to leave out some parts of the total range.

The IB forward interaction path must use an MPEG-2 transport stream with a modulated QAM channel as defined in [EN 429] and described in chapter 10. The interaction channels have a bandwidth of 1 or 2 MHz for the downstream signals and of 0.2, 1, 2 or 4 MHz for the upstream signals.

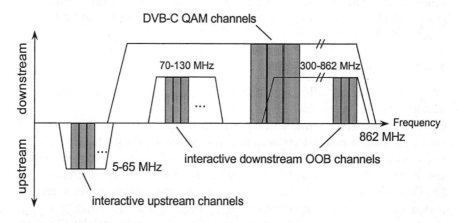

Fig. 13.5. Frequency spectrum assignment of interactive channels

Table 13.2 shows the possible data rates depending on the bandwidth and the chosen modulation scheme. With a QPSK modulation the OOB downstream channel has a transmission bit rate of 1.544 Mbit/s or 3.088 Mbit/s. The support of 3.088 Mbit/s is mandatory, 1.544 Mbit/s is optional for both INA and NIU. For upstream transmission, the INA can use a QPSK or a 16-QAM modulation. Therefore, six types of transmission rates are possible, specifically 12.352 Mbit/s, 6.176 Mbit/s, 3.088 Mbit/s, 1.544 Mbit/s, 512 kbit/s and 256 kbit/s. The support of 3.088 Mbit/s is mandatory, other rates are optional for INA and NIU. The INA is responsible of indicating which rate NIUs may use.

Table 13.2. Supported bit-rates on interactive channels

| | Bandwidth | | | |
Modulation	200 kHz	1 MHz	2 MHz	4 MHz
QPSK	256 kbit/s	1.544 Mbit/s	3.088 Mbit/s	6.176 Mbit/s
16-QAM	512 kbit/s	3.088 Mbit/s	6.176 Mbit/s	12.352 Mbit/s

The system supports the use of several upstream channels operated in parallel. The MAC protocol uses one downstream channel for the control of up to 8 upstream channels. In instances where multiple downstream channels are used at least one channel must support the provisioning functionality. Within this channel, designated as the MAC control channel, the NIU receives basic parameters for sign-on and link management. For the communication during the initialisation and provisioning procedures a dedicated service channel is used in upstream direction.

In a cable TV network downstream information is sent in a broadcast mode to all connected users. Therefore, an address assignment is needed to identify different users. The address scheme of the cable return channel defined by DVB uses a MAC address. The MAC address is a 48-bit value, which may be hard-coded in the device or may be provided by external sources.

Figure 13.6 explains the signal processing for OOB downstream transmission in the INA. For data transport ATM cells with a size of 53 bytes are used. For the forward error correction two Reed-Solomon bytes are added. An interleaver prevents the occurrence of burst errors that would not be correctable at reception. Ten ATM cells are combined with additional signalling information to a structure called Signalling Link Extended Superframe. The signalling information allows the user terminals to adapt to the network and send synchronised information back via the upstream. After the creation of a frame, the bytes are mapped to a serial bit stream, randomised, differentially encoded and QPSK modulated. The receiver within the modem has to apply the inverse procedure.

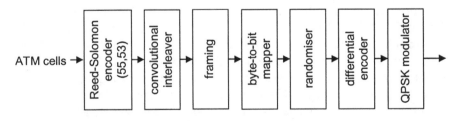

Fig. 13.6. Signal processing for OOB downstream transmission in the INA

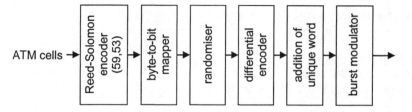

Fig. 13.7. Signal processing for upstream transmission in the NIU

The upstream channel is split into time slots, which can be assigned to different users. The technique is called time-division multiple access (TDMA). Figure 13.7 depicts the block diagram for the signal processing of upstream transmission in the NIU. The transmission is based on ATM cells. Due to higher interference risk in the upstream, 6 bytes Reed-Solomon forward error correction is used for each ATM cell. Interleaving is not suitable since the transmission takes place in bursts originating from different users. The bytes are mapped to a serial bit stream, randomised and differentially encoded. Before the modulator a 4 bytes unique word is inserted. This enables the INA to synchronise the receiver on the bursts. To increase the number of available upstream slots it is possible to replace an ATM cell by 3 minislots. A minislot contains 20 bytes including the 4 bytes unique word and 2 bytes Reed-Solomon forward error correction.

To synchronise the transmission of different users the downstream provides a common time base transmitted every 3 ms. At a data rate of 1.544 Mbit/s it describes the beginning of a new superframe. In the IB case this marker must be embedded every 3 ms into the MPEG-2 transport stream. The upstream is also divided into a 3 ms raster. Depending on the data rate a fixed number of slots is available. From the individual receiving time of the marker the receiver is able to derive the exact point in time of when to start the transmission of a burst.

13.4.1.3 MAC Layer Protocol

As mentioned in section 13.2 the MAC protocol provides tools for higher layer protocols in order to transmit data transparently and independently of the physical layer. The main objective is to control the access to the transmission channel. Other tasks are initialisation, provisioning and sign-on management, connection management, link management and security.

The MAC protocol exchanges information among the functional components using MAC messages. These messages can be sent in a broadcast and a unicast mode. MAC messages are embedded in the packet structure of the channels.

The MAC protocol provides three different access modes and a mode for power and timing calibration of the user terminals. These modes are called contention, reservation, fixed-rate and ranging. Within upstream channels users send packets with a TDMA-type access scheme. This means that each channel is shared by many different users, that can either send packets at an arbitrary time with a resulting possibility of collisions or request the allocation of exclusive transmission opportunities and use the slots assigned by the INA. The distribution of time slots assigned to a particular access mode is organised in access regions. The location of each access region is provided to the NIUs by information contained in the slot boundary fields of the downstream signalling messages. Knowledge about the limits between access regions allows users to decide when to send data in contention slots with a risk of collision or in individually allocated reservation or fixed-rate regions.

- **Contention** based techniques are used for multiple subscribers that have equal access to the signalling channel. It is possible that simultaneous transmissions occur in a single slot, which then results in a so-called collision. The specification has been designed to support cable round trip delays of up to 800 μs, which corresponds to a cable length of approximately 80 km. It is not possible for a NIU to detect a collision by listening to the channel like on an Ethernet network. The INA utilises a reception indicator to inform the NIUs whether ATM cells were received successfully. To reduce the risk of new collisions, contention resolution algorithms are employed.
- **Fixed-rate** access requires data to be sent in slots assigned to the fixed-rate access region in the upstream channel. These slots are uniquely allocated to a connection by the INA.
- **Reservation** access implies that data is sent in the slots assigned to the reservation access region in the upstream channel. These slots are uniquely allocated to a connection for one-time usage by the INA. This assignment is made at the request of the NIU for a given connection.
- **Ranging** access means that the data is sent in a slot preceded and followed by slots not used by other users. These slots allow users to adjust their clock depending on their distance to the INA so that their slot timing is correctly synchronised to the timing determined by the INA.

13.4.1.4 Mid Layer Protocols

To provide an interface between the MAC layer and the higher layers, [ES 800] defines three ways to encapsulate IP packets: Direct IP, Ethernet MAC Bridging and PPP. Support of Direct IP is mandatory for INA and NIU and the other two methods are optional. In the IB downstream channel the IP datagrams are carried using DVB multiprotocol encapsulation. In the OOB downstream and upstream channels the IP datagrams are carried in ATM AAL-5.

13.4.1.5 DOCSIS® - the Alternative Cable Return Channel

The considerable interest of the market in providing interactive services across cable TV networks and the large potential they offer in terms of infrastructure and available transmission capacity led to the parallel development of alternative solutions for the interaction channel for cable. The Data-Over-Cable Service Interface Specification (DOCSIS®) developed particularly in the United States has reached a significant market penetration. With some adaptations to fit into the European system of cable networks it also became an ETSI standard [ES 488].

DOCSIS and its European variant known as Euro-DOCSIS offer a transparent bi-directional transmission platform for IP based data traffic across the cable television network. The system does not inter-operate with the DVB Return Channel for Cable, i.e. a user terminal compliant to DOCSIS can not communicate with an interactive network adapter running the DVB-RCC MAC protocol and vice versa. The reference architecture for the DOCSIS system is depicted in figure 13.8.

The transmission path extends from the cable modem termination system (CMTS) equivalent to the interactive network adapter in the DVB reference model as shown in figure 13.2 to the cable modem (CM) in the user's premises incorporating a specific type of network interface unit. The downstream from the CMTS to the CM consists of a continuous flow of MPEG-2 packets.

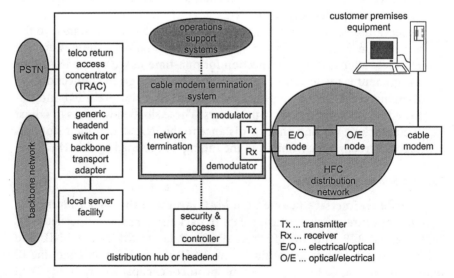

Fig. 13.8. Reference architecture of the DOCSIS system

The PID in the header identifies the payload as DOCSIS (see also chapter 5 for details on multiplexing an MPEG-2 transport stream). Modulation, error correction and physical channel parameters accord to the ITU-T Recommendation J.83 Annex B [ITU J.83B] with a channel width of 6 MHz. The Euro-DOCSIS variant adopts the 8 MHz wide modulated QAM channel as defined in [EN 429] and described in chapter 10. The downstream channel has a gross transmission rate between 30.3 and 41.4 Mbit/s with 64-QAM and between 42.9 and 55.2 Mbit/s with 256-QAM depending on the type of modulation, the symbol rate and the channel width. The upstream uses a combination of frequency-division multiple access (FDMA) and time-division multiple access (TDMA). In the upstream frequency range of 5 to 42 MHz (DOCSIS) or 5 to 65 MHz (Euro-DOCSIS) several channels are allocated, each of which is either 200, 400, 800, 1600 or 3200 kHz wide. Within one frequency channel bursts of minislots with variable length are modulated with QPSK or 16-QAM. Most of the transmission parameters such as modulation scheme, error correction parameters and burst length are configurable burst by burst through the CMTS. This results in a considerable flexibility in available data rates. With QPSK they range from 320 kbit/s to 5.12 Mbit/s and with 16-QAM a data rate of up to 10.24 Mbit/s is possible.

Similar to the DVB solution the DOCSIS MAC protocol is responsible for the dynamic configuration of the communication system and the allocation of upstream resources. Other tasks include functions such as timing and synchronisation, security of the data link and support of Quality of Service. The upstream is divided into minislots, which are time-units with variable length. By transmitting MAC management messages in the downstream the CMTS assigns each interval of minislots to a particular purpose and/or to a particular cable modem. Six types of usage for an interval of minislots are defined:

- **Initial Ranging:** Cable modems send their first message in this interval to establish the initial contact to the CMTS. All new cable modems are allowed to use that transmission opportunity concurrently. It is not expected that the messages are sent synchronously, and the transmission parameters are chosen to be very robust.
- **Periodic Ranging:** Intervals of this usage type are assigned to individual cable modems to fine tune synchronisation and power levels.
- **Request:** All modems requiring transmission opportunities can request them in minislots allocated to this type of usage. Requests are subject to collisions.
- **Request/data:** Modems can send requests for the allocation of bandwidth and data in this interval. All transmissions may suffer from collisions.

- **Short data:** An interval of minislots assigned to this usage type is reserved for a single cable modem and is intended for the transmission of a small amount of payload data.
- **Long data:** This interval has the same purpose as intervals of type 'short data'. However, certain transmission parameters are optimised for a longer duration of the burst.

The basic data transport service provided by the MAC protocol is 'best effort'. Simple mechanisms allow the assignment of some fundamental classes of service to particular cable modems and, thus, enhance their performance. In DOCSIS 1.1 sophisticated upstream scheduling services were included that enable dynamic guarantees of Quality of Service for individual flows of packets. Packet fragmentation and payload header suppression complement these measures to increase the efficiency of the upstream bandwidth utilisation. For further details the reader is referred to [ES 488].

13.4.2 Interaction Channel for LMDS

LMDS is the acronym for Local Multipoint Distribution System and describes a wireless technology, which is able to transmit a large amount of data at very high frequencies using microwave radio. LMDS solutions operate in frequency bands between 10 and 66 GHz. The usable spectrum is subject to licensing and must be allocated by the local regulatory bodies.

The term 'local' in the name of the system denotes the fact that propagation characteristics of signals in this frequency range limit the potential coverage area to a single cell site. Recent systems have bridged up to 10 km at lower frequencies, while at higher frequencies the distance is smaller. LMDS is a 'multipoint' distribution system, which indicates that signals are transmitted in a point-to-multipoint or broadcast method; the wireless return path, from an individual subscriber to the base station, is a point-to-point transmission. A microwave transceiver is for example installed onto a building at the subscriber site and another one at the LMDS base station. LMDS is used to distribute data, which may consist of simultaneous voice, data, Internet, and video traffic. The great advantage in comparison to the cable TV network is the simple installation of wireless links.

DVB defined an LMDS interaction system [EN 199], which is based on the specification of the return channel for cable [ES 800]. The main difference between cable and LMDS is the usage of different frequency bands. The intermediate frequencies could for compatibility reasons even be the same as on cable. These frequencies are then modulated and transformed to the higher frequency band. Alternatively, intermediate frequencies of up to 2 GHz are defined.

13.4.3 Interaction Channel for Satellite Distribution Systems

The specification [EN 790] documents the baseline for the provision of the interaction channel for geostationary satellite interactive networks with fixed return channel satellite terminals (DVB Return Channel Satellite (DVB-RCS)). Currently, the scope of such services is in a business-to-business environment, but it is envisaged that with growth of the market the satellite interactive terminal will become viable as a consumer product in a domestic environment.

The system model for an interactive satellite network is shown in figure 13.9. It is an extension to the DVB reference model for interactive services, which is described in chapter 13.2. Three main functional blocks can be identified, the feeder station, the gateway station and the network control centre (NCC). The feeder station transmits the forward link signal, which is the standard DVB-S signal [EN 421], provided by a broadcast service provider. On this link the user data, and the control and timing signals, which are needed for the synchronisation between uplink and downlink, are multiplexed. The reference model shows a clear separation between the downlink transmission (SAT FW) and the return channel uplink transmission (SAT

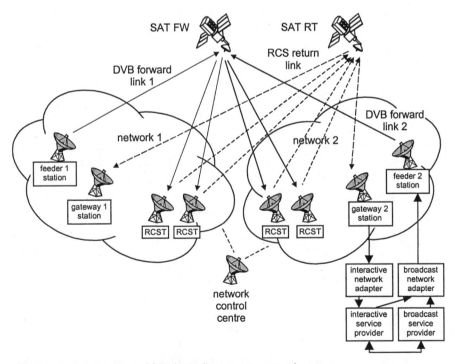

Fig. 13.9. System reference model for the satellite interactive network

RT). In reality, this separation does not need to exist. Instead, the same satellite may be used for uplink and downlink. Then for example, the downlink may use the 12 GHz band (Ku) whereas the uplink is installed using the 20 GHz band (Ka). The gateway station receives the return signals from the return channel satellite terminal (RCST) and provides access to other networks. The network control centre generates control and timing signals for the operation of the satellite interactive network to be transmitted by one or several feeder stations, and monitoring functions are also provided.

For the forward link the RCST is able to receive digital signals conforming to [EN 421]. The return channel uses specific physical layer and medium access control functions. The signal processing for each RCST transmitter contains burst formatting, energy dispersal, channel coding and burst modulation. An important feature is the synchronisation of the RCST to the satellite interactive network. The synchronisation scheme is based on a network clock reference (NCR) and on signalling messages in DVB/MPEG-2-TS private sections. The NCR is distributed with a specific PID within the MPEG-2 transport stream. The RCST reconstructs the reference clock from the received NCR and compensates the carrier frequency.

Data is transmitted in bursts on the upstream. Four types of bursts are specified. They create a traffic, acquisition, synchronisation or common signalling channel. The traffic bursts are used for carrying payload data from the RCST to the gateway. These bursts may be based on ATM cells or MPEG-2-TS. Synchronisation and acquisition bursts are required to accurately position RCST's burst transmissions during and after logon. Common signalling channel bursts are only used by an RCST to identify itself during logon. Each burst is followed by a guard time allowing the RCST to ignore transients and system timing errors. The length of the guard time depends on the network. No general value has been specified.

The serial data stream of a burst is randomised by a specific pseudo-random sequence. Forward error correction is applied for all burst types. A turbo coding and a concatenated coding scheme are described. Both schemes are mandatory for the RCST. Modulation is done by a Gray-coded QPSK with absolute mapping and the signals shall be square root raised cosine filtered.

A set of MAC messages has been defined to enable RCSTs to request transmission capacity and to support synchronisation, control and management. The transmission capacity needs to be managed by the NCC since the satellite link is a shared medium that is used concurrently by all RCSTs. The satellite access scheme is a multi-frequency time-division multiple access (MF-TDMA) technique. It allows a group of RCSTs to communicate with a gateway using a set of multiple carrier frequencies, each of which is divided into time slots. The allocation of transmission capacity to an individual active RCST contains a series of bursts each characterised by a carrier frequency, a bandwidth, a start time and a duration in terms of the number of subsequent time slots the burst may occupy in this particular frequency channel. There are two mechanisms to distribute the available transmission capacity among the active return channel terminals. In the Fixed-Slot MF-TDMA the bandwidth and duration of successive traffic bursts assigned to an RCST is fixed. Thus, the only variable from burst to burst is the frequency. Optionally, a Dynamic-Slot MF-TDMA scheme can be implemented using additional RCST flexibility to vary the bandwidth and duration of successive slots allocated to an RCST. The RCST may also change the rate of transmission and coding between successive bursts. A resulting advantage is the more efficient adaptation to the widely varying transmission requirements of multimedia services.

The return link of a satellite interactive network is organised in time slots as the smallest unit of resource allocation. A collection of time slots in different frequency channels with possibly varying characteristics forms a frame. Frames are, in turn, combined as superframes. The time slot allocation process supports five categories of transmission capacity. In general, transmission capacity is requested by the RCST based on a desired data rate, i.e. a certain volume of data to be transmitted in a defined amount of time, or based on a desired data volume without indication of time constraints. With continuous rate assignment, a number of time slots is allocated to a particular RCST in each superframe without individual requests. The size of the assignment is sufficient to accommodate a constant data rate that has been previously negotiated between the RCST and the NCC. Rate based dynamic capacity is transmission capacity which is requested dynamically by the RCST indicating the desired data rate to be transmitted. Similarly resulting from individual requests by the RCST is the assignment of volume based dynamic capacity. In this case, however, the RCST indicates its required transmission capacity as a cumulative or absolute data volume. A free capacity assignment is performed to allocate transmission capacity to RCSTs that would otherwise remain unused. Such a capacity assignment starts automatically and does not involve any exchange of signalling messages.

Control and management purposes are addressed in the MAC protocol to allow an RCST to log on to the network. This involves identification, power control and a logon procedure. An extensive set of messages and descriptors is defined to support this.

13.4.4 Interaction Channel for Satellite Master Antenna TV

The Satellite Master Antenna TV (SMATV) system is defined as a system intended to distribute television, sound and multimedia signals to users located in one or more adjacent buildings by sharing the reception resources. The signals received by the satellite antenna and possibly combined with terrestrial signals are further distributed through a coaxial network. As a result, the network connecting a user to the service provider consists of a satellite section most likely based on DVB-S and a coaxial section that may among other alternatives be based on DVB-C. The primary consideration for the specification of an SMATV system is the transparency of the interface between the satellite and the cable section for the transmitted signals. In that way, the SMATV headend would remain simple and cost effective as required for the targeted consumer electronics market segment.

To enable interactive services across the SMATV infrastructure, [TR 201] defines an interaction channel system through the seamless concatenation of the return channel systems in the satellite and the coaxial section. The recommended solutions to enable the satellite and the coaxial network for bi-directional data communication are DVB-RCS and DVB-RCC, respectively, as described in the previous sections. However, the interface between the two sections allows inter-operability irrespective of which combination of interactive solutions is implemented. Figure 13.10 shows an adaptation of the general reference architecture for DVB interactive services as described in section 13.2 as it would be used for an interaction channel implemented with an SMATV infrastructure. The forward interaction path may or may not be embedded in the broadcasting channel of the satellite and cable distribution systems. Depending on the SMATV system approach the signals are trans-modulated from the satellite QPSK modulation scheme to the QAM modulation scheme used in DVB-C (system A), or the satellite signal is directly transmitted across the cable network after a simple frequency conversion (system B). Traffic in the reverse interaction path of the coaxial section is collected at the SMATV head-end by the grouping terminal (GT). The satellite interactive terminal (SIT) enables forwarding of the signals to the service provider across the satellite section.

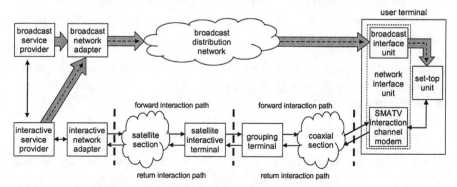

Fig. 13.10. System reference model with an interaction channel for SMATV systems

It is expected that the SMATV coaxial distribution network is smaller than general cable TV systems in orders of magnitude, both in size and in the number of users. Therefore, some significant simplifications can be applied to DVB-RCC. Those simplifications can reduce the cost of the SMATV headend, while the possibility to use off-the-shelf DVB-RCC set-top boxes and cable modems as user terminals can be retained. The simplifications are:

- Use of the out-of-band option defined in DVB-RCC with a unique carrier in each direction, thus, reducing the complexity of the radio frequency interface of the headend.
- Restriction of the upstream to a single carrier shared among all users.
- Reduction of the frequency range for the reverse interaction path (e.g. 15-35 MHz).
- Due to the short distances and the rather constant conditions in the small coaxial network, omission of power and time ranging, which can lead to a vastly simplified MAC protocol.
- Use of a fixed slot allocation to each terminal further simplifying the MAC protocol.

The crucial factor for seamless integration of satellite and coaxial return channel systems is an interface that provides interoperability between the sub-systems and transparency for the transmitted signals. A very low-cost, serial interface is recommended as a generic solution. The definition is based on RS-232 with the grouping terminal being designated as a data terminal equipment and the SIT as a data control equipment. The data rate is configurable up to 119 200 kbit/s. Alternative types of interfaces including Ethernet or X.25 based solutions have been proposed as well.

13.4.5 Interaction Channel for Digital Terrestrial Television

Among the broadcast systems that are defined in the DVB family of specifications the terrestrial transmission system was the last to be equipped with a return channel. The emergence of commercial requirements for bi-directional data communication across terrestrial networks led to the development of specification [EN 958] and, thus, to the most current member of the set of DVB return channel systems [FARIA].

The DVB Return Channel Terrestrial (DVB-RCT) system is able to provide interactive services using the existing infrastructure for digital terrestrial TV. It is based on an interaction channel that connects return channel terrestrial terminals (RCTT) to an interactive network adapter (INA). The forward interaction path is always embedded in the broadcast channel. As a consequence, the terrestrial interactive network consists of two unidirectional physical layers with the downstream direction comprising the broadcast as well as the forward interaction paths and the upstream direction providing the return interaction path. The downstream makes use of the terrestrial digital broadcast system DVB-T as specified in [EN 744] and described in chapter 11 of this book. It is based on an MPEG-2 transport stream. A specific PID was assigned to indicate signalling messages for interactive services. Application data and signalling messages in the reverse interaction path are encapsulated into ATM cells that, in turn, are mapped onto physical bursts of a VHF/UHF transmission system. A typical DVB-RCT system is depicted in figure 13.11.

The downstream transmission from the base station (INA) provides synchronisation and management information to all RCTTs. With that information RCTTs are able to access the upstream channel synchronously and transmit information to the base station. A single antenna is sufficient for receiving the broadcast and the forward interaction channels and for transmitting in the reverse interaction channel. To allow access by multiple users, the VHF/UHF frequency band used for reverse path transmission is partitioned into sub-channels in the frequency domain and into slots in the time domain. The frequency synchronisation is derived from global DVB-T timing signals in the broadcast channel, while time synchronisation results from the transmission of MAC management packets in the forward interaction path.

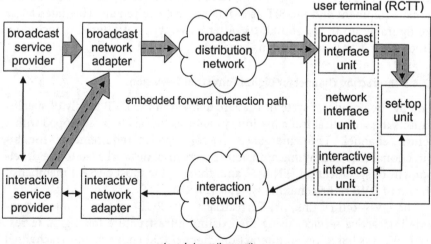

Fig. 13.11. System reference model with an interaction channel for digital terrestrial television

13.4.5.1 Physical Layer and Framing

The physical transmission channel for the return interaction path of the DVB-RCT system is formed by a dedicated radio frequency channel in the VHF/UHF band that is organised to allow concurrent access from many individual RCTTs. For that purpose, the available transmission channel is split in time and frequency to form a grid of time-frequency slots with each slot usable by an individual RCTT. The partition method was inspired by the DVB-T system and is based on OFDM. A slot in time contains an OFDM symbol which is made up of 1024 (1K mode) or 2048 (2K mode) individually modulated sub-carriers. DVB-RCT defines three targeted inter-carrier distances of approximately 1, 2 and 4 kHz. The inter-carrier distance determines the guard space between individual sub-carriers in the frequency domain and governs the robustness of the transmission system with regard to possible misalignment of a RCTT. Each value implies a maximum size of the area that can be served by a single base station as well as the resistance to the Doppler shift resulting from a RCTT in motion. The bandwidth of the DVB-RCT channel is a function of the number of sub-carriers and the carrier spacing. If the DVB-RCT system is used in conjunction with an 8 MHz DVB-T system, it is in the range of approximately 900 kHz for 1K mode and 1 kHz inter-carrier spacing to 7.6 MHz in 2K mode with 4 kHz sub-carrier spacing. To provide immunity against inter-carrier and inter-symbol interference as well as protection against other disturbing effects, each sub-carrier is either Nyquist or rectangularly shaped in the time domain. The number of sub-carriers and a specific inter-carrier distance define one of a total of six possible transmission modes. Transmission mode and sub-carrier shaping are to be exclusively chosen for an individual radio transmission channel in a particular time slot.

The time-frequency slots provided by the DVB-RCT radio frequency channel are organised in transmission frames defining a repetitive structure in which each slot is assigned a specific purpose. In addition to data symbols carrying the payload sub-carriers are dedicated to null symbols, ranging symbols and pilot symbols. These dedicated symbols provide the means for synchronisation and transmission management. There are two types of transmission frames (TF). TF1 contains a single null symbol, a series of 6, 12, 24 or 48 ranging symbols and a number of data symbols. It is used to carry one burst structure of type 1 (BS1) or four consecutive burst structures of type 2 (BS2). TF2 is made up of a set of general purpose OFDM symbols, which can contain either data or ranging symbols. It is suitable to carry eight burst structures of type 3 (BS3) or one BS2 stuffed with 4 null symbols to equal the duration of eight BS3. Figure 13.12 shows an example of the allocation of time-frequency slots in a TF1 in 1K mode.

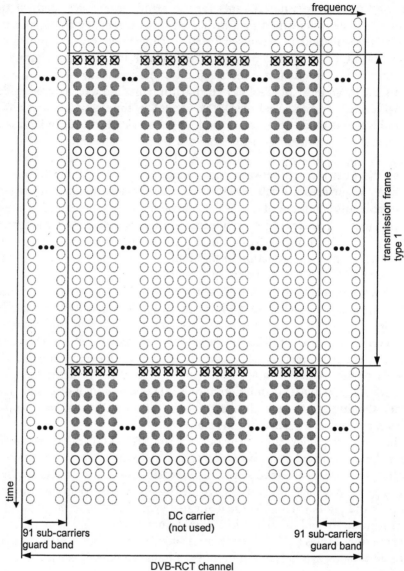

Fig. 13.12. Structure of a transmission frame type 1 (TF1) in 1K mode

An important feature of the DVB-RCT system is synchronisation. Constraints are imposed on the RCTTs to minimise interference between users. A two-step synchronisation scheme is employed comprising a coarse (initial) synchronisation based on timing information contained in the downstream and a subsequent fine synchronisation based on the ranging procedure defined by the MAC protocol.

Before the data can be transmitted the physical DVB-RCT signal has to be constructed by the RCTT. The following functions are involved:

- a variable data randomisation procedure,
- an error correction encoding using a Turbo code or the concatenation of a Reed-Solomon and a punctured convolutional code (under the control of the base station, a given RCTT can produce successive bursts with different encoding rates; this variable encoding capability provides flexible adaptation to the individual transmission channel conditions for the RCTT),
- bit interleaving,
- formatting, where the data bursts are organised into one of three predefined burst structures and where the pilot symbols are inserted,
- a QPSK, 16-QAM or 64-QAM modulation with a subsequent mapping of the modulated symbols to the allocated sub-carriers,
- a Nyquist or rectangular shaping in time.

The basic transmission slot allocated to a RCTT is called a burst. Independent from the encoding scheme and pulse shaping a burst carries 144 data symbols. The three burst structures defined in DVB-RCT define the allocation of modulated data and pilot symbols to time-frequency slots in the transmission frames. They offer various combinations of time and frequency diversity, thereby providing various degrees of robustness, burst duration and a wide range of bit-rate capacity of the system. The set of sub-carriers occupied by a single burst is called a sub-channel. Sub-channels can be allocated to an individual RCTT by the MAC process.

- Burst structure 1 (BS1) uses a unique sub-carrier in each OFDM symbol to carry the total burst over time. An optional frequency hopping law may be applied within the duration of the burst.
- Burst structure 2 (BS2) simultaneously occupies four sub-carriers in a single OFDM symbol each carrying a quarter of the total burst over time.
- Burst structure 3 (BS3) carries 1/29 of the total data burst in one OFDM symbol using 29 sub-carriers.

The definition of the burst structures includes a pilot allocation scheme to enable coherent detection in the base station. The pilot insertion ratio is approximately 1:6, which means that approximately one carrier in six is assigned a pilot symbol and not a data symbol.

With the three burst structures and the two types of transmission frames a mapping scheme is required to define the allocation of one or more bursts to a particular transmission frame. In DVB-RCT three medium access schemes (MAS) are defined.

- MAS1 is used exclusively to map BS1 onto TF1. When rectangular shaping is used, there are 144 data symbols and 36 pilot symbols in a burst. This total of 180 symbols including a null symbol and 6, 12, 24 or 48 ranging symbols is mapped onto a single sub-carrier in 187, 193, 205 or 229 subsequent OFDM symbols.
- MAS2 is used exclusively to map BS2 onto TF1.
- MAS3 is used to map BS3 and, optionally, BS2 onto TF2.

With the various options for transmitting data across the DVB-RCT return interaction path a wide range of transmission capacity from about 500 bit/s up to about 15 kbit/s is available. The bit rate depends on the number of sub-carriers in use, the order of the modulation scheme and the code rate. The length of the guard interval in rectangular shaping and the roll-off factor in Nyquist shaping also affect the bit rate.

13.4.5.2 MAC Layer Protocol

Since all RCTTs use the same transmission media, a MAC protocol is required to control access to the network and to ensure fair distribution of resources according to individual demand. The MAC protocol of DVB-RCT is based on DVB-RCC [ES 800] with specific modifications to suit the particular characteristics of the DVB-RCT physical layer. The MAC layer of DVB-RCC is described in detail in section 13.4.1. Thus, this section is restricted to give an overview of modifications required for DVB-RCT.

As described in the previous section the upstream physical layer divides the radio frequency channel into slots in the time domain. Each time slot is then split in the frequency domain into several sub-carriers. The group of sub-carriers that is used to transport a single burst according to the definition of the burst structure (BS1, BS2, and BS3) is called a sub-channel. The MAC layer controls the assignment of sub-channels and time slots to the individual RCTT. The central control instance is located in the base station (INA). Similar to the mechanism in DVB-RCC RCTTs are able to request resources at the INA and have them granted. The signalling takes place using a dedicated bi-directional connection that is initially set up when a RCTT becomes active. All connections of a particular RCTT are maintained on the same pair of upstream and downstream frequencies. For the exchange of MAC messages and data contention, reservation, fixed-rate and ranging access modes are available. They are defined in [ES 800] and their functions are described in section 13.4.1. Sign-on and ranging procedures are based on concepts developed for DVB-RCC as well.

14 The Multimedia Home Platform (MHP)

The previous chapters of this book primarily dealt with various DVB specifications describing source coding and transmission. In the time of converging media the DVB Project determined that the specification of a transport platform alone was not sufficient for the next generation of services and interoperable user terminals. Based on the experience in the Internet world, the demand grew for interactive services offered to TV users. With the Internet various platforms and user terminals render services differently and provide varying functionality although all main functions are based on widely accepted and deployed standards – one example would be browsers. Such behaviour is not acceptable for TV based services where viewers expect a consistent look, feel and behaviour of applications. The word "application" stands for something, which would be called a software "program" in the PC world. Since the word programme is used for TV content in the world of broadcasting a new term needed to be defined: application. Therefore, the DVB consortium started the work with the goal to specify a technical solution for the user terminal enabling the reception and presentation of applications in an environment that is independent of specific equipment vendors, application authors and broadcast service providers.

14.1 The Role of Software Platforms in the Receiver

To enable interactive services at all, it is necessary to define a common execution environment (middleware) for the applications that are received by the user terminals via a broadcast channel or via an interaction channel. The term "Middleware" was introduced to denote a new software layer located between the classical operating system and the application – see figure 14.1. A so-called API (Application Programming Interface) offers a set of functions that is available to each application. The interfaces hide all underlying components like hardware or the operating system from the application and offer functionality only via standardised method calls. However, a middleware specification needs to go far beyond the definition of an API. General aspects such as the presentation of colours, fonts, graphics and pictures as well as mechanisms for downloading, synchronisation, application life cycling and resource management need to be specified.

The advantage of such a middleware specification is the interoperability of various applications and all compliant user terminals. A service developer using an API function to create an application can be sure that this application will be executed in the same way on every user terminal implementing this API.

An API implementation consists roughly of three layers (figure 14.1).

Hardware layer: Besides the set of components that are required to enable the fundamental functionality of a set-top box, like an MPEG-decoder and RF front-end, additional components like CPU, hard disc and memory are necessary to implement an API. Modern set-top boxes and integrated digital television sets (iDTV) usually contain those components and are, thus, ready in principal to provide interactive TV services.

Operating system layer: As in personal computers, this software layer makes hardware functions available to applications. Software programs in the PC world are usually optimised for a special operating system and are, at least, aware of the underlying hardware.

Middleware: The middleware decouples services and executing hardware/ system software. This means that an application designed for a particular API will run in the same way on every implementation of this API. The software layer is implemented specifically for the operating system running on the receiver. It presents a consistent interface (API) to the applications that is completely independent from hardware and operating system. The middleware controls the incoming and outgoing events as well as life cycling and resource management.

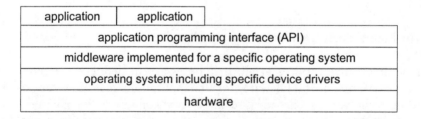

Fig. 14.1. General structure of a middleware software stack

14.2 Some Non-MHP Solutions

The introduction of digital transmission standards for TV services in the middle of the 1990s enabled the broadcasters to add interactive services to video and audio content. At the beginning, electronic program guides, or so-called EPGs, were the main services added to the classic (video, audio and possibly teletext) broadcast stream. In most cases, an EPG evaluates the data contained in the service information (SI) tables of the DVB/MPEG 2 transport streams. Information related to the TV service like bouquet and channel name as well as details about the current content can be displayed.

To enable the TV viewer to use such EPGs as well as other interactive applications, appropriate hardware and software is required in the receiver. The first interactive TV platforms using DVB transmission standards were all offered in vertical market operations, where one institution had control of the whole chain, including the set-top boxes, the software and the conditional access (CA) systems. A few years later DVB members became more and more aware of the potential benefits of a universal API offered under the same terms that pertain to other DVB specifications. But what was later becoming available under the DVB name Multimedia Home Platform (MHP) took several years to develop. Only in 2001 were the first MHP applications launched (in Finland). Many markets, for example the U.K., could not wait that long and therefore decided to introduce other proprietary or publicly available systems (e.g. MHEG-5). Some of these systems will be described in the following sections.

This listing (in alphabetic order) does not claim to be complete, but it is intended to show the history and the progress in the middleware development.

14.2.1 ATVEF

The Advanced Television Enhancement Forum (ATVEF) is a cross-industry group formed to specify a standard for delivering interactive television applications that can be authored one time using a variety of tools and deployed to a variety of television sets, set-top boxes and PC-based receivers. The Enhanced Content Specification defines the fundamentals necessary to enable a creation of hypertext mark-up language (HTML)-enhanced television content so that it can be reliably broadcasted across different networks to different compliant receivers.

The ATVEF specification basically describes HTML 4.0 and delivers TV programming over both analogue and digital video systems. The ATVEF specification consists of three parts:

1. Content specifications to establish minimum requirements for receivers.
2. Delivery specifications for transport of enhanced TV content.
3. A set of specific bindings.

Up to now just a few broadcasters have worked with experimental ATVEF prototypes (applications and set-top boxes). Although for the time being a migration to MHP markets has not been envisioned, creating some kind of an HTML browser as an MHP plug-in to execute ATVEF applications on MHP boxes would be a reasonable option.

14.2.2 Betanova

Betanova 1.xx
This middleware was developed by the German company Beta Research. It opened the software environment of the so-called d-box to C/C++ programmers. With Betanova 1.xx, it is possible to develop interactive applications.

Betanova 2.xx
Betanova 2.xx is a Java-compliant software platform that simplifies the adaptation of existing applications to the d-box. In addition, the use of Java simplifies the development of applications for the d-box. With the implementation of Java, Betanova took its first step towards a possible migration path to MHP. At the time of writing it is unclear whether or not Betanova 2.xx will be introduced by the German Pay-TV broadcaster Premiere.

14.2.3 Liberate

Liberate Technologies is a provider of platforms for delivering enhanced content and services to television viewers. Interactive digital services include enhanced TV broadcasts, video-on-demand, personalised content, TV chat, instant messaging, digital video recording, and others.
The Java based Liberate software engine is called TV Navigator Standard. TV Navigator is customisable to reflect the individual identity of the network operator, the capabilities of the set-top box and the available interactive TV services.

14.2.4 Mediahighway

Established in 1993, Canal+ R & D began to develop a digital interactive pay-TV system called Mediahighway, which first debuted in 1995.
In its current version, Mediahighway supports Java as a programming language. The Mediahighway Virtual Machine is hardware independent and

implements the Mediahighway API in compliance with the specifications provided by Canal+ Technologies.

Since the Mediahighway Virtual Machine has the capability of incorporating various interpreters to read applications created in different languages, such as Java, HTML, MHEG-5, and so on, it can make use of a wide variety of existing applications.

The latest release of Mediahighway, marketed under the brand name Mediahighway+ is said to comply with the open specifications for a Multimedia Home Platform (MHP). Based on the MHEG-5 and DAVIC specifications the platform allows the consumer to connect digital devices such as DVD, DVHS and PC to the set-top box.

14.2.5 MHEG

MHEG-5

MHEG [ISO 13522-5] is an acronym for the Multimedia and Hypermedia information coding Experts Group. This group developed, within the International Standardisation Organisation (ISO), several standards, which deal with the coded representation of multimedia/hypermedia information.

MHEG defined the term multimedia as a representation of information by several media types, such as audio, video, text, and graphics. MHEG uses specialised links that enhance multimedia services by allowing digital TV subscribers to navigate through a screen of objects. Besides defining hypermedia and multimedia objects, the MHEG standard is also concerned with the interchange of these objects between storage devices, telecommunication and broadcast networks.

The set of MHEG specifications is composed of a total of seven parts.

MHEG-5 was designed to be supported by systems with minimal resources, which makes it a suitable middleware product for low-end digital set-top boxes. An MHEG-5 application, which is basically a set of multimedia and hypermedia objects that reside on a computer at the TV operators' head-end, is converted into a bit stream format and is broadcasted over a broadband network. On the receiving end, a digital set-top box with a software component called a MHEG-5 engine extracts the multimedia/hypermedia from the incoming digital stream, interprets the data, and displays it on the viewers television screen. In addition to handling incoming data, the MHEG-5 engine is also responsible for synchronisation and support of local interactivity with the subscriber.

During the definition phase of the MHP specification, the MHEG consortium provided considerable input. Actually, DAVIC (Digital Audio Visual Council) proposed that MHEG-5 be chosen as the presentation engine with an optional Java virtual machine for advanced applications. After a serious discussion the DVB Steering Board made the decision to develop a Java-only solution, but nevertheless MHEG-5 and MHP share many common technologies, such as:

- DSMCC
- "Plug-in" concept
- Font & graphics formats

MHEG-5 was published as a European standard by ETSI (European Telecommunication Standard Institute) in 1997 [ETS 777].

Currently MHEG-5 content is provided in the UK via DVB-T by several British broadcasters – the most prominent among them is the BBC (British Broadcasting Corporation). Applications like EPGs, tickers, Superteletext and interactive games are on air.

Using the plug-in mechanism of the MHP, an MHEG-5 browser can potentially be integrated into an MHP software stack, which enables a set-top box to execute MHEG-5 services as well as MHP services.

MHEG-6

MHEG-6 [ISO 13522-6] extends the functionality of MHEG-5 by providing the necessary components of a Java environment to allow complex processing of multimedia/hypermedia information in a relatively efficient manner. The users of MHEG-5 requested such a capability, especially for sophisticated set-top boxes, near video-on-demand offerings, distributed network games, and other highly interactive applications.

MHEG-7

MHEG-7 [ISO 13522-7] defines a test suite that can be used to test an MHEG-5 engine's conformance to a specific application domain. It also defines a format for test cases that can be used to extend the test suite either for more detailed testing or for extensions defined by the application domain.

MHEG-8 (MHEG-XML)

MHEG-8 is an ISO/IEC Standard [ISO 13522-8], which provides XML encoding for MHEG-5.

14.2.6 OpenTV

The American Company OpenTV Inc. is a leading supplier of middleware solutions for broadcasters in many countries.

The core middleware architecture developed by OpenTV is said to be hardware-independent, modular, and extensible. The execution layer provides compatibility with applications authored in C, HTML and Java code. The C-code execution engine is provided as part of the basic middleware package, and allows for execution of C-code applications. HTML and Java execution engines are offered as options.

14.2.7 Migration Concepts

Broadcasters and network operators who started broadcasting interactive content before the MHP became available may want to migrate to the MHP in the future. A migration concept has to solve the problems of different transport protocols, content formats, application code and receiver capabilities. The migration concepts can be coarsely distinguished into four options, which all have advantages as well as disadvantages.

Simulcasting means that the service provider broadcasts a legacy as well as an MHP-based service. This allows for a smooth transition with a variety of receivers in the market but it is necessary to design and broadcast the same application for the two systems in parallel.

Shared broadcast using MHP content formats is similar to the above scenario with the advantage that content (pictures, sounds, etc.) is shared. The legacy system has to be able to understand the MHP content formats and has to have access to the transport protocols.

Plug-in for the legacy system operating in a MHP receiver:
During the development of the MHP, "plug-ins" for legacy systems were often discussed. The concept of plug-ins is described in 14.4.1. A MHP implementation is enhanced by software running on top of the MHP API or directly on top of the operating system which is able to interpret legacy applications. This unavoidably adds to the complexity of the receiver. The idea promises that legacy content can be broadcast, which is used even by the population of MHP receivers. Eventually, though, migration from the legacy system to the MHP will have to be managed.

A *Clean-cut* is a hard switch from the legacy system to the MHP system at a determined date. After this deadline all legacy API boxes will be unable to execute applications because there will no longer be any corresponding applications on air.

In some cases it may be possible to update the software in the legacy receivers "over night" using the system software update (SSU) mechanisms described in chapter 12.6.

14.3 MHP 1.0

The MHP specification version 1.0 was finalised by the DVB Project in February 2000 and was published by ETSI in July 2000. A thorough corrigenda process removing many errors and ambiguities resulted in version MHP 1.0.3., which was finalised in June 2003 [ES 812].

To enable different terminal classes, the MHP specification defines three profiles. MHP 1.0 includes the enhanced television and the interactive television profiles. They allow for local interactivity as well as the use of a return or interaction channel. The interaction channel opens the door for personalised services on the TV screen.

The most enhanced third MHP profile – Internet access profile – is covered by the MHP version 1.1 which is described in chapter 14.4.

The MHP serves as a neatly defined Java-based middleware layer that hides the specifics of the hardware and the operating system from the application. The MHP specification describes an API. The functions of this API are so generally designed that they can be implemented on a large set of different hardware devices and operating systems. This abstraction has the consequence that an MHP-compliant application runs on all kinds of hardware devices, as long as these devices fully implement the MHP specification.

14.3.1 Basic Architecture

The basic MHP architecture is shown in figure 14.2. It is very similar to the general structure of a middleware stack as shown in figure 14.1. The operating system layer resides on top of the hardware layer. One of the main tasks of the operating system is the integration of all hardware drivers (e.g. for the DVB front end). All implementation specific software components which are necessary to connect the lower layers, e.g. the supported transport protocols or file access with the MHP API, can be realised within the operating system. On top of the operating system, a Java Virtual Machine is responsible for the abstraction of the hardware and operating system layer. The MHP API layer is implemented on top of the Java Virtual Machine and includes for example the "core" Java APIs (Personal Java), Digital TV APIs

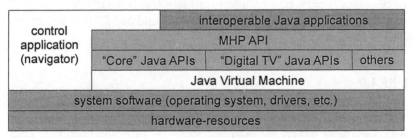

Fig. 14.2. Basic MHP architecture

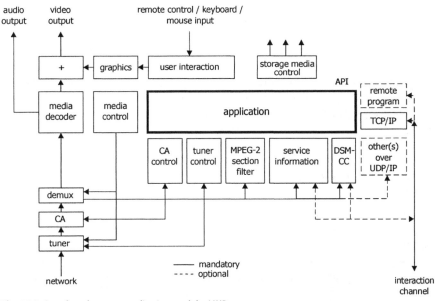

Fig. 14.3. Interfaces between applications and the MHP system

or user input/output APIs. The sum of all MHP Java functions will be referred to in the following as DVB-Java (DVB-J). The interoperable applications running on the DVB-J are called DVB-J applications. In contrast to DVB-J applications, also DVB-HTML applications are defined within MHP, which do not need to run on Java, but on an HTML-engine.

The control application (also referred to as navigator) serves as the basic interface for the user. Due to the fact that this is a pre-installed, vendor specific software it can work directly on the operating system. The main task of the control application is to enable the service selection and the monitoring and control of the services. The implementation of an additional control application as an interoperable Java application is possible.

In figure 14.3 the relations between an MHP application and the main functions implemented in the terminal are shown. These functions provide access to the digital broadcast network and the interaction channel, to the audio- and video-output and to the user input. The application can access all these functions as if the hardware executing the requested task was within one single physical device. This means that even if the hardware resources are physically spread over different devices, it is the task of the MHP implementation to logically bundle them for the application.

In the next sections some of these functions will be described in greater detail. That description includes the access of the MHP terminal to the broadcast and interaction channels, the way an application is transported into the terminal (using Service Information and the data carousel of Digital Storage Media- Command and Control (DSM-CC)) as well as the signalling of applications and the control of their life-cycles. Media control, graphics and user interaction are reflected in the section about content formats and graphics capabilities. Finally, the security architecture of the MHP is explained. Other elements like protocols or hardware modules, which are needed by an MHP implementation, are considered in other parts of this book.

14.3.2 Transport Protocols

The MHP supports a large number of different transport protocols in order to access the DVB broadcasting channel as well as the interaction channel. In MHP version 1.0 the only way to download an application together with its accompanying data into the terminal is through the broadcasting channel. The interaction channel is optional for the enhanced broadcasting profile and mandatory for all higher profiles. If available, it can be used by an application for all kinds of remote server interaction or Internet access.

14.3.2.1 Broadcast Channel Protocols

Figure 14.4 shows the protocol stack of the broadcast channel. The MPEG-2-transport stream is the transport medium for different kinds of digital data such as video, audio and applications. In order to distinguish between the different data types the first information that a terminal evaluates is the service information (SI). SI also describes the applications contained (e.g. application name, signalling of the application). MHP applications as well as the related resources are transmitted via an object carousel. This object carousel is an extension of the MPEG-2 DSM-CC user to user object carousel which supports nested directory structures like on a PC-file system. The content of the object carousel is cached within the receiver, and an application can be executed as soon as the relevant files are available.

An alternative protocol stack, mandatory only for the Internet access profile (described in MHP 1.1), builds on the MPEG-2 layer with DVB multiprotocol encapsulation which can be used for various kinds of protocols, but in this case it is used for the transport of IP packets. The user datagram protocol (UDP) is selected as the upper layer of the stack.

The interested reader is referred to chapter 12 for further details on DVB data broadcasting.

application				
application programming interface				
DVB object carousel		UDP	DVB SI	
DSM-CC user-to user object carousel		IP		
DSM-CC data carousel	DVB multiprotocol encapsulation			
MPEG-2 sections (DSM-CC sections)				
MPEG-2 transport stream				

broadcast channel

Fig. 14.4. Broadcast channel protocol stack

14.3.2.2 Interaction Channel Protocols

As shown in figure 14.5 the interaction channel protocol stack is mainly based on the Internet Protocol (IP) which may be used to transmit data across various kinds of network dependent protocols. The reason for choosing IP is that it is widely accepted and that its application is independent of the available return channel, as for example the cable return channel or the return channel via the telephone line. On top of IP, UDP (User Datagram Protocol) and TCP (Transmission Control Protocol) packets can be transmitted to an MHP terminal. As a further option the application can access the DSM-CC user-to-user carousel which is encapsulated using the Universal Networked Object-Remote Procedure Call

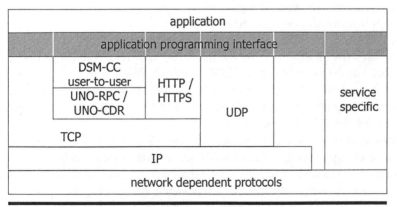

network connection

Fig. 14.5. Interaction channel protocol stack

(UNO-RPC) and the Universal Networked Object-Common Data Representation (UNO-CDR) protocols on top of the TCP stack. TCP has proved to be robust against packet loss in bad transmission channels and allows the easy routing of data between complex and changing networks. The hypertext transfer protocol (HTTP) and the hypertext transfer protocol over secure socket layer (HTTPS) allow the transport of content from standard Internet servers. In parallel to the IP stack, service specific protocols are possible. These allow applications, which may come from the network provider, to use other network-related protocols.

14.3.3 Application Model and Signalling

One of the core mechanisms of the MHP is the application model. When the user selects a TV channel in an MHP environment, he may be offered a bouquet of MHP applications, available together with video and audio programmes within the channel. Applications are announced and described using signalling mechanisms within a DVB-SI table called AIT (Application Information Table). Two different types of MHP applications are defined. DVB-Xlets are Java applications implementing the "Xlet" interface as defined within the "JAVAX.TV" package. DVB-HTML applications, on the other hand, are sets of XHTML pages.

Signalling is the remote control of the application's state by the broadcasting service provider. An application can be signalled as being of the "autostart" type, which means that the application will be started immediately as soon as the service containing the application is selected. An example for an automatically started application would be an EPG, which gives the user an overview of all available services as soon as a programme bouquet is entered.

The application can also be signalled as being "present", which means that the application will not be started without user interaction. To stop the execution of an application the signalling "destroy" or "kill" is possible. Using the "destroy" method allows the initiation of a clean shut down by the application, whereas using the "kill" method deletes the application immediately and completely. For the case of DVB-HTML applications, also the "prefetch" signalling is defined which informs the terminal to load but not display a DVB-HTML application.

Another kind of signalling ("service binding") is required to co-ordinate the execution of applications when the TV service being watched is changed. If the user leaves a TV programme to select a new one, two things may happen to the running applications, depending on the signalling in the originating TV service. In the default case (that means for all applications bound to one TV service), the applications will be stopped immediately by the MHP terminal. This behaviour is achieved by signalling the applications as "service bound". The applications will continue to run if they are signalled as "service unbound". However, if the applications are not indicated to be available in the new programme, they will be stopped, anyway.

The feature of unbundling applications and TV services is particularly helpful to TV operators that control multiple TV services in a bouquet. By using applications indicated as "service unbound" (maybe a bouquet-wide EPG application), execution of the applications is enabled on all channels without interruption by stopping and restarting, which otherwise may occur when new programmes in the bouquet are selected.

The external signalling by the broadcaster is reflected in the well-defined application states within the MHP terminal. A broadcaster can start, stop and destroy applications. The terminal needs a way to internally store the different application states. For a DVB-Xlet there are four states defined (loaded, paused, active and destroyed). In figure 14.6 the transitions between the states are depicted in the form of a state diagram. For a DVB-HTML application, there are five defined states (loading, paused, active, killed and destroyed) – the state transitions are quite similar to the Xlet transitions. In the following, the term application will always refer to DVB-Xlets.

The application manager of the MHP in the receiver is responsible for the controlling and monitoring of the application states. Following a user selection (e.g. in the Navigator), the application manager will load the application. In the next step the application manager calls the initXlet() function and the application enters the pause mode. Consequently the application will carry out its basic initialisation. At this point, it can already use some limited resources. In the next step, the application enters the active state if the application manager calls the method startXlet(). Now the application is running and can for example present its service to the user.

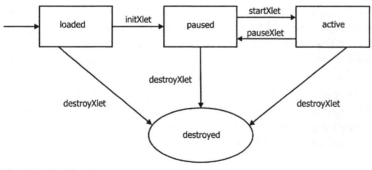

Fig. 14.6. Xlet life cycle

If one of the described state changes failed, the application enters the destroyed state. When the application has finished the presentation to the user, it can return to the pause state. In that state the service should try to minimise its resource consumption in order to raise the probability not to be destroyed if the terminal needs to provide resources to other services.

This may for instance happen, if the terminal is requested to start a second application. Even though it is possible to have multiple applications running at the same time, the number of simultaneous applications is always limited by the availability of resources. Basic rules of resource management employed by the resource manager are defined within the MHP. However, specific regulations are to a large extent left to the implementation. As a consequence, an MHP implementer has considerable freedom to decide in which complexity the resources are managed. Some elements of the MHP have their own resource management mechanisms, for other components a general resource management mechanism exists. The latter approach helps DVB-J applications to share resources among each other. For applications which run in the same service context all resource conflicts are solved with the help of the application priority field set in the AIT. The application with the highest priority first gains access to the resources.

14.3.4 Content Formats

It is no surprise that MHP supports all important DVB video and audio formats. In addition, various content formats have to be supported by the MHP implementations in order to establish true multimedia capabilities.

In general, applications may present pictures, movies, music and text to the user. Streaming formats (audio and video) are not yet supported and will follow in future MHP versions. In addition to running video based on MPEG-2, applications have the possibility to play short video "drips" (slideshow-like) which are sequences of MPEG-2 I- and P- frames.

All supported content formats are listed in table 14.1. They may be dependent on the MHP profile. "M" means mandatory and "-" means not required by the platform. The GIF-image format does not need to be supported due to the license reasons. Additional content formats which are not supported by MHP, like new formats developing in the Internet, can be enabled via the plug-in mechanism. For details refer to chapter 14.4.1.

To allow the application to gain access to content formats listed in table 14.1, the MHP includes a set of different APIs such as the Java Media Framework (control of the display of video and audio streams), the API for font downloading or the HAVi API. This is explained in the following section.

Table 14.1. Content formats

Content Type	Content Format	Enhanced Broadcast (M = mandatory)	Interactive Broadcast (M = mandatory)
Bitmap Pictures	PNG with restrictions	M	M
	PNG without restrictions	-	-
	GIF	-	-
	MPEG-2 I-Frames	M	M
	JPEG with restrictions	M	-
	JPEG without restrictions	-	M
Audio clips	Monomedia format for audio clips	M	M
Video drips	MPEG-2 Video "drips"	M	M
Text encoding	Monomedia format for text	M	M
Video	MPEG-2 Video [TR 154]	M	M
Audio	MPEG-1 Layer 2 Audio [TR 154]	M	M
Subtitles	DVB Subtitles and Teletext	M	M
Fonts	"Tiresias" as the resident font	M	M
	Downloadable fonts	M	M

14.3.5 Graphics Model

The MHP graphics model offers a large number of different tools to display video, graphics, buttons or text. The various elements of a displayed picture are located in three graphics layers which are, from back to front, a background plane, a video plane and a graphics plane. The graphics plane is used to visualise all interactive components or graphics; the background layer can be applied to present a still (background-)image and the video layer shows the running video signal. The layers can be combined with a set of different paint rules. The most important paint rule includes the possibility to draw parts of layers transparent or semi-transparent, which allows the presentation of applications together with the running video in the background. In figure 14.7 an example for the combination of the different layers is given. All three layers can be controlled by a DVB-J application using different function calls.

The MHP supports the so-called lightweight components of the Java AWT (Abstract Windowing Toolkit) graphics interface. The attribute "lightweight" indicates that in this API only abstract functions are defined, which do not allow the access to operating-system dependent paint functions. That means, elements painted with lightweight components look and behave the same way on all possible MHP implementations.

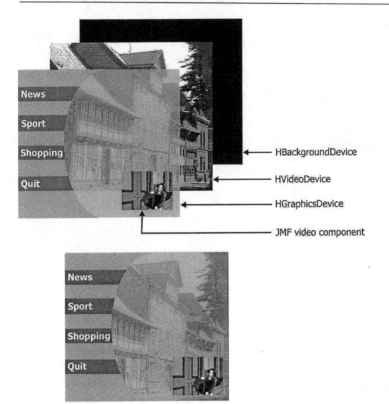

HBackgroundDevice

HVideoDevice

HGraphicsDevice

JMF video component

Fig. 14.7. MHP display stack / graphics reference model

As a substitute for the Java AWT heavyweight components, the graphics components of HAVi (Home Audio Video Interoperability) are supported. They allow to determine the "Look & Feel" of an application in a predictable way, so that the application appears in the same make-up on all displays. HAVi is also required since it allows the control of an application via a classical remote control avoiding the necessity of a mouse device in the TV environment. For the control of video and audio signals the Java Media Framework (JMF) is part of the specification.

The MHP supports a number of different screen modes. The minimal screen resolution is 720x576 (702 pixels, are nominally visible per row) supporting the display aspect rations of 4:3 and 16:9. Optionally MHP terminals may also support square pixel resolutions of 768x576 for 4:3, and 1024x576 for 16:9 displays. Because two different aspect ratios can be selected using the same minimum screen resolution, a small distortion of the application-user interface will be visible if the application itself does not take measures against it.

14.3.6 User Interface

The MHP is designed to work within a TV set, which in terms of its user interface is very different compared to a "normal" PC. In a PC environment the most important input devices are the mouse and the keyboard. These devices are not common in the living room environment of the MHP. Instead, it has to be possible to control applications by a remote control. A minimum set of input events is defined, which has to be present e.g. on a remote control in order to control the applications. The set of events consists of four coloured keys (which may also be labelled A to D), four direction buttons, number keys 0..9, a teletext key and an enter key. This allows a limited user interaction, which is sufficient for many types of MHP applications. Of course it is also possible (but not necessary) to add further input devices to an MHP terminal such as a minute keyboard hidden in the remote control.

14.3.7 Security Architecture

One of the most important features of a software platform, connected to the World Wide Web or to other open networks, is its security. With regard to the MHP, the security has three main aspects: the link security over different networks, the application security within the terminal and the certificate management to verify the owner of an application.

The security of the DVB broadcast link is already covered through DVB specifications (see chapter 8). The return channel security is built on standard technologies as known from the Internet. The basic protocol used here is the transport layer security (TLS) protocol. The minimal cipher algorithms that are supported are Rivest/Shamir/Adelmann (RSA), message digest (version 5) (MD5), secure hash algorithm (SHA-1), and the data encryption standard (DES).

The application security consists of two key elements, the security policies for applications and the authentication of applications. The security policy for applications refers mainly to DVB-J applications, as they have access to a variety of security relevant functions like access to private user data, local storage or the return channel. The basic idea of the MHP application security architecture is the so-called sandbox. The principle of the sandbox is to describe an execution environment which is safe in a way, that the executed application cannot access any resource (hardware or software) which is not released, and so cannot cause any harm to the MHP platform. An MHP application runs within a sandbox as long as it is unsigned. The application is unsigned, as long as it is not authorized by any kind of security key. From within the sandbox it cannot access any security relevant functions. However, the sandbox supports features sufficient for a number of typical MHP applications including local interactivity.

An unsigned MHP application may overcome the limitations of the sandbox by requesting additional rights.

A permission request file, which is an XML-file that comes along with the application, contains this request e.g. access to specific MHP interfaces. It is also possible to ask for access to files which were transmitted by other broadcasters. The rights of the application, which are requested by the broadcaster, can be furthermore limited by the end-user of the terminal. That may be necessary if the user has, for example, security concerns and wants to prevent the application from accessing a remote server.

After the application is loaded into the MHP system and after all security checks have been passed successfully the application may be started. The first step in starting an application is the activation of a class loader mechanism, which loads pieces of Java programs (class-files) into the memory enabling the execution. The class loader mechanism prevents communication between different applications, which has the consequence that two applications will never be able to access the same copy of any application-defined static variable. Nevertheless, an inter-application communication API which is based on Java RMI (Remote Method Invocation) is used to make communication between applications possible.

The class unloading during runtime is supported as well, which allows the withdrawal of application rights. That might become necessary if, for example, a security key expires or has been revoked.

The second element of application security is the authentication of applications, which is required to identify the origin of the application. Three types of security messages are used by the MHP security framework, cryptographic hash codes, signature files and certificate files. Applications are authenticated with the help of certificates, which clearly disclose the owner of the application to the receiver.

As shown in figure 14.8, the application and all the associated data are transmitted in a file tree. This file tree contains compiled JAVA classes (*.class), a permission request file (*.perm), signature files, certificates, hash files and application specific data files. For each directory, a hash file is calculated containing the hash codes of all files in that directory and of all hash files from subdirectories. The hash codes are computed only from the content of the files and are independent from all transport medium dependent data. A signature file is placed in the root tree of the directory

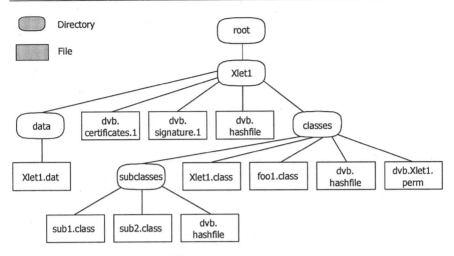

Fig. 14.8. Example of an authenticated file tree

structure or within the top directory of a sub-tree. It holds a reference to a certificate that contains the public key (the public part of an asymmetric security key) to decode the signature, the used hash algorithm and the value of the signature. The certificate file contains the public key that is used to decode a hash code contained in a signature and so enables a tree/sub tree to be verified. A trusted certification authority signs the certificate.

The certificate management is the third part of the MHP security architecture. It has to assure, that certificates can be removed from MHP terminals (revoked) or possibly replaced by newer editions. This will be required if the broadcaster's private key (the secret counterpart of the public key) is assumed to be compromised or if the certification authority no longer certifies the broadcaster. Two mechanisms exist for the revocation of two different types of certificates, root and non-root certificates (which are used for application authentication). Certificate revocation lists (CRLs) are compiled and published by every certification authority. The lists can be distributed either via the broadcast channel or (if present) via the return channel. To achieve a high degree of security against filter attacks (i.e. the filtering of CRLs with the goal of misusing revoked certificates), it is important that the terminal stores all received CRLs for a long period of time. The root certificate management is used to replace root certificates, if necessary. In every terminal, a set of root certificates is stored during production in the factory. The messages used for certificate management are called Root Certificate Management Messages (RCMMs). Multiple signatures belonging to the root certificates stored in the terminal authenticate

RCMMs. An MHP terminal will accept a RCMM message if and only if it has at least two signatures. The use of multiple signatures guarantees that the set of root certificates can be updated securely even if one of the root certificates has been compromised.

14.3.8 Minimum Platform Capabilities

The MHP runs on a large variety of different hardware platforms. Within the MHP specification itself, the only specified minimum platform capabilities refer to security parameters. Hardware resources like the size of the memory and the CPU performance are not covered by the specification. Due to the complexity of the implementation these parameters may vary over a wide range. With a growing number of implementations, the marketplace will assess the capabilities of commercially available terminals.

MHP terminals which received the MHP logo after successful passing of the MHP test suite during December 2002/January 2003 are said to include the following hardware resources:

	CPU	CPU type	Flash memory	RAM
Type 1	121 MHz	Special purpose	16 MB	40 MB
Type 2	166 MHz	32 Bit RISC	8 MB	32 MB
Type 3	160 MHz	32 Bit RISC	8 MB	32 MB

RISC: Reduced Instruction Set Computer

14.4 MHP 1.1

The MHP 1.1 defines the next generation of the Multimedia Home Platform enhancing it to a truly interactive digital platform. MHP 1.1 builds upon MHP 1.0 and adds certain functionality. Since both MHP 1.0 and MHP 1.1 are undergoing a thorough corrigenda process, new versions are being created. For example MHP 1.0.3 was published in Summer 2003. That version of the MHP 1.0 is the basis of MHP 1.1.1 [TS 812].

Of the three MHP profiles the most advanced is the Internet access profile, and is only described in MHP 1.1. This specification in addition includes DVB-HTML and the possibilities of using Internet driven technologies such as ECMAScript, CSS, DOM etc.

The following sections present the important enhancements of MHP 1.1 from MHP 1.0 [Sieburg].

14.4.1 Enhancements in MHP 1.1

Internet Access Profile
A first approach enabling access to the Internet from TV based services was already outlined in version 1.0 of the specification. However, MHP 1.1 includes its description in greater detail. Classical Internet services like web browsing or email exchange are based on a client-server architecture requiring specific software in the user terminal. The high dynamics in the market of Internet applications (e.g. a new browser release every few month) make it impossible to predict the future development, which is why no particular client software is specified for the MHP user terminal. It is left to the implementer to choose it. The Internet access profile defines only an API to access resident Internet client applications. Via this API it is possible, for instance, to start a browser with a URL, to send emails and to register news groups.

DVB-HTML
The most significant enhancement in MHP 1.1 compared to MHP version 1.0.x is DVB's definition of HTML. This feature is intended to be used in all three MHP profiles (enhanced, interactive, Internet access profile). It is an optional feature and, thus, it is up to the MHP device manufacture to implement it. One reason for making it optional is the amount of CPU performance and memory size required for decoding and presenting DVB-HTML which has a great impact on the price of an MHP box. One must remember that MHP 1.0 provides all the tools offering content on the TV screen and therefore, DVB-HTML offers much redundancy.

To enable Internet access and to avoid the well-known interoperability problems with HTML which are known from standard PC-based browsers, DVB created a completely defined mark-up language called DVB-HTML. It is based on XHTML (Extensible Hypertext Mark-up Language) which is in turn based on XML. The advantages of XHTML as compared to HTML 4.x (based on SGML - Standard Generalized Mark-up Language) as the standard language of the World Wide Web are the following:

- Syntax is 100% compatible to all important XML standard languages like SVG (Scalable Vector Graphics), WML (Wireless Mark-up Language), SMIL (Synchronised Multimedia Integration Language), etc.
- Support of embedded XML compatible content
- Support of standardised script languages / access to elements of the HTML document via DOM

To promote interoperability, both between different MHP vendors and with the Internet community, a lot of commonly used Internet standards are

Table 14.2. Content types accessible in DVB-HTML

MIME media type	common name
text/XML	XML
application/XML	
text/CSS	CSS
text/plain	monomedia format for text
text/dvb.utf8	
audio/mpeg	Monomedia format for audio clips or audio
image/jpeg	JPEG
image/png	PNG
image/gif	GIF
image/mpeg	MPEG-2 I-frames
video/dvb.mpeg.drip	MPEG-2 video drips
video/mpeg	MPEG-2 video
multipart/dvb.service	multipart DVB service
application/dvb.pfr	downloadable fonts
application/dvbj	initial class file of an Xlet
text/ecmascript	ECMAScript

supported. As an example, the table 14.2 describes a specific subset of the data formats defined for use in the World Wide Web, which was adopted by the MHP 1.1.

In cases where media types seem to be identical to these listed in table 14.1 a detailed analysis will show differences with respect to some restrictions in MHP 1.0 which no longer exist in MHP 1.1.

In the "classical" Internet world all HTML pages are stored individually on a web server. The only relation between them is established by the hyper links contained in one page and pointing to another one.

A DVB-HTML application in the context of MHP is a set of resources selected from the DVB-HTML family of content formats as defined in table 14.2. The interpretation of a DVB-HTML application is handled by a support application (commonly called a user agent in W3C (World Wide Web Consortium) terminology).

For the purposes of exposition, DVB defines an "actor" which is the runtime context associated with a DVB-HTML application maintained by the user agent, in which there is a one to one correspondence between actors and running applications. A user agent is an application that interprets a content format (in this case a DVB-HTML document) which could be implemented as a plug-in. There may be many to one relationship between actors and user agents.

During the lifetime of a DVB-HTML application the user agent performs three types of processing on resources to enable the desired behaviour of the DVB-HTML application:

- Decoding
- Presentation
- Interaction

Decoding is the process of reading the bit representation of the resource and constructing a runtime representation in the actor context. The nature of the runtime representation is not specified, but it shall respect the required semantics of the resource type. Most resource types have a visible or audible representation.

Presentation is the process of rendering the runtime representation of a resource to the relevant device. Many resources can be partially presented while decoding takes place. It is an implementation option as to whether this occurs in any given user agent. Some resource types have a behavioural model.

Interaction is the generic term used for the processing required by the user agent to correctly exhibit the behaviour model defined by the semantics of the resource. Similarly it is possible that the interaction processing may overlap with decoding and presentation.

Theoretically it is possible to create an HTML service once and to adapt it for a specific terminal environment using stylesheets. But in practice it is quite different to display content on a PC monitor compared to presenting it on a TV screen. A DVB-HTML application must be controllable with a conventional remote control – with the basic input events (colour keys, up, down, etc.). Also the physical parameters of the display devices diverge. As a typical computer monitor works with a resolution higher than 800x600 @ 75Hz (progressive), the classic TV set displays the content with 720x576 @ 50 Hz (interlaced). A larger font size, a special anti-flickering filter and specific Java graphic components can solve these problems.

Two of the DVB-HTML modules that mostly enable dynamic, interoperability (between Xlets and HTML) as well as a proper layout on every MHP terminal are CSS and ECMA-script.

CSS (Cascading Stylesheets) is a W3C standard, which allows a flexible/adaptable layout of content for different platforms. With the DVB-HTML-stylesheets, which can be preinstalled in every box, a common default layout for all applications is defined. Nevertheless, the application can use another layout rather than the default layout by using stylesheet information, which can be sent together with the application. If the stylesheets require another font type, which is not installed at the MHP box the respective font has to be downloaded from a remote server.

The *ECMA-Script (European Computer Manufacturers Association)* Standard is based on several contributing technologies, the most well known being JavaScript (Netscape) and JScript (Microsoft). ECMA-Script is an object-oriented programming language for performing computations and manipulating computational objects within a host environment. It was originally designed to be a web scripting language, providing a mechanism to bring life to web pages in browsers and to perform server computation as part of a web-based client-server architecture.

In MHP, the ECMA-Script can access elements of the HTML page via DOM (Document Object Model) and change them dynamically. Furthermore the script can access Cookies (short pieces of information stored as a string on the local box) as well as the CSS information and can manipulate them. Via a special API, the scripts can use all features of the DVB-J API.

The interaction of DVB-HTML with embedded Xlets, ECMA-Script and CSS enables the creation of highly complex interactive services in a very modular way.

Downloading and Signalling of Applications via an Interaction Channel

MHP 1.0 offers the download of applications – and the associated signalling – exclusively via the broadcast channel. Additional resources like images or audio files are downloadable via the interaction channel – also known as return channel. With MHP 1.1 this restriction is no longer valid. Specialised or personalised applications (or part of them) can be loaded via any interaction channel. The benefits of its use are the reduction of the "boot" time of the application and the possibility to create more complex and personalised services.

Storing of Applications

MHP 1.0 only allows the writing of data (images, text, etc.) to local storage devices. MHP 1.1 also provides the tools to store pieces or the whole of the application code. This option is completely controlled by the broadcaster. Once all parts of the application program code are received and stored, the stored application can be started directly from the box. The broadcaster provides information in the AIT concerning the application e.g. the version number. This information for example allows the MHP terminal to start the locally stored application if there is no newer version of the application in the broadcast stream instead of waiting for a new one to be downloaded. This feature speeds up the starting time of applications. This is especially interesting for frequently used services like EPGs.

Stored Services

A new class of applications – called stand-alone applications – enables services, which can be executed independently from a broadcast service. By setting a tag in the "application storage descriptor" to the type "stand alone" (in contrast to "broadcast related"), a broadcaster can request the MHP box to store an application locally. The option to execute a stored application when receiving different channels and bouquets provides the possibility to create more complex services e.g. EPGs covering multiple bouquets. The stored service feature does not include the storage of audio and video data.

Plug-ins

Plug-ins enable a MHP-box to decode and present data formats which the standard itself does not support. The concept of MHP plug-ins was first conceived to enable the migration of applications written for non-MHP solutions in MHP terminals (see section 14.2.7). But of course plug-ins may be also used for Internet standards like Flash.

A special API was created which allows the plug-in to be stored in the terminal. Thereby the need to download the plug-in every time an application requires such plug-in is eliminated.

Incoming applications are then checked for the type of the data and the appropriate plug-in is started. The communication between non-MHP applications running within the plug-in and other MHP applications is also supported. The plug-in API also enables the execution of very complex formats like HTML.

The standard defines two types of plug-ins:

An interoperable plug-in is executed on top of the MHP API. This means that when using the MHP API the plug-in emulates the non-MHP conforming runtime environment. From the point of view of the MHP implementation, the plug-in is a normal Xlet, which is independently responsible for the life cycle, resource management, etc. of the application running on top of it.

An implementation specific plug-in is executed directly using the functionality of the operating system of the set-top box. This implies that it is impossible to install such a plug-in via a broadcast channel to run interoperable on all receiving boxes. In practice, this type of plug-in runs independently from the MHP runtime environment.

Open Card Framework (OFC)

Based on the CA-API (Conditional Access), defined in MHP 1.0, which mainly focuses on the conditional access via the common interface (CI), DVB opened this API towards smart cards. The embedded Open Card Framework enables access to all kinds of smart cards, like money or phone cards. Implementation of this API is optional.

14.5 The Future of MHP

14.5.1 Ongoing Developments

The MHP is a software environment, which has become quite complex and therefore requires permanent maintenance. In addition, the market players are requiring new features. One example is the incorporation of all software tools required to support Personal Digital Recorders (PDR). A second example is the addition of an "IP-Tuner" which will replace the current tuner API in cases where an MHP terminal is connected to a broadband Internet network. Many more new requirements are known and will be incorporated in future versions of the MHP.

The MHP concept of an open interoperable software platform has attracted the interest of organisations not typically involved in broadcasting and consumer electronics. The MHP therefore may be on its way into new environments.

14.5.2 Aspects of a mobile MHP

The MHP was designed originally with the clear goal to be deployed in a "living room" environment. In this environment power supply of the terminals is not an issue, the conditions of the broadcast channel are stable and the user interfaces are clearly defined.

In the "mobile world" the terminal conditions are fundamentally different [Kli/Schi]. The broadcast channels as well as the interaction channels are both radio channels. Especially if the user moves, it might appear that channel conditions change or that the connection even gets lost. This means that the received application can be incomplete (missing class-files or resources). A version of the MHP for use in mobile terminals should be able to handle such problems employing strategies such as downloading missing parts via the interaction channel. Furthermore the application should be much more tailored towards non-perfect channels compared to those required for the "stationary" MHP system. To provide more robust and predictable operation it might be useful to provide the application with information on the conditions of the radio and about the status of the power resources (e.g. accumulator status). This information could help the application to adapt its behaviour like storing specific information in the case of decreasing power.

The usage of MHP-based platforms on mobile devices like PDAs or in-car systems leads to new demands on the user interfaces which differ from those of the "classical MHP". The introduction of speech control for applications

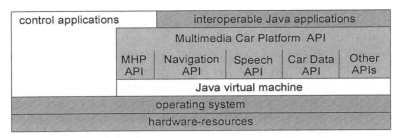

Fig. 14.9. MCP API extensions

in the car environment as an alternative input method is an obvious new requirement. The great variety of possible terminal classes requires much more flexibility concerning the user interface. This includes on the one hand the control of input events (e.g. via touch screen, speech control, car specific input devices) and on the other hand the output options (from the small low resolution screens in mobile phones – to the full sized TV screens for passengers in buses).

Figure 14.9 shows a software stack developed as an extension of the MHP by the European research consortium Multimedia Car Platform (MCP). The MCP API shown includes the MHP API and additionally a navigation API, a car data API, a speech API, etc. The new software stack enables the use of location based services. This ranges from services, which guide the MCP user through unknown cities to services which select content displayed to the user depending on his current location.

14.6 The MHP Test Suite

The MHP is a solution addressing what is called a "horizontal market". This term implies that a customer can buy a receiver, which is MHP compliant and use it to run all the MHP applications on air – irrespective of where these applications originate. Considering the enormous complexity of the MHP specification it becomes clear that such interoperability will be difficult to achieve. The key tools to achieve interoperability in practice are the MHP test suite and the related self-certification program.

The DVB Project founded an MHP Experts Group assigned with the task of compiling a sound and "waterproof" set of MHP tests which would be able to guarantee that only interoperable MHP receivers and applications be offered to the public. The group drew tests from providers of the original technology (Sun Microsystems for much of the Java part) and from a significant number of DVB member companies and compiled a first version of the test suite which could be released in the summer of 2002. This first version included about 10,600 tests. A second version which included additional tests for certain critical areas was made available to the implementers of the MHP in January 2003.

The test suite is delivered to implementers on request by a custodian organisation (ETSI) in the form of a DVD. Upon successful testing of the implementation against each individual test, the implementer sends the custodian a document in which he certifies that all tests have been conducted. The custodian will then inform the DVB Project about the receipt of the certificate. The DVB Project consequently will release the MHP logo to the implementer.

The ideas underlying this procedure are the following. First it is assumed that the successful passing of the test suite will guarantee the required degree of interoperability between MHP applications and receivers. Implicitly this means that MHP applications are assumed to be MHP compliant. In reality the authors of such applications will have to test them on a certain number of MHP compliant receivers before they can start broadcasting them. The second underlying idea is that the MHP logo is able to signal the interoperability to a consumer buying a receiver. The third idea is that – like with all DVB specifications – the process of self-testing the compatibility by a receiver manufacturer will be sufficient to guarantee stability in a mass market.

14.7 The Globally Executable MHP

Not surprisingly, the development of the MHP was carried out with those countries and broadcasting organisations in mind which use the complete DVB line-up of technologies and standards. The MHP therefore relies, for example, on the broadcast channel protocol stack shown in figure 14.4. When the specification was finished and, very importantly, when it could be demonstrated to the world in 2002 that DVB had created an open standard and had solved the many surrounding problems related to intellectual property, testing, self-certification etc. organisations which were not part of the original activities showed an interest in the MHP. At this point the problem arose that the DVB protocol stack has not been used in its entirety in countries such as the U.S.. Instead, transport layer protocols had been developed and successfully put in use, which included specific elements for which the MHP had no interfacing mechanisms available.

The DVB Project created an activity aimed at the joint development of a specification now known as the Globally Executable MHP (GEM) [TS 819] upon the request of organisations, who offer services not fully in line with the DVB system but are interested in the use of the MHP in their networks. This document in conjunction with the MHP specification itself creates a complete software platform which can be incorporated in, for example, cable networks in the U.S. Where required, GEM defines functional equivalents to what the DVB stack includes and therefore "glues" the MHP into non-DVB systems.

At the time of writing this text in the summer of 2003 the MHP has via versions of the GEM specification been accepted in Japan and in the U.S. both for cable and digital terrestrial television.

The selection of technologies focuses on open Java-based software platform standards established for mobile and/or multimedia applications.

14.8 Other Java-Based Software Platforms

This section is intended to complete the picture of Java based software platforms. Like the MHP the following described systems are aiming at the abstraction of special hardware devices and/or the underlying transport protocols.

The selection of technologie focuses on open Java-based software platform standards established for mobile and/or multimedia applications.

14.8.1 OSGi

The Open Service Gateway Initiative is a group that has the goal of defining a software gateway, which has the capability to connect all kinds of different networks using just one single middleware platform. The Java-based platform is a resident server, which is able to store, control and run so-called OSGi-bundles. These bundles are software packages consisting of either OSGi-compatible Java programs or native code, which can only run on manufacturer specific gateways [OSGi 2].

The main difference between OSGi and MHP is, that OSGi is not targeting an end-user terminal but a server-system. That means, OSGi does not handle e.g. user input or graphical output; it rather concentrates on the handling of the Java-bundles. Some bundle types are already defined, such as an HTTP-bundle.

OSGi implementations can not only be found in the home networking area, but also more and more in the automotive environment.

14.8.2 DAB Java Specification

The World DAB (Digital Audio Broadcast) consortium is working on a software platform called DAB Java, which has been standardised in its first version 1.1.1. by ETSI [TS 993].

This Personal Java based platform is somewhat similar to the MHP (e.g. it also supports Java-Xlets, but has not such a complete API to control the application lifecycle). The DAB Java interface allows compliant services to access directly the underlying DAB network. The basic idea is to broadcast applications via DAB and to have a small and modular platform to run these applications on. One possible user terminal class could be in-car radios (even with a small display). However, some important elements like e.g. authentication of applications are not yet standardised.

14.8.3 Mobile Information Device Profile – MIDP

Since the beginning of the development of the Java technology, Sun Microsystems always pursued the idea of putting Java inside small devices. The first initiative to achieve that was PersonalJava. PersonalJava defines a subset of the J2SE (Java 2 Standard Edition) APIs that can be run in devices with limited memory and processing power.

At the end of 2000, Sun proposed a new architecture for the Java APIs. Java platforms now are defined for four main "worlds": Server, Desktop, "Other devices", and SmartCards. Each of these worlds is characterised by a VM and an appropriate set of APIs. Within each world, still some variations exist according to the device capabilities (especially in the "other devices" world).

The "other devices", or Java 2 Micro Edition (J2ME) world targets a diversity of terminals, like stationary devices such as STBs and mobile devices such as cell phones. Consequently, it was broken into two main sub-worlds: the limited and the non-limited devices. To address this distinction, the J2ME API set was divided into three main layers that are presented in figure 14.10 and described below.

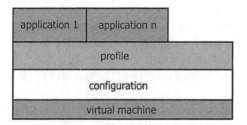

Fig. 14.10. J2ME Layers

VM: Two different virtual machines are commonly used: KVM (K Virtual Machine) and JVM (Java Virtual Machine). The KVM is a limited VM which does not include some of the well-known features of a full JVM implementation, such as floating point support and thread groups.

Configuration: The Configuration layer defines the minimum set of Java core APIs that are available to applications. Two configurations were defined: CDC (Connected Device Configuration) and CLDC (Connected

Limited Device Configuration). The CLDC should be used with limited-resource devices, like cell phones, while the CDC addresses more capable devices, like STBs.

Profile: The profile defines the high level APIs that are specific to a particular class of devices. The profile usually includes user interface, communication and application management APIs. Some profiles were defined, but the only one which is being widely used at the present time is the MIDP.

Optional packages, like RMI or Bluetooth (described in JSR 82 - Java Specification Request 82), can also be included in a device, if this device has the minimum set of APIs necessary to run the package. For instance, RMI cannot run in standard MIDP, since it requires some APIs that are not defined in CDLC and MIDP.

As already described, MIDP is one of the J2ME profiles that can be run on CLDC. The MIDP specification encompasses:

- Execution environment
- Application lifecycle
- Application provisioning
- Security environment
- API sets for I/O, storage and UI

Although MIDP is a Java-based middleware, such as MHP, a list of restrictions caused by the limited hardware resources has to be considered. Especially, the support of for example the presentation of audio and video, datagram services, and TCP/IP is not included in the basic version of MIDP.

The fact that parts of the MIDP specification covering important aspects are not mandatory and that each manufacturer may add optional/proprietary packages implies that two different MIDP implementations may provide different functionalities to the application. Thus, at the moment there is no way to guarantee that a MIDP application will run on all MIDP devices. To solve this API fragmentation problem, JSR 185 (Java Technology for the Wireless Industry) was created to define some of the optional parts as mandatory in this specification. A JTWI compliant device must have CLDC 1.0, MIDP 2.0, JSR 120 (Wireless Messaging API) and JSR 135 (Mobile Media API).

15 Measurement Techniques for Digital Television

The measurement of characteristic parameters in digital television can be divided into two areas. First of all, the source signals in the baseband are processed to form compressed data streams. These are subsequently furnished with an error protection and then digitally modulated for transmission. The measurement techniques for the processing steps in both areas will be introduced in this chapter.

When compressing audio and video signals, noticeable changes can occur in the output signal (see chapters 3 and 4). The MPEG-2 standard offers extensive possibilities for video coding to influence the coding result. Depending on the parameters chosen and the sequential image content, the resulting image quality is subject to considerable variation. A test sequence was developed at Braunschweig Technical University with the aid of which the image quality provided by an encoder can be visually assessed. This is described in section 15.1.1.

The data streams resulting from the source coding of several programmes are combined to form a "transport stream" (TS) in the time multiplex (see chapter 5). In order to check the correctness and standard conformity of the coding of the information in the data streams, special protocol testers are required which are able to interpret and detect errors in the data.

The functionality of decoders can be checked with test data streams, which can be audio or video elementary streams or transport streams. For checking various aspects of the decoding, data streams are required which contain specific features critical for a decoder.

Digital transmission technique predominantly employs measuring methods known from other fields of communication technology. Section 15.2 will introduce several measurements obtained from transmitters and receivers, as well as measurements of the bit-error rate.

Within the DVB Project the DVB Measurement Group (DVB-MG) developed the DVB "measurement guidelines" [ETR 290]. This document can be considered the fundamental handbook to be used for all sorts of baseband and RF measurements. The guidelines include sections covering measurement and analysis of the MPEG-2 Transport Stream, measurements specific for DVB-C, DVB-S and DVB-T, recommendations for the measurement of delays and various annexes describing measurement methods etc..

15.1   Measurement Techniques for Source-Signal Processing in the Baseband

15.1.1 Quality Evaluation of Video Source Coding

The image quality which can be obtained with hybrid coding (as, for instance, with MPEG-2) cannot at the moment be satisfactorily assessed with quantitative measurement parameters. The assessment of image quality is therefore only possible using video artefacts of a coded image sequence. This statement is also true for the comparison of encoders and decoders.

A sequence which facilitates such an assessment, and whose luminance signal can be seen in figure 15.1 [ROHDE], was developed at the Institute for Communications Technology of Braunschweig Technical University. All typical artefacts of a hybrid coding can be clearly demonstrated with this sequence. Furthermore, it enables statements to be made about the pre-processing, the structure of the "group of pictures" (GOP, see section 4.2.2) and the value of the global quantisation factor. It consists of various numbered image zones, which will be briefly described below.

Image zone 1 consists of finely shaded bar patterns and is divided into ten fields. On the left-hand side there are five fields with horizontal bars, two of which consist of black-and-white patterns, and adjacent, on the right, there are five fields with vertical bars, two of which again comprise black-and-white patterns. The remaining fields consist of green-magenta patterns, whose chrominance components C_B and C_R represent the lowest permissible values (green) and the highest permissible values (magenta) respectively. The same luminance value has been chosen for both colours in order to obtain a complete separation between luminance and chrominance for the analysis of the pre-processing. Hence the coloured bar patterns in figure 15.1 are not discernible. With the help of the above fields, one can see at a glance the steps that preceded the data compression, such as subsampling and filtering.

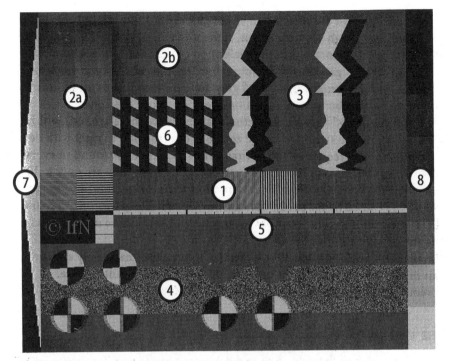

Fig. 15.1. Test sequence for the evaluation of the image quality

The vertical and horizontal grey wedges (image zones 2a and 2b) show such errors as occur due to the quantisation of the DCT coefficients of low order. Due to the brightness modulation of the grey wedges, which results in a shifting of the grey values, even negligible quantisation errors become obvious.

Image zone 3 is composed of various edges. In the left part these are immobile, while in the right part they move up and down. As before, the green and magenta coloured zones, which are adjacent to the right of the black zones, are not visible. Since the representation of the steep skirts of the edges specifically calls for DCT coefficients of a high order, but as these are only coarsely quantised at the encoder, quantisation errors are particularly prominent at the edges.

Further critical picture contents are the moving objects in image zone 4. At the edges of these objects errors are still noticeable even if – in the case of a generally good image quality – hardly any visible errors occur in the remaining picture. The sector wheels at the top on the left rotate around their centre. All others move evenly in a horizontal direction. The sector wheel comprising green and magenta at the top on the right is not identifiable for the known reasons.

Image zone 5 is for the estimation of the global quantisation factor, which can be used, by a rough approximation, as the measure of image quality and is therefore of great importance in the evaluation of the coded test sequences. It can be read off by regarding the structure border between the even grey zone on the left and the weakly structured grey zone on the right as an indicator pointing to a value between 1 and 31 on the scale above the zone.

The oblique "grey" bars (in the actual test sequence these are red) within the six vertical white stripes of image zone 6 occur from left to right with decreasing temporal frequency. In the first stripe they are to be found in every second image, whereas in the stripe furthest to the right they are only to be found in every seventh image. In conjunction with the geometric arrangement, errors caused by the prediction can be made visible in this way. In so far as these prediction errors occur obviously enough, they can be used to determine the organisation of the GOP.

Image zone 7 indicates whether, instead of a complete image, only a section of an image has been coded in order to decrease the amount of data.

Colour errors can be identified with the EBU colour bar (image zone 8).

15.1.2 Checking Compressed Audio and Video Signals

The conformance of the video and audio elementary streams (see chapters 3 and 4) with the MPEG-2 standard can be verified by means of protocol analyses. A selection of the possibilities provided by the MPEG-2 standard has been chosen by the DVB Project for audio and video coding. New, DVB-specific elements were not introduced for this level [ETR 154]. The DVB Measurement Group has therefore not dealt with this topic. Further information on testing the conformance of audio and video streams is to be found in part 4 of the MPEG-2 standard [ISO 13818], where the individual stream elements are discussed in great detail.

15.1.3 Checking the MPEG-2 Transport Stream

The MPEG-2 "transport stream" (TS) is of particular interest for protocol testing because all the useful information to be transmitted is integrated in it (see chapter 5). It is also easily accessible because the DVB Physical Interface Group (DVB-PI) has specified three different interfaces for it [EN 50083-9]:

- the data are transmitted with an 8-bit width via the synchronous parallel interface. In addition there is a clock signal, a data-valid signal and a sync signal;
- the data format of the "synchronous serial interface" (SSI) corresponds to that of the parallel interface after a parallel/serial conversion;

- the "asynchronous serial interface" (ASI) is operated with a constant data rate of 270 Mbit/s. The jitter occurring as a result of the mapping of the useful bytes of the TS onto the high-rate stream must be equalised by the first-in first-out (FIFO) memory at the receiver.

Errors in the source coding of the individual audio and video signals, errors of the multiplex and some of the errors occurring during transmission can be identified by analysing the transport stream.

A great number of tests have been recommended by the DVB Measurement Group for the MPEG-2 transport stream. These were developed for in-service testing and can either be used for continuous or for periodic monitoring of the most important elements [ETR 290]. During the development of the TS tests great care was taken to strike a reasonable balance between the importance of a parameter and the implementation requirements for the testing, thus making continuous monitoring of the transport stream possible without an excessive hardware requirement.

These tests should be seen solely as a "health check" of a TS and are not intended as complete conformance tests as described in part 4 of the MPEG-2 standard [ISO 13818]. A more thorough analysis of a TS by means of further tests is of course possible. Owing to the complexity of the TS and the higher data rate, this is, however, not possible in real time and can therefore only be carried out on a part of the TS. The representation in this section is limited to the tests recommended by the Measurement Group.

Many of the tests listed below refer to the information in the TS header. They can also be carried out with a scrambled TS (see chapters 5 and 8) since the header always remains unscrambled. One of the prerequisites for the tests is a quasi error-free TS as should be available under normal operating conditions at the output of the Reed-Solomon decoder. If a transmission error is detected, or if the measuring instrument is no longer locked to the TS, further error messages are suppressed.

The tests are conceived in such a way that particular errors are assigned to an indicating element so that the respective protocol tester or measuring decoder can indicate the operational error to the operator in a clearly visible form. A sequence of LEDs, labelled with the names of the errors, will suffice to carry out a comprehensive monitoring.

The tests are divided into three groups according to their importance. The tests in the first group check whether the TS can be synchronised and whether no errors have occurred during transmission. In addition, errors in the compilation of the TS can be detected. Tests in the second and third groups are for the conformance of the repetition rates and the existence of various elements of the TS as well as for deviations of the clock references.

The following pages refer to various syntactical elements of the MPEG-2 TS and to errors defined by the MG. In order to avoid confusion, the spelling of these elements and errors adopted here is that used in the corresponding documents [ISO 13818], [ETR 290].

Group 1: High-priority Tests (Basic Monitoring)

TS_sync_loss

The measuring device for demultiplexing and carrying out further testing must first be synchronised with the TS, for instance by means of a hysteresis function. The device can be regarded as synchronised if a certain number of correct sync bytes following upon each other at packet intervals have been received. This state of being synchronised is only changed when several consecutive sync bytes do not have the value 47_{HEX}. The MG recommends that a device be synchronised to the TS after five correct consecutive sync bytes. When there have been two or more consecutive erroneous sync bytes the synchronisation is lost.

Synchronism between the measuring device and the TS is an essential condition for all further testing. If there is no synchronism, the results of the remaining tests become irrelevant. All other error messages must therefore be suppressed or ignored.

Sync_Byte_error

If the already synchronised condition of an individual sync byte takes a value different from 47_{HEX}, then this will be indicated as Sync_Byte_error.

PAT_error

By means of the "program association table" (PAT) the demultiplexer is informed which programmes are contained in the TS and with which "packet identification" (PID) those packets are labelled in which, by means of the "program map table" (PMT), more detailed information is to be found about the compilation of the programme. PAT and PMT, therefore, have the function of an index, and it is only through these that access to the audio, video and auxiliary data is made possible.

Errors are indicated

- if packets with PID 0000_{HEX}, as specified for the PAT, do not contain any PAT data,
- if the time interval between two packets with PID 0000_{HEX} is greater than 0.5 seconds, or
- if the scrambling_control_field of the header indicates that the packet is scrambled. PAT data must not be scrambled.

Continuity_count_error

In each TS packet header there is a 4-bit field, the value of which is incremented in each packet of a particular PID. Hence, there exists one independent continuity_counter per PID. Under certain conditions a packet may be transmitted twice.

An error is identified

- if the order of the packets is not correct,
- if a packet is triply or multiply transmitted, or
- if a packet is missing.

A continuity_count_error can occur, for example, when a non-correctable transmission error is identified by the Reed-Solomon decoder. This can be indicated by the setting of the transport_error_indicator in the header of an altered packet. The packet can then no longer be unambiguously assigned to a PID because the PID itself might have been altered. It is most probable that a jump in the continuity_counter of this PID will occur in the next packet with the same PID.

Generally it must be observed that, if they are indicated by the discontinuity_indicator in the adaptation field, even discontinuities of the continuity_counter are permitted.

PMT_error

Together with the PAT described above, the "program map table" (PMT) takes care of allocating the audio, video and auxiliary data of a programme to the PIDs.

An error is indicated

- if packets with the PID of a PMT are repeated at a time interval greater than 0.5 seconds, or
- if the scrambling_control_field of the header indicates that the packet is scrambled (scrambling not being permissible in packets with PMT).

To check the conditions, the PAT must be completely decoded to start with, because it is only then that those PIDs become known with the aid of which the PMTs can be decoded.

PID_error

When the PAT and all the PMTs have been decoded, one can check, using the list of all the PIDs referenced in the tables, whether packets were actually transmitted with each PID stated in the list. A PID_error is identified when there is a PID in which no packet is transmitted within a user-defined time limit. This type of error points to an inconsistency between the PIDs of the transmitted packets, the contents of the PAT, and the PMTs. The opposite case, i.e. that of a PID not listed in the tables but appearing in the TS, is permitted and does not cause a PID_error.

Group 2: Lower-priority Tests (Recommended for Continuous or Periodic Monitoring)

Many of the following tests are only possible if the useful data of the TS are not scrambled. They sometimes require considerable computation (e.g. CRC_error), which must be balanced against the usefulness of the information gathered.

Transport_error

One bit of the TS packet header is used as the transport_error_indicator, which indicates whether uncorrected errors occurred in the packet in the course of the transmission or processing. If a bit has been set it may not be set back. The payload of the packet is unusable for decoding purposes.

If the bit is set, there is a Transport_error. Further tests can no longer deliver any reliable results. Therefore erroneous packets must not be tested.

CRC_error

A certain amount of information is transmitted in the TS in the form of tables. These include the already mentioned "program association table" (PAT) and the "program map tables" (PMTs) as well as further service information (SI) in the "network information table" (NIT), the "event information table" (EIT), the "bouquet association table" (BAT) and the "service description table" (SDT) (see section 5.4). For the transmission, the individual sections that a table is made up of are provided with a cyclic redundancy-check code. The decoder can then carry out the check and determine whether the data have been correctly received.

PCR_error

The "program clock references" (PCRs) serve to synchronise the "system time clocks" (STCs) of the multiplexer and the demultiplexer. For each programme, separate PCRs can be transmitted in adaptation fields (see section 5.3). The MPEG-2 standard stipulates that the time interval between two consecutive PCR values of the same programme must not exceed 100 ms. The DVB Implementation Guidelines [ETR 154] recommend that this interval should not be greater than 40 ms.

There is a PCR_error

- if the interval between the PCRs of a programme is greater than 40 ms, or
- if the difference between two subsequent PCRs of a programme is greater than 100 ms, without a temporal discontinuity being indicated.

To check the second condition, the coded PCRs themselves are compared, as opposed to the first condition. If the difference between them is greater than the maximum permissible interval of 100 ms, there is bound to be discontinuity in the STC. This calls for a signalling by the discontinuity_indicator in the adaptation field. If the discontinuity of the coded time is not indicated there is a PCR_error.

PCR_accuracy_error

The MPEG-2 standard [ISO 13818] specifies the tolerances for the STC. A deviation of ±810 Hz from the nominal frequency of 27 MHz is permissible. The frequency drift may be ±0.075 Hz/s at the most. In addition, an error of ±500 ns is allowed for the PCRs received. Possible causes can be inaccuracies in the generation of the PCRs. A further source of errors is the remultiplex. When new programmes are compiled for a transport stream, the PCRs must be recalculated ("restamping") due to the fact that their position in the data stream can be changed. If the data of a programme are repeatedly subjected to a remultiplex, care should be taken that the PCRs do not exceed the maximum permissible deviations.

Discrepancies in the reception time of the PCRs due to jitter on the transmission path must not be registered for these tolerances. Thus, carrying out these tests only requires knowledge of the PCRs and of their position in the TS. The reception time is of no importance here.

As each programme can have its own time base, the PCRs of several programmes may have to be checked separately. Alternatively, the test can also be carried out for one chosen programme.

An inequality for checking the conformance with the specifications is to be found in part 4 of the MPEG-2 standard. However, this method can only be applied if the data rate in the TS is constant.

PTS_error

The "presentation time stamps" (PTSs) supply the synchronisation of the video and audio signals of a programme. The stipulated maximum time interval, which should not exceed 700 ms in any elementary stream, is to be checked. Due to the fact that the PTS values are in the header of the "packetised elementary stream" (PES) they are only available with an unscrambled data stream.

CAT_error

The information required for decoding scrambled transport streams is coded in the "conditional-access table" (CAT).

A CAT_error is indicated

– if a TS contains packets whose transport_scrambling_control fields signal scrambled packets although there is no CAT, or
– if a section of another table is coded in a packet with PID 0001_{HEX}, reserved for the CAT.

Group 3: Optional Tests (Application-dependent Monitoring)

In this group the majority of the tests refer to further tables with service information [ETS 468]. These include the Network Information Table_error (NIT_error), the Service Description Table_error (SDT_error), the Event Information Table_error (EIT_error), the Running Status Table_error (RST_error), and the Time and Date Table_error (TDT_error), each of which is checked as to whether the packets of the PID defined by the DVB Project contain other tables than those specified. The point to be taken into consideration in this context is that the DVB Project has defined a stuffing table which may be contained in service information packets beside the useful information. In addition to the above procedure, adherence to the specified minimum repetition rates in the tables is tested. These rates vary, depending on the tables [ETR 211].

For the Service Information_repetition_error (SI_repetition_error), in addition to checks being carried out on the minimum repetition rates already included in the above individual tests, the sections of the SI are tested as to whether they are not repeated too fast.

For the Buffer_error the course of the buffer occupancy of the ideal decoder described in the MPEG-2 standard ("transport stream system target decoder" [T-STD]) has to be reconstructed. An overflow of one buffer, also an underflow of several buffers, will cause a Buffer_error.

A TS may also contain streams with PIDs which are not listed in the PAT or in a PMT. If this is the case, it is indicated as Unreferenced_PID.

15.1.4 Checking the Functionality of the Decoder

A further field of measurement techniques is the checking of the conformity of decoders with the standards. Here, there can be problems for two reasons. One reason is the great flexibility of the MPEG-2 standard. For example, it is acceptable, but not recommended, to code a data stream with plenty of overhead in the form of headers and time stamps which a decoder might not be able to decode with sufficient speed. The second reason lies in the fact that the MPEG-2 standard [ISO 13818] describes an ideal decoder, the T-STD, which, in the form described, cannot be implemented because, among other things, in the T-STD the data are removed from the interim stores instantaneously and not over a period of time. Consequently, manufacturers of decoders are bound to choose an implementation which varies from the ideal decoder, but one which nevertheless enables a decoding conforming with the standard.

For checking the correct functioning of decoders it is appropriate to use special test streams, by means of which different coding aspects can be realised for different purposes. For example, a change in the resolution and in the aspect ratio can be coded in a video elementary stream, or all the possible motion vectors can be coded in a data stream (cf. section 4.2.2). The same applies to the audio coding and the multiplex.

In the development stage of the MPEG-2 standard, video and audio elementary streams were coded and exchanged by the participating companies and organisations in order to verify the standard.

Transport streams for testing purposes can contain special video and audio streams, or they show peculiarities of the multiplex. Examples in this field are:

– a change of the programme definition by the coding of new tables (PAT and PMT),
– a jump of the STC (cf. section 5.3),
– the coding of the PCR with the maximum permissible deviations.

For production and service, data streams with which the tasks of contemporary resolution-test-pattern generators can be accomplished are also of interest. For even in the future the test patterns familiar from analogue television will be required for controlling cathode ray tubes. They can be coded to form video streams and can then be multiplexed to form a transport stream. The test patterns fed with such a TS could be shown on a DVB receiver and one could switch between images with programme selection buttons.

The supply of a DVB receiver with the necessary data streams will require a generator. This is composed of a storage medium for the data streams, an interface for the output of the data streams, and a device for the control of the operational run and for the internal data transport.

15.2 Measurements for Digital Transmission Technology

An overview of some important measurement quantities and their respective measuring instruments for the transmission technology of digital television is given in the following table:

Measurement quantities	Measuring instruments
eye diagram, ISI	oscilloscope
vector diagram (constellation)	vector signal generator or vector modulation analyser
power distribution	power analyser
linearity	power analyser, vector modulation analyser
interference; S/N, C/N ratio	spectral analyser, special measuring instruments
error rate	transmission analyser, additional decoder for on-line measurement
regenerator phase noise	phase detector, spectrum analyser

Interference, for instance expressed as S/N or C/N power ratio, considerably affects the achievable error rate of the transmission. As is sufficiently known, the said powers can be determined by a spectrum analyser. For in-service measurements it is possible to use a channel-state analyser, which can carry out an analysis on the basis of the behaviour of the forward error correction (FEC, see chapter 6) when the source signals are known.

The following sections describe the measurement of the error rate and the power distribution as well as the eye diagram and vector diagram, followed by the measurement of the regenerator phase noise.

15.2.1 Representation of the Eye Diagram

The eye diagram presents a very clear measure of the quality of a digital transmission [SÖDER]. Generally, this measure is the size of the eye-opening area, although the amplitude and phase noise as well as the clock jitter complicate the practical evaluation. The eye diagram usually only presents an overall evaluation of the system which allows no conclusions about individual errors.

The eye diagram is very easy to represent by means of an oscilloscope as shown by the block diagram of the measurement set-up in figure 15.2. In a binary system the decision threshold and its sampling instant correspond with the symmetry axes of the eye diagram.

Intersymbol interference (ISI, see section 7.1) occurs in the transmitted signal when the previous or subsequent signals affect the actual signal in its eye opening. Examples are shown in figure 15.2. When the impulse response of the entire transmission channel shows zeros for integral multiples of the clock period the system is free from intersymbol interference (first Nyquist criterion), not, however, from overshoots which could cause overdrive effects and consequently lead to a non-linear behaviour of the channel. This point was discussed in detail in section 7.1.

For multilevel systems the eye opening decreases in proportion to the number of decisions. Figure 15.2 shows a quaternary system with four ampli-

Representation of an eye diagram

Examples

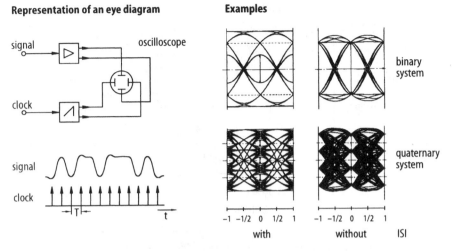

Fig. 15.2. Eye diagrams: representation and examples [SÖDER]

tude thresholds and three eye openings. The decision levels are at the centre line of the respective eye level (dash-dot lines).

15.2.2 Measurements Carried out at Modulators and Demodulators

The principle of quadrature amplitude modulation (QAM) has already been explained in detail in section 7.5. Figure 15.3 shows as an example the block diagram of a QAM transmission. The mapper (byte-to-symbol-word converter) divides the incoming serial data stream into two parallel signals I and Q (in-phase and quadrature phase), which, subsequent to passing through low-pass filters, modulate in quadrature (90° phase shift to each other) a carrier signal in its amplitude. The D/A converters required for multilevel I- and Q-signals are assumed to be included in the mapper.

Subsequent to running through the transmitter amplifier, the transmission path and the receiver input stages, the QAM signal is demodulated in the reverse order. For this purpose, a reference carrier signal has to be regenerated from the received signal and is then fed to both the synchronous demodulators. Another requirement is that the phase position of this reference signal be known in relation to the modulation. A full account of this procedure has been given in section 10.4. The stages of the signal processing, as described here, are assumed to be integrated in the carrier regenerator of the block diagram.

The block diagrams of the transmitter and the receiver, as illustrated here, are also valid for a vector signal generator or for a vector modulation analyser [SCHERER]. Both measuring instruments contain the same structural com-

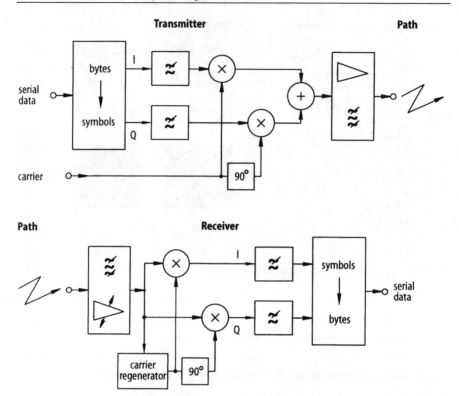

Fig. 15.3. Principle of the quadrature modulation

ponents as represented here, that is, with reference to their functions. However, the components are designed as universally usable sets with very close tolerances. With these measuring sets it is possible to carry out detailed investigations not only of whole systems, but also of individual components. This is the subject of the following considerations, which generally pertain to other types of modulation as well, e.g. the modulation with quadrature phase shift keying (QPSK) which can be represented as QAM with binary I- and Q-signals.

15.2.2.1 Measurements at a QAM Transmitter

Figure 15.4 shows a measurement set-up. The eye diagrams of the baseband signals can be represented by means of a vector modulation analyser, for instance in order to investigate the effect of the band-limiting low-pass filters. In the carrier-frequency domain the vector representation of the signal components is useful in the (complex) I,Q-plane since it facilitates the measuring of static or dynamic amplitude- and phase deviations. In this case, three rep-

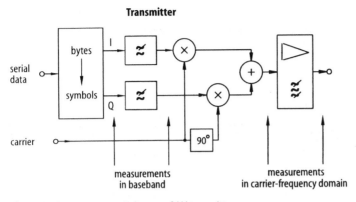

Fig. 15.4. Measurements carried out at a QAM transmitter

resentations are possible: firstly, one which continuously shows the whole temporal course with all transients but which often has an unclear effect; secondly, a temporally discrete representation of the vector diagram which only indicates the optimal signal-sampling instants ("constellation", see section 7.5); and, thirdly, the temporal representation of the signal according to its amplitude and phase course.

The power analyser measures the average and the maximum power of the modulated signal [SCHERER, RHODES]. Theoretically there is a relationship between the two values which is given by the type of modulation (peak-to-average power ratio); however, in practice this relationship can be changed by non-linearities in the transmitter. The display of a histogram of the output power for the various vectors of the QAM makes a further investigation into this effect possible.

15.2.2.2 Measurements at a QAM Receiver

Figure 15.5 shows the construction of a suitable measurement set-up. For the purpose of measuring the characteristics of the QAM receiver by means of a universal, practically ideal transmitter it is recommendable to use a vector signal generator for the input. This generator enables definite changes of various system parameters, e.g. of the I/Q amplitude ratio, the quadrature angle, or the characteristics of the I/Q pre-filters. Depending on these parameter variations, the respective deviations in the receiver can be measured with a vector modulation analyser as described above. Sometimes the answer required is the error rate, the measurement of which is explained in section 15.2.3.

The simulation of the transmission path comprises its partition into individual parallel channels, which can be equipped independently of each other with varying attenuation, delay time, and Doppler shift [SCHERER]. By com-

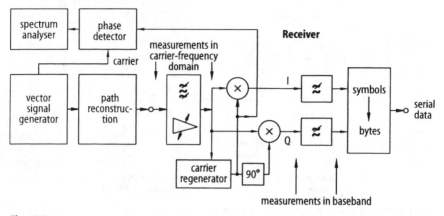

Fig. 15.5. Measurements carried out at a QAM receiver

bining these individual channels a transmission path of considerable complexity is created which, controlled by programme, is also dynamically variable. In this way both multipath propagation and fading effects can be simulated; the fading effects may, in addition, show a correlation between the individual simulation channels.

For measuring the properties of the carrier regenerator, especially its phase noise, a phase detector is provided which reads out the phase difference between the very stable carrier signal of the vector signal generator and the regenerated reference signal of the QAM receiver. The phase noise and other interference in the reference signal can thus be detected with a spectrum analyser. The above-mentioned properties have a direct effect on the width of the opening in the eye diagram at the receiving end.

15.2.2.3 Measurements at an OFDM Transmission

The transmission of digital signals by means of an orthogonal frequency division multiplex (OFDM), which is also referred to as multiple-carrier technique, has already been described in section 7.6. Firstly, the measurements at an OFDM transmission system in the carrier-frequency domain include the testing of the individual subcarriers for their correct frequency and modulation. This is possible by means of a spectrum analyser with suitable resolution if a blockwise periodically recurring data pattern is transmitted (cf. section 11.3.3). Secondly, it is also possible to measure, with the same equipment, the complete power density in the channel. In order to determine the power and the overdriving limits of the transmitter, the respective distribution of the output signal is also important. This can be ascertained by means of a power analyser, as described in section 15.2.2.1. What was stated in section 15.2.2.2 is also valid for the simulation of the transmission path.

15.2.3 Measurement of Error Rate

The error rate in a digital transmission is the most frequently used quality criterion. However, this is only an overall measure of quality, which barely expresses anything about the individual causes of errors. Contrary to the error probability which was often used as operand in the preceding sections, the error rate is a measured value whose determination in its turn will lead to further errors. Therefore there can only be an approximate conformity between the two quantities.

15.2.3.1 Direct Measurement of Error Rate

The measuring arrangement shown in figure 15.6 comprises a generator which creates a pseudo-random binary sequence (PRBS) that is fed into the

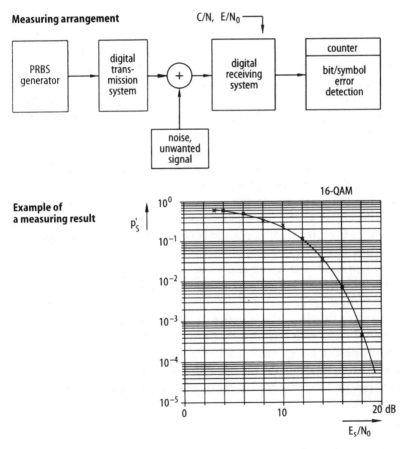

Fig. 15.6. Direct measurement of error rate: measuring arrangement and example

digital transmission system as a data stream. The interference which is super-imposed on the transmission path is represented as the addition of noise and interference signals in the block diagram. In the digital reception system the decisions as to which signs or symbols were transmitted are thus corrupted, the corruption being the result of a random process. These decisions are checked at the bit/symbol-error identification stage and the errors found are counted, either for a predetermined duration or for a specified number of de-cisions [WELLHSN].

The duration of such a measurement is therefore dependent on these base values, on the error rate itself, and on the permissible measuring error. It can be very long for a statistically relevant number of errors if their occurrence is infrequent. Furthermore, this measuring technique requires either a free channel or implementation during non-operational times, so that it is mainly used in laboratories or when putting a system into operation.

The length of the pseudo-random sequence should be chosen to be as long as possible, so that the shape of its spectrum is almost uniform for all data rates used. To recognise errors at the receiving end, in the case of link tests, it is necessary to synchronise the receiver with the known sequence of the transmitter by cross-correlation. However, the duration of this procedure is directly dependent on the length of the sequence, so that a compromise be-tween both requirements is necessary. For the known hierarchies of the digi-tal transmission systems with high data rates, sequences of the length of $2^{15} - 1$ and $2^{23} - 1$ are given.

The error rate is usually represented as dependent on the signal-to-noise ratio. In the carrier-frequency domain of the system this quantity is the ratio of the carrier power to the noise power C/N, which is relatively easy to deter-mine with a spectrum analyser. In the baseband, the quantity is S/N, which is the ratio of the signal to the noise power; or else the ratio of the energies E_b/N_o or E_s/N_o is used, for which the signal energy refers to a bit or a symbol respectively (see chapter 7).

The curve represented in figure 12.6 serves as an example of a measuring result in a 16-QAM transmission (16 possible modulation vectors). The meas-ured values are shown as crosses, and the curve calculated in accordance with (7.46) is shown as a thin line. The interdependence between the bit-error rate and the symbol-error rate has already been described in section 7.4. The symbol-error rate can be approximated as a product of the symbol length (here, for instance, 4 bits) and the bit-error rate, as long as the last-mentioned does not exceed 10^{-3} and double bit errors within a symbol are practically negligible as a consequence.

Measuring arrangement

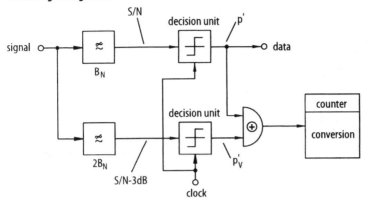

Fig. 15.7. Comparison measurement of error rate

15.2.3.2 Comparison Measurement of Error Rate

Figure 15.7 shows the measuring arrangement for a method that permits the comparison measurement of the error rate even during operation [FEHER 1, FEHER 2]. This method is valid on the assumption that the impairing interference and its effect on the transmission system are known, as indicated by a curve like the one shown in figure 15.6.

For this method there are two decoders of the same type present in the receiver and both are fed with the same signal. The only difference between the two branches is the noise bandwidth of the pre-switched channel filters, the latter are symbolised here by low-pass filters. The noise bandwidth of the comparison channel is to be chosen double that of the data channel. It follows from this that the S/N ratio in the comparison channel – assuming a spectrally uniformly distributed interference – will be 3 dB less than in the data channel.

On the further assumption that the error rate in the data channel is much smaller than that in the comparison channel, the data channel can be used as an "ideal" reference for the evaluation of the errors in the comparison channel. This evaluation takes place by means of an EX-OR gate, at the output of which the discrepancies between the two decision units are counted. By conversion, using the predetermined error-rate characteristic of the system, the corresponding value for the data channel is obtained.

Due to the considerably greater error rate in the comparison channel the measuring speed increases accordingly. Variations of this method which have become known include the feeding of an additional interference signal into the comparison channel and the changing of the threshold value belonging to that channel.

Symbols in Chapter 15

B_N noise bandwidth
E_b energy/bit
E_s energy/symbol
I in-phase component
N_o noise-power density
p' error rate
p_S' symbol-error rate
p_V' comparative error rate
Q quadrature component
T clock period
t time in general

16 Bibliography

[Agraw]   Agrawal, B.P.; Shenoi, K.: Design Methodology for $\Sigma\Delta M$. IEEE Trans.
             on Comm., COM-31, No. 3, March 1983, pp. 360 - 370.
[Appelq]   Appelquist, P.: HD-DIVINE, a Scandinavian terrestrial HDTV pro-
             ject. EBU Technical Review, No. 256, Summer 1993, pp. 16 - 19.
[Benven]   Benveniste, A.; Goursat, M.: Blind Equalizers. IEEE Trans. on
             Comm. COM-32, No. 8, August 1984, pp. 871 - 883.
[Beyers]   Beyers, B.W.; Christofer, L.; Saint Girons, R.; Teichner, D.: Di-
             recTV/USSB/DSS - das erste Multi-Kanalsystem für die digitale Sa-
             tellitenübertragung von Fernsehprogrammen. 16. Jah-restagung der
             Fernseh- u. Kinotechn. Gesellschaft (FKTG), Nürnberg 1994, pro-
             ceedings, p. 227.
[Biaesch]   Biaesch-Wiebke, C.: CD-Player und R-DAT-Recorder. 3rd edition.
             Vogel Buchverlag, Würzburg 1992.
[Bidet]   Bidet, E.; Joanblanc, C.; Senn, P.: A Fast Single Chip Implementation
             of 8192 Points Complex FFT. Proc. of the CMOS Integrated Circuits
             Conference (CICC), San Diego 1994.
[Bronst]   Bronstein, I.N.; Semendjajew, K.A.: Taschenbuch der Mathematik.
             Verlag Harri Deutsch, Thun 1984.
[Candy]   Candy, J.C.: A Use of Double Integration in Sigma Delta Modulation.
             IEEE Trans. on Comm., COM-33, No. 3, March 1985, pp. 249 - 258.
[CCIR 10]   Multi-channel stereophonic sound system with and without
             accompanying picture. CCIR Draft New Recommendation (Doc.
             10/11), Doc. 10/BL/3. March 1992.
[CCIR 563]   Radiometeorological Data. CCIR Report 563-4, 1990.
[CCIR 564]   Propagation Data and Prediction Methods Required for Earth-Space
             Telecommunication Systems. CCIR Report 564-4, 1990.
[Chester]   CEPT: The Chester 1997 Multilateral Coordination Agreement relat-
             ing to Technical Criteria, Coordination Principles and Procedures
             for the introduction of Terrestrial Digital Video Broadcasting (DVB-
             T), Chester 1997.
[Clark]   Clark, G.C.; Cain, J.B.: Error-Correction Coding for Digital
             Communications.   Plenum Press, New York 1988.
[Cost 207]   COST 207: Digital Land Mobile Radio Communications. Final re-
             port, Commission of the European Communications, Information
             Technology and Sciences, 1989.
[Costas]   Costas, J.P.: Synchronous Communications. Proc. IRE 44 (1956), pp.
             1713 - 1718.

[de Bot]   de Bot, P.G.M.; Le Floch, B.; Mignone, V.; Schütte, H.D.: An Over-
                view of the Modulation and Channel Coding Schemes Developed for
                Digital Terrestrial Television Broadcasting within the dTTb Project.
                Proc. of the Internat. Broadcasting Convention, Amsterdam 1994, p.
                569.

[DIGISMATV]   Satellite Digital TV in Collective Antenna Systems. SMATV Refer-
                ence Channel Model for Digital TV. Race Digital Satellite Master An-
                tenna Television (DIGISMATV) Project, Doc. DIGISMATV-NT-
                PART (A-D)-031-HSA, June 1994.

[DVBTNORD]   Modellversuch DVB-T Norddeutschland. Interim Report, Braun-
                schweig, March 2001.

[Engels]   Engels, V.; Rohling, H.; Breide, S.: OFDM-Übertragungsverfahren
                für den Digitalen Fernsehrundfunk. Rundfunktech. Mitt. 37 (1993),
                No. 6, pp. 260 - 270.

[EN 001]   Attachments to the Public Switched Telephone Network (PSTN);
                General technical requirements for equipment connected to an ana-
                logue subscriber interface in the PSTN. European Standard (EN) 300
                001 v1.5.1, European Telecommunications Standards Institute
                (ETSI), October 1998.

[EN 012]   Integrated Services Digital Network (ISDN); Basic User-Network In-
                terface (UNI); Part 1: Layer 1 specification. European Standard (EN)
                300 012-1 v1.2.2, European Telecommunications Standards Institute
                (ETSI), Mai 2000.

[EN 192]   Digital Video Broadcasting (DVB); DVB specification for data
                broadcasting. European Standard (EN) 301 192 v1.3.1, European Tele-
                communications Standards Institute (ETSI), January 2003.

[EN 193]   Digital Video Broadcasting (DVB); Interaction channel through the
                Digital Enhanced Cordless Telecommunications (DECT). European
                Standard (EN) 301 193 v1.1.1, European Telecommunications Stan-
                dards Institute (ETSI), July 1998.

[EN 195]   Digital Video Broadcasting (DVB); Interaction channel through the
                Global System for Mobile communications (GSM). European Stan-
                dard (EN) 301 195 v1.1.1, European Telecommunications Standards
                Institute (ETSI), February 1999.

[EN 199]   Digital Video Broadcasting (DVB); Interaction channel for Local
                Multi-point Distribution Systems (LMDS). European Standard (EN)
                301 199 v1.2.1, European Telecommunications Standards Institute
                (ETSI), June 1999.

[EN 240]   Digital Enhanced Cordless Telecommunications (DECT); Data Ser-
                vices Profile (DSP); Point-to-Point Protocol (PPP) interworking for
                internet access and general multi-protocol datagram transport. Eu-
                ropean Standard (EN) 301 240 v1.1.3, European Telecommunications
                Standards Institute (ETSI), June 1998.

[EN 403]   Integrated Services Digital Network (ISDN); Digital Subscriber Sig-
                nalling System No. 1 (DSS1) protocol; Signalling network layer for
                circuit-mode basic call control; Part 1: Protocol specification. Euro-
                pean Standard (EN) 300 403-1 v1.3.2; European Telecommunications
                Standards Institute (ETSI), November 1999.

[EN 421]   Digital Video Broadcasting (DVB); Framing structure, channel cod-
              ing and modulation for 11/12 GHz satellite services. European Stan-
              dard (EN) 300 421 v1.1.2 (formerly named ETS 421), European Tele-
              communications Standards Institute (ETSI), August 1997.

[EN 429]   Digital Video Broadcasting (DVB); Framing structure, channel cod-
              ing and modulation for cable systems. European Standard (EN) 300
              429 v1.2.1(formerly named ETS 429), European Telecommunications
              Standards Institute (ETSI), April 1998.

[EN 744]   Digital Video Broadcasting (DVB); Framing structure, channel cod-
              ing and modulation for digital terrestrial television. European Stan-
              dard (EN) 300 744 v1.4.1 (formerly named ETS 744), European Tele-
              communications Standards Institute ETSI, January 2001.

[EN 748]   Digital Video Broadcasting (DVB); Multipoint Video distribution
              Systems (MVDS) at 10 GHz and above. European Norm EN 300 748,
              European Telecommunications Standards Institute ETSI, August
              1997.

[EN 749]   Digital Video Broadcasting (DVB); Framing structure, channelcod-
              ing and modulation for Multipoint Video Distribution Systems
              (MVDS) below 10 GHz. European Norm EN 300 749, European Tele-
              communications Standards Institute ETSI, August 1997.

[EN 765]   Digital Enhanced Cordless Telecommunications (DECT); Radio in
              the Local Loop (RLL) Access Profile (RAP); Part 1: Basic telephony
              services. European Standard (EN) 300 765-1 v1.3.1, European Tele-
              communications Standards Institute (ETSI), April 2001.

[EN 790]   Digital Video Broadcasting (DVB); Interaction channel for satellite
              distribution systems. European Standard (EN) 301 790 v1.3.1, Euro-
              pean Telecommunications Standards Institute (ETSI), November
              2002.

[EN 958]   Digital Video Broadcasting (DVB); Interaction channel for Digital
              Terrestrial Television (RCT) incorporating Multiple Access OFDM.
              European Standard (EN) 301 958 v1.1.1, European Telecommunica-
              tions Standards Institute (ETSI), March 2002.

[EN 50083-9] Cabled distribution systems for television, sound and interactive
              multimedia signals; Part 9: Interfaces for CATV/SMATV headends
              and similar professioal equipment for DVB/MPEG-2 transport
              streams. European Norm EN 50083-9, Comité Européen de Normali-
              sation Electrotechnique CENELEC, June 1998.

[EN 50221]   Common Interface Specification for Conditional Access and other
              Digital Video Broadcasting Decoder Applications. European
              Norm EN 50221, Comité Européen de Normalisation Électrotech-
              nique CENELEC, February 1997.

[ES 488]   Access and Terminals (AT); Data-Over-Cable Systems. ETSI Stan-
              dard (ES) 201 488 v1.2.2, European Telecommunications Standards
              Institute (ETSI), October 2003.

[ES 800]   Digital Video Broadcasting (DVB); DVB interaction channel for Ca-
              ble TV distribution systems (CATV). ETSI Standard (ES) 200 800
              v1.3.1, European Telecommunications Standards Institute (ETSI),
              October 2001.

[ES 812]   Digital Video Broadcasting (DVB); Multimedia Home Platform
                (MHP) Specification 1.0.3. ETSI Specification ES 101 812, European
                Telecommunications Standards Institute (ETSI), December 2003.

[ETR 154]   Digital Video Broadcasting (DVB); Implementation guidelines for
                the use of MPEG-2 systems, Video and Audio in satellite, cable and
                terrestrial broadcasting applications. ETSI Technical Report TR 101
                154 (formerly named ETR 154), European Telecommunications
                Standards Institute ETSI, July 2000.

[ETR 211]   Digital Video Broadcasting (DVB); Guidelines on implementation
                and usage of Service Information (SI). ETSI Technical Report TR 101
                211 (formerly named ETR 211), European Telecommunications Stan-
                dards Institute ETSI, January 2003.

[ETR 289]   Digital Video Broadcasting (DVB); Support for use of scrambling
                and Conditional Access (CA) within digital broadcasting systems.
                ETSI Technical Report ETR 289, European Telecommunications
                Standards Institute ETSI, October 1996.

[ETR 290]   Digital Video Broadcasting (DVB); Measurement guidelines for DVB
                systems. ETSI Technical Report ETR 290, European Telecommuni-
                cations Standards Institute ETSI, August 1997.

[ETS 401]   Radio broadcasting systems: Digital Audio Broadcasting (DAB) to
                mobile, portable and fixed receivers. European Telecommunication
                Standard ETS 300 401, European Telecommunications Standards In-
                stitute (ETSI), May 1997.

[ETS 402]   Integrated Services Digital Network (ISDN); Digital Subscriber Sig-
                nalling System No. 1 (DSS1) protocol; Data link layer; Part 1: General
                aspects. European Telecommunication Standard (ETS) 300 402-1
                ed.1, European Telecommunications Standards Institute (ETSI), No-
                vember 1995.

[ETS 421]   Digital Video Broadcasting (DVB); DVB framing structure, channel
                coding and modulation for 11/12 GHz satellite services. European
                Norm EN 300 421 (formerly named ETS 421), European Telecom-
                munications Standards Institute ETSI, August 1997.

[ETS 429]   Digital Video Broadcasting (DVB); DVB framing structure, channel
                coding and modulation for cable systems. European Norm EN 300
                429 (formerly named ETS 429), European Telecommunications
                Standards Institute ETSI, April 1998.

[ETS 468]   Digital Video Broadcasting (DVB); Specification for Service Infor-
                mation (SI) in Digital Video Broadcasting (DVB) Systems. European
                Norm EN 300 468 (formerly named ETS 468), European Telecom-
                munications Standards Institute ETSI, May 2003.

[ETS 472]   Digital Video Broadcasting (DVB); Specification for conveying ITU-
                R System B Teletext in Digital Video Broadcasting (DVB) bitstreams.
                European Norm EN 300 472 (formerly named ETS 472), European
                Telecommunications Standards Institute ETSI, May 2003.

[ETS 473]   Digital Video Broadcasting (DVB); Satellite Master Antenna Televi-
                sion (SMATV) distribution systems. European Norm EN 300 473
                (formerly named ETS 473), European Telecommunications Stan-
                dards Institute ETSI, August 1997.

[ETS 500]   Digital cellular telecommunications system (Phase 2) (GSM); Principles of telecommunication services supported by a GSM Public Land Mobile Network (PLMN) (GSM 02.01). European Telecommunication Standard ETS 300 500 ed.2, European Telecommunications Standards Institute (ETSI), January 1996.

[ETS 743]   Digital Video Broadcasting (DVB); Subtitling systems. European Technical standard ETS 300 743, European Telecommunications Standards Institute ETSI, September 1997.

[ETS 744]   Digital Video Broadcasting (DVB); Framing structure, channel coding and modulation for digital terrestrial television. European Norm EN 300 744 (formerly named ETS 744), European Telecommunications Standards Institute ETSI, January 2001.

[ETS 777].   Terminal Equipment (TE); End-to-end protocols for multimedia information retrieval services; Part 1: Coding of multimedia and hypermedia information for basic multimedia applications (MHEG-5); Part 3: Application Programmable Interface (API) for MHEG-5. European Telecommunication Standard ETS 300 777, European Telecommunications Standards Institute (ETSI), 1997.

[ETS 801]   Digital Video Broadcasting (DVB); Interaction channel through Public Switched Telecommunications Network (PSTN) / Integrated Services Digital Networks (ISDN). European Telecommunication Standard (ETS) 300 801 ed.1, European Telecommunications Standards Institute (ETSI), August 1997.

[ETS 802]   Digital Video Broadcasting (DVB); Network-independent protocols for DVB interactive services. European Telecommunication Standard ETS 300 802 ed.1, European Telecommunications Standards Institute (ETSI), November 1997.

[Faria]   Faria, G.; Scalise, F.: DVB-RCT: A New Standard for Interactive TV. SMPTE Journal 111 (2002), No 1, pp 34-44.

[FCC]   Federal Communications Commission: ATV System Recommendation. IEEE Trans. on Broadcasting 39 (1993), No. 1, pp. 2 - 208.

[Feher 1]   Feher, K.: Digital Communications: Microwave Applications. Prentice-Hall Inc., Englewood Cliffs, N.J. 1981.

[Feher 2]   Feher, K. u.a.: Telecommunications Measurements, Analysis and Instrumentation. Prentice- Hall Inc., Englewood Cliffs, N.J. 1987.

[Flaherty]   Flaherty, J.A.: Digital Television and HDTV in America - A progress report. EBU Technical Review, No. 260, Summer 1994, pp. 30 - 35.

[Forney]   Forney, G.D.: Burst-Correcting Codes for the Classic Bursty Channel. IEEE Trans. on Comm. Technol. COM-10, No. 5, October 1971, pp. 772-781.

[Friedr]   Friedrich, J.; Herbig, P.; Reuber, H.J.: Theorie und Praxis bandbreiteneffizienter Modulations- und Codierungverfahren. ANT Nachrichtentechn. Berichte, No. 11, March 1994, pp. 23 - 40.

[FTZ 1]   Bezugskette für die Übertragung von Fernseh- und Tonrundfunksignalen im nationalen Breitbandkommunikations (BK)-Verteilnetz. FTZ-Richtlinie 151R8, Fernmeldetechnisches Zentralamt, Darmstadt 1985.

[FTZ 2]   Bedingungen und Empfehlungen für den Anschluß privater Breit-
               band/Rundfunk-Empfangsantennenanlagen. FTZ-Richtlinie 1R8-15,
               Fermeldetechnisches Zentralamt, Darmstadt 1985.
[Furrer]   Furrer, F.J.: Fehlerkorrigierende Block-Codierung für die Daten-
               übertragung. Birkhäuser Verlag, Basel 1981.
[Gauss]   Gauß, C.F.: Theorie der den kleinsten Fehlern unterworfenen Com-
               binationen der Beobachtungen. In: Boersch, A.; Simon, P. (Hrsg.):
               Abhandlungen zur Methode der kleinsten Quadrate. Stankiewicz
               Buchdruckerei, Berlin 1887, pp. 1 - 27.
[Gerdsen]   Gerdsen, P.; Kröger, P.: Digitale Signalverarbeitung in der Nachrich-
               tenübertragung. Springer-Verlag, Berlin - Heidelberg - New York
               1993.
[Gralli]   Grand Alliance HDTV System: Specification Version 2.0. Chapter 6:
               Transmission System. Chapter 7: Grand Alliance System Summary. 7
               December 1994.
[Gray]   Gray, Paul R.; Meyer, Robert G.: Analysis and Design of Analog Inte-
               grated Circuits, University of California, Berkeley, 1993.
[Guillou]   Guillou, L.C.; Giachetti, J.L.: Encipherment and Conditional Access.
               SMPTE Journal 103 (1994), No. 6, pp. 398 - 406.
[Hagenauer]   Hagenauer, J.: Rate-Compatible Punctured Convolutional Codes
               (RCPC Codes) and their Applications. IEEE Trans. on Comm. COM-
               36, No. 4, April 1988, pp. 389-400.
[Hessenm]   Hessenmüller, H.; Jakubowski, H.: Die Signalqualität bei digitaler
               Tonübertragung - subjektive Testergebnisse, objektive Meßverfah-
               ren. Rundfunktech. Mitt. 27 (1983), No. 1, pp. 3 - 16.
[Hopkins]   Hopkins, R.: Digital terrestrial TV for North America - The Grand
               Alliance HDTV System. EBU Technical Review, No. 260, Summer
               1994, pp. 36 - 50.
[Hung]   Hung, A.C.: PVRG-MPEG CODEC 1.1. Portable Video Research
               Group (PVRG), Stanford University, June 1993.
[IdSmitten]   In der Smitten, F.J.: Digital Video Broadcasting - Feldversuche zur
               digitalen Übertragung OFDM-codierter Farbbildsignale in einem
               terrestrischen Fernsehkanal. Rundfunktech. Mitt. 37 (1993), No. 3,
               pp. 130 - 141.
[IEC 958]   Interface audionumérique/Digital audio interface. Norme Internati-
               onale/International Standard CEI/IEC 958, 1989-03. Commission E-
               lectrotechnique Internationale/International Electrotechnical Com-
               mission, Geneva 1989.
[IRT]   Richtlinie für die Beurteilung der Fernsehversorgung bei ARD/ZDF
               und DBP. Institut für Rundfunktechnik, Richtlinie No. 5R10, Munich
               1981.
[Irwin]   Irwin, D.W.; Cox, N.R.: Hybrid Clamping in NTSC Digital Video
               Equipment. IEEE Trans. on Broadcasting 38 (1992), No. 1, pp. 19 - 26.
[ISO 7498]   Information processing systems - Open Systems Interconnection -
               Basic Reference Model. ISO International Standard IS 7498, 1984.
[ISO 10918]   Information technology - Digital compression and coding of con-
               tinuous-tone still images. ISO/IEC International Standard IS 10918,
               July 1992.

[ISO 11172]   Information technology - Coding of moving pictures and associated
                 audio for digital storage media up to about 1.5 Mbit/s. ISO/IEC In-
                 ternational Standard IS 11172, November 1992.
[ISO 13522-5]   Information technology -- Coding of multimedia and hypermedia
                 information -- Part 5: Support for base-level interactive applications.
                 ISO/IEC International Standard ISO/IEC 13522-5, 1997.
[ISO 13522-6]   Information technology -- Coding of multimedia and hypermedia
                 information -- Part 6: Support for enhanced interactive applications.
                 ISO/IEC International Standard ISO/IEC 13522-6, 1998.
[ISO 13522-7]   Information technology -- Coding of multimedia and hypermedia
                 information -- Part 7: Interoperability and conformance testing for
                 ISO/IEC 13522-5. ISO/IEC International Standard ISO/IEC 13522-7,
[ISO 13522-8]   2001.
                 Information technology -- Coding of multimedia and hypermedia
                 information -- Part 8: XML notation for ISO/IEC 13522-5. ISO/IEC
                 International Standard ISO/IEC 13522-8, 2001.

[ISO 13818]   Information technology - Generic coding of moving pictures and as-
                 sociated audio. ISO/IEC International Standard IS 13818, November
                 1994.
[ISO 13818-6]   Information technology - Generic coding of moving pictures and as-
                 sociated audio information - Part 6: Extensions for Digital Storage
                 Media - Command and Control (DSM-CC). International Standard
                 ISO/IEC 13818-6, International Organisation for Standardisation
                 (ISO), 1998.
[ISO 8802-2]   Information technology -- Telecommunications and information
                 exchange between systems -- Local and metropolitan area networks
                 -- Specific requirements -- Part 2: Logical link control. ISO/IEC In-
                 ternational Standard ISO/IEC 8802-2, 1998.
[ITU 601]   Studio Encoding Parameters of Digital Television for Standard 4:3
                 and Wide-Screen 16:9 Aspect Ratios. Draft Revision of Recommen-
                 dation ITU-R BT.-601-4. International Telecommunication Union,
                 Geneva 1994.
[ITU 656]   Interfaces for Digital Component Video Signals in 525-line and 625-
                 line Television Systems Operating at the 4:2:2 level of Recommenda-
                 tion 601. ITU-R Recommendation BT.656-1 (originally CCIR Rec.
                 656-1, Geneva 1986).
[ITU H.261]   Video codec for audiovisual services at p x 64 kbit/s. ITU-T Recom-
                 mendation H.261, March 1993.
[ITU H.263]   Video coding for low bit rate communication. ITU-T Recommenda-
                 tion H.263, February 1998.
[ITU J.83B]   Digital multi-programme systems for television, sound and data
                 services for cable distribution - Annex B. ITU-T Recommendation
                 J.83, International Telecommunications Union (ITU), April 1997.
[Jakubow]   Jakubowski, H.: Quantisierungsverzerrungen in digital arbeitenden
                 Tonsignalübertragungs- und -verarbeitungssystemen. Rundfunk-
                 tech. Mitt. 24 (1980), No. 2, pp. 91 - 92.

[Jansky]   Jansky, D.M.: Methods for Accommodation of HDTV Terrestrial
                Broadcasting. IEEE Trans. on Broadcasting 37 (1991), No. 4, pp. 152 -
                157.
[Johann]   Johann, J.: Modulationsverfahren. Nachrichtentechnik, vol. 22.
                Springer-Verlag, Berlin - Heidelberg - New York 1992.
[Kammeyer]   Kammeyer, K.D.: Nachrichtenübertragung. Verlag B.G. Teubner,
                Stuttgart 1992.
[Kenter]   Kenter, H. (Hrsg.): Ton- und Fernsehübertragungstechnik und
                Technik leitergebundener BK-Anlagen. R.v.Decker's Verlag, Heidel-
                berg 1988.
[Klinken]   van Klinken, N.; Renirie, W.: Receiving DVB-T: Technical Challen-
                ges. Proceedings of the International Broadcasting Convention, Am-
                sterdam 2000.
[Kli/Schi]   Klinkenberg, F.; Schiek, U.: Das MCP-Terminal als zukünftige SW-
                Plattform für mobile DVB Dienste. 20. Jahrestagung der Fernseh-
                und Kinotechnischen Gesellschaft, Zürich 2002, proceedings, pp
                932-948.
[Knoll]   Knoll, A.: Der MPEG-2-Standard zur digitalen Codierung von Fern-
                sehsignalen. Der Fernmeldeingenieur, Verlag f. Wissenschaft u. Le-
                ben G. Heidecker, Erlangen, July 1992.

[Ladebusch]   Ladebusch, U.: Optimierung von Datenstrukturen im digitalen
                Rundfunk. Doctoral Theses at Technical University of Braunschweig
                2000.
[Laflin]   Laflin, N.J.; Wright, D.T.: Planning and Field Testing for Digital Ter-
                restrial Television. Proc. of the Internat. Broadcasting Convention,
                Amsterdam 1994, p. 527.
[Lee]   Lee, E.A.; Messerschmitt, D.G.: Digital Communication. 2nd edition.
                Kluwer Academic Publishers, Boston-Dordrecht-London 1994.
[Liss]   Liss, C.: Diversity-Empfang für mobile DVB-T-Systeme. Fernseh-
                und Kinotechnik 56 (2002), No 1-2, pp 425-228
[Lohsch]   Lohscheller, H.: A Subjectively Adapted Image Communication Sys-
                tem. IEEE Trans. on Comm. COM-32, No. 12, December 1984, pp.
                1316 - 1322.
[Lüke 1]   Lüke, H.D.: Signalübertragung - Grundlagen der digitalen und ana-
                logen Nachrichtenübertragungssysteme. 5th edition. Springer-Verlag,
                Berlin - Heidelberg - New York 1992.
[Lüke 2]   Lüke, H.D.: Korrelationssignale. Springer-Verlag, Berlin - Heidel-
                berg - New York 1992.
[Mäusl]   Mäusl, R.: Digitale Modulationsverfahren. 3. Auflage. Telekommu-
                nikation, Bd. 2. Hüthig Buch-Verlag, Heidelberg 1991.
[Morgenst]   Morgenstern, G.: Zur Berechnung der spektralen Leistungsdichte
                von digitalen Basisband-Signalen. Der Fernmelde-Ingenieur 33
                (1979), No. 12, pp. 1 - 39.
[Muschall]   Muschallik, C.: Einfluß der Oszillatoren im Frontend auf ein OFDM-
                Signal. Fernseh- und Kino-Technik 49 (1995), No. 4, pp. 196 - 205.
[Muschall2]   Muschallik, C.: Ein Beitrag zur Optimierung der Empfangbarkeit
                von Orthogonal-Frequency-Division-Multiplexing-(OFDM) Signa-
                len. Doctoral Theses at Technical University of Braunschweig 2000.

[Musmann]   Musmann, H.G.; Pirsch, P.; Grallert, H.D.: Advances in Picture Coding. Proc. IEEE 73 (1985), No. 4, pp. 523 - 548.

[NHK]   Nippon Hoso Kyokai (NHK): Digital Modulation Scheme for ISDB (Integrated Services Digital Broadcasting) in the 12 GHz Band. NHK Laboratories Note No. 428, September 1994.

[North]   North, D.O.: An Analysis of the Factors which Determine Signal/Noise Discrimination in Pulsed-Carrier Systems. Proc. IRE 51 (1963), No. 7. pp. 1016 - 1027.

[Nyquist]   Nyquist, H.: Certain Topics in Telegraph Transmission Theory. Trans. of the American Institute of Electrical Engineers 47, February 1928, pp. 617 - 644.

[Oppenhm]   Oppenheim, A.V.; Willsky, A.S.: Signale und Systeme. VCH-Verlag, Weinheim 1989.

[Park]   Park, S.: Principles of Sigma-Delta Modulation for Analog-to-Digital Converters. Application Note APR8/D, Motorola Inc. 1990.

[Pennebk]   Pennebaker, W.B.; Mitchell, J.L.: JPEG still image data compression standard. Van Nostrand Reinhold, New York 1993.

[Petke]   Petke, G.: Planungsaspekte für digitales terrestrisches Fernsehen. Rundfunktech. Mitt. 37 (1993), No. 3, pp. 119 - 129.

[Proakis]   Proakis, J.G.: Digital Communications. Second Edition. McGraw-Hill Book Company, New York 1989.

[Reed]   Reed, S.; Solomon, G.: Polynomial Codes over Certain Finite Fields. Journal of the Society for Industrial and Applied Mathematics 8, June 1960, pp. 300-304.

[Reimers 1]   Reimers, U.: European perspectives on digital television broadcasting - Conclusions of the Working Group on Digital Television Broadcasting (WGDTB). EBU Technical Review, No. 256, Summer 1993, pp. 3 - 8.

[Reimers 2]   Reimers, U.: Systemkonzepte für das Digitale Fernsehen in Europa. Fernseh- und Kino-Technik 47 (1993), No. 7-8, pp. 451 - 461.

[Reimers 3]   Reimers, U.: Das europäische Systemkonzept für die Übertragung digitalisierter Fernsehsignale per Satellit. Fernseh- und Kino-Technik 48 (1994), No. 3, pp. 115 - 123.

[Reimers 4]   Reimers, U.: Concept of a European System for the Transmission of Digitized Television Signals via Satellite. SMPTE Journal 103 (1994), No. 11, pp. 741 - 747.

[Reimers 5]   Reimers, U.: Digitales Fernsehen für Europa - Ein Statusbericht. Fernseh- und Kino-Technik 48 (1994), No. 10, pp. 517 - 519.

[Reimers 6]   Reimers, U.; Behnke, J.; Ladebusch, U.; Liss, C.; Petke, G.; Wächter, Th.: Die Kosten von DVB-T Sendernetzen. Fernseh- und Kinotechnik 55 (2001), No 55, pp 271-278

[Rhodes]   Rhodes, C.W.: Measuring Peak and Average Power of Digitally Modulated Advanced Television Systems. IEEE Trans. on Broadcasting 38, No. 4, December 1992, pp. 197 - 201.

[Roddy]   Roddy, D.: Satellitenkommunikation. Hanser Verlag, München-Wien / Prentice Hall, London 1991.

[Rohde]   Encoder Test-Sequenz. Produktinformation DVTS PD 757.1790.21, Rohde & Schwarz GmbH & Co. KG, Munich, June 1995.

[Roy]   Roy, A.; Reimers, U.: FDMA-Betrieb eines Satelliten-Transponders
                für die Zuführung mehrerer TV-Programme zu terrestrischen PAL-
                Sendern. Rundfunktechn. Mitt. 39 (1995), No. 2, pp. 63 - 70.

[Ruelbg]   Ruelberg, K.D.: Kanalcodierung und Modulation für digitale Fern-
                sehübertragung über Satellit, Kabel und terrestrische Sendernetze.
                16. Jahrestagung der Fernseh- und Kinotechn. Gesellschaft (FKTG),
                Nürnberg 1994, proceedings, pp. 139 - 162.

[Schaaf]   Schaaf, C.: Digital Delivery by Cable. IEE Symposium on Emerging
                Broadcast Technology, Cambridge 1994.

[Scherer]   Scherer, K.: Measurement Tools for Digital Video Transmission.
                IEEE Trans. on Broadcasting 39, No. 4, December 1993, pp. 350 - 363.

[Schönfd 1]   Schönfelder, H. (Hrsg.): Digitale Filter in der Videotechnik. Drei-R-
                Verlag, Berlin 1988.

[Schönfd 2]   Schönfelder, H.: Fernsehtechnik. Teil 2. Justus von Liebig-Verlag,
                Darmstadt 1973.

[Schöps]   Schöps, G.: Bandbreiteneffiziente Modulationsverfahren für die Ka-
                belübertragung von Digital-HDTV. Rundfunktechn. Mitt. 38 (1994),
                No. 4, pp. 60 - 67.

[Schüssler]   Schüßler, H.W.: Digitale Signalverarbeitung. Band I: Analyse diskre-
                ter Signale und Systeme. Springer-Verlag, Berlin - Heidelberg - New
[Sieburg]   York 1988.

                Sieburg, M.; Vergleich von MHP 1.0 mit MHP 1.1. Fernseh- und Ki-
                notechnik 55. Jahrgang (2001), pp 604-606.

[Skritek]   Skritek, P.: Handbuch der Audio-Schaltungstechnik. Franzis-Verlag,
                Munich 1988.

[Söder]   Söder, G.; Tröndle, K.: Digitale Übertragungssysteme. Theorie, Op-
                timierung und Dimensionierung der Basisbandsysteme. Springer-
                Verlag, Berlin - Heidelberg - New York - Tokyo 1985.

[Stenger]   Stenger, L.: Das Systemkonzept für die Übertragung digitaler Fern-
                sehsignale im Kabel. 16. Jahrestagung der Fernseh- und Kinotechn.
                Gesellschaft (FKTG), Nürnberg 1994, proceedings, pp. 67 - 80.

[Stoll 1]   Stoll, G.; Theile, G.: MASCAM: Minimale Datenrate durch Berück-
                sichtigung der Gehöreigenschaften bei der Codierung hochwertiger
                Tonsignale. Fernseh- und Kino-Technik 42 (1988), pp. 551 -558.

[Stoll 2]   Stoll, G.: MPEG-2 Layer 2 "Musicam-Surround": The new mul-
                tichannel audio coding standard for broadcast, telecommunication
                and multimedia. 16. Jahrestagung der Fernseh- und Kinotechn. Ge-
                sellschaft (FKTG), Nürnberg 1994, proceedings, pp. 101 - 113.

[Strubenh]   Strubenhoff: Synchronisationsverfahren für die OFDM-Übertragung
                beim digitalen terrestrischen Fernsehen. Doctoral Theses at Univer-
                sity of Wuppertal, VDI-Verlag, Düsseldorf 1997.

[Sweeney]   Sweeney, P.: Codierung zur Fehlererkennung und Fehlerkorrektur.
                Hanser Verlag, München-Wien / Prentice Hall, London 1992.

[Teichner]   Teichner, D.; Herpel, C.; Schröder, E.F.; Spille, J.; Riemann, U.: Der
                MPEG-2-Standard - Generische Codierung für Bewegtbilder und zu-
                gehörige Audio-Information. Fernseh- und Kino-Technik 48 (1994),
                Nos. 4 - 10.

[Telekom]   Preliminary Criteria for Planning Digital Terrestrial Television Ser-
                vices. Deutsche Bundespost Telekom, ITU-R Document 11C/28-E, 28
                July 1993.

[Theile]   Theile, G.; Stoll, G.; Link, M.: Low bitrate coding of high quality au-
                dio signals - An introduction to the MASCAM system. EBU Techni-
                cal Review, No. 230, August 1988, pp. 158 - 181.

[Tietze]   Tietze, U.; Schenk, Ch.: Halbleiter-Schaltungstechnik. 10. Auflage.
                Springer-Verlag, Berlin - Heidelberg - New York 1993.

[TR 154]   Digital Video Broadcasting (DVB); Implementation guidelines for
                the use of MPEG-2 Systems, Video and Audio in satellite, cable and
                terrestrial broadcasting applications. Technical Report TR 101 154,
                European Telecommunications Standards Institute (ETSI), July
                2000.

[TR 194]   Digital Video Broadcasting (DVB); Guidelines for implementation
                and usage of the specification of network independent protocols for
                DVB interactive services. Technical Report TR 101 194 V1.1.1, Euro-
                pean Telecommunications Standards Institute (ETSI), June 1997.

[TR 196]   Digital Video Broadcasting (DVB); Interaction channel for Cable TV
                distribution systems (CATV); Guidelines for the use of ETS 300 800.
                Technical Report TR 101 196 V1.1.1, European Telecommunications
                Standards Institute (ETSI), December 1997.

[TR 201]   Digital Video Broadcasting (DVB); Interaction channel for Satellite
                Master Antenna TV (SMATV) distribution systems; Guidelines for
                versions based on satellite and coaxial sections. Technical Report TR
                101 201 V1.1.1, European Telecommunications Standards Institute
                (ETSI), October 1997.

[TR 205]   Digital Video Broadcasting (DVB); Guidelines for the Implementa-
                tion and Usage of the DVB Interaction Channel for Local Multipoint
                Distribution Systems (LMDS). Technical Report TR 101 205 V1.1.2,
                European Telecommunications Standards Institute (ETSI), July
                2001.

[TR 790]   Digital Video Broadcasting (DVB); Interaction channel for Satellite
                Distribution Systems; Guidelines for the use of EN 301 790. Techni-
                cal Report TR 101 790 V1.1.1, European Telecommunications Stan-
                dards Institute (ETSI), September 2001.

[TS 006]   Digital Video Broadcasting (DVB); Specification for System Software
                Update in DVB Systems. Technical Specification ETSI TS 102 006
                V1.2.1 (2002-10). European Telecommunications Standards Institute
                ETSI, October 2002.

[TS 197]   Digital Video Broadcasting (DVB); DVB Simulcrypt; Part 1:
                Head-end architecture and synchronization. Technical Specifica-
                tion TS 101 197-1. European Telecommunications Standards Insti-
                tute ETSI, June 1997.

[TS 812]   Digital Video Broadcasting (DVB); Multimedia Home Platform
                (MHP) Specification 1.1.1. Technical Specification TS 102 812, Euro-
                pean Telecommunications Standards Institute (ETSI), June 2003.

[TS 819]   Digital Video Broadcasting (DVB); Globally Executable MHP (GEM)
                Specification 1.0.0. Technical Specification TS 102 819, European
                Telecommunications Standards Institute (ETSI), January 2003.

[TS 993]   Digital Audio Broadcasting (DAB); A Virtual Machine for DAB: DAB
                Java Specification. Technical Specification TS 101 993, European Te-
                lecommunications Standards Institute (ETSI), March 2002.

[Unger]   Unger, H.G.: Elektromagnetische Wellen auf Leitungen. 2nd edition.
                Hüthig-Verlag, Heidelberg 1986.

[Unger 2]   Unger, H.-G.: Hochfrequenztechnik in Funk und Radar. 3rd rev. ed.,
                Teubner Studienskripten No. 18, Stuttgart 1988.

[Velders]   Velders, A.M.; Tichelaar, J.Y.: Measurement on PALplus Signals in
                Dutch CATV Networks. Dutch National Platform HDTV, Hilversum
                1993.

[Viterbi]   Viterbi, A.J.: Error Bounds for Convolutional Codes and an Asymp-
                totically Optimum   Decoding Algorithm. IEEE Trans. on Informa-
                tion Theory IT-13 (1967), No. 2, pp. 260-269.

[Wallace]   Wallace, G.K.: The JPEG Still Picture Compression Standard. IEEE
                Trans. on Consumer Electronics CE-38, No. 1, February 1992.

[Wellhsn]   Wellhausen, H.W.; Martin, D.: Fehlerhäufigkeitsmessungen. Nach-
                richtentechn. Zeitschrift 24, Heft 11, November 1971, pp. 553 - 558.

[WGDTB]   Reimers, U. et al.: Report to the European Launching Group on the
                Prospects for Digital Terrestrial Television. Europäisches DVB-
                Projekt, Document WGDTB 1063, Geneva, November 1992.

[Zander]   Zander, H.: Die digitale Audiotechnik. Grundlagen und Verfahren.
                1st edition. Drei-R-Verlag, Berlin 1987.

[Zwicker]   Zwicker, E.: Psychoakustik. Springer-Verlag, Berlin 1982.

17 Acronyms and Abbreviations

AAL	ATM adaptation layer
ABSOC	Advanced Broadcasting Systems of Canada
AC	alternating-current (component)
ACI	adjacent channel interference
A/D	analogue-to-digital
ADC	analogue-to-digital converter
ADSL	Asymmetrical Digital Subscriber Line
ADVEF	Advanced Television Enhancement Forum
AES	Audio Engineering Society
AF	adaptation field
AFC	automatic frequency control
AGC	automatic gain control
AIT	application information table
AM	amplitude modulation
API	application programming interface
ARD	Arbeitsgemeinschaft der öffentlich-rechtlichen Rundfunkanstalten der Bundesrepublik Deutschland (one of the German public broadcasters)
ASK	amplitude shift keying
ATM	asynchronous transfer mode
ATV	advanced television
AWGN	additive white Gaussian noise
AWT	abstract window toolkit
BAT	bouquet association table
BCH	Bose-Chaudhuri-Hocquenghem (code)
BER	bit-error rate
BOM	bill of material
BP	band-pass filter
BS	burst structure
BSS	broadcasting-satellite service
BTA	Broadcasting Technology Association (Japanese industrial organisation)
CA	conditional access
CAT	conditional-access table
CATV	community-antenna television
CCI	co-channel interference
CCIR	Comité Consultatif International de Radiodiffusion (now ITU-R)

CCITT	Comité Consultatif International du Télégraphe et du Téléphone (now ITU-T)
CD, CD-I	compact disc, compact disc – interactive
CDC	connected device configuration
CEC	Commission of the European Communities
CEI/IEC	Commission Électrotechnique Internationale/International Electrotechnical Commission
CENELEC	Comité Européen de Normalisation Électrotechnique
CEPT	Conférence Européenne des Postes et des Télécommunications
C/I	carrier-to-interference ratio (in dB)
CLDC	connected limited device configuration
CLK	clock (timing signal)
CM	cable modem
CMOS	complementary metal-oxide semiconductor
CMTS	cable modem termination system
C/N	carrier-to-noise ratio (in dB)
CNR	carrier-to-noise ratio = C/N
COFDM	coded orthogonal frequency-division multiplex
CRC	cyclic redundancy check
CRL	certificate revocation list
CSS	cascading stylesheets
CVBS	colour-video blanking synchronisation
D2-MAC	television transmission standard (sound and data: digital [duobinary]; image:multiplexed analogue component)
DA	distribution amplifier in broadband cable plants
D/A	digital-to-analogue
DAB	Digital Audio Broadcasting
DAC	digital-to-analogue converter
DAPSK	differential amplitude phase-shift keying
DAVIC	Digital Audio Visual Council
DBP	Deutsche Bundespost (former Federal German postal authority)
DC	direct-current (component)
DCO	digitally controlled oscillator
DCT	discrete cosine transform
DECT	Digital Enhanced Cordless Telecommunication
DES	data encryption standard
DFT	discrete Fourier transform
DOCSIS	Data-Over-Cable Service Interface Specification
DOM	document object model
DPCM	differential pulse-code modulation
DSC	digital serial components
DSM-CC	digital storage media- command and control
DSM-CC U-N	DSM-CC user-to-network interface
DSP	data services profile
DSR	digital satellite radio
DSS	digital satellite system
DTMF	Dual Tone Multi-Frequency

DTS	decoding time stamp
DTVB	Digital Television Broadcasting
DVB	Digital Video Broadcasting
DVB-C	DVB Cable
DVB-CM	DVB Commercial Module
DVB-J	DVB Java
DVB-MC	DVB Microwave Cable-based
DVB-MG	DVB Measurement Group
DVB-MS	DVB Microwave Satellite-based
DVB-NIP	DVB Network-Independent Protocols
DVB-PI	DVB Physical Interface
DVB-RCC	DVB Return Channel for Cable
DVB-RCCS	DVB Return Channel for SMATV
DVB-RCD	DVB Return Channel for DECT
DVB-RCG	DVB Return Channel for GSM
DVB-RCL	DVB Return Channel for LMDS
DVB-RCP	DVB Return Channel for PSTN
DVB-RCS	DVB Return Channel for Satellite
DVB-RCT	DVB Return Channel Terrestrial
DVB-S	DVB Satellite
DVB-T	DVB Terrestrial
DVC	digital video cassette
DVD	Digital Versatile Disc
EAV	end of active video
EBU	European Broadcasting Union = EUR
ECM	entitlement control message
ECMA	European Computer Manufacturers Association
EDTV	enhanced-definition television
EIRP	equivalent isotropically radiated power
EIT	event information table
EMC	electromagnetic compatibility
EMM	entitlement management message
EN	European Norm
EPG	electronic program guide
ERO	European Radiocommunications Office
ES	elementary stream
ETR	ETSI Technical Report
ETS	European Telecommunication Standard
ETSI	European Telecommunications Standards Institute
EX-OR	exclusive-or (function) = XOR
FCC	Federal Communications Commission
FDMA	frequency-division multiple access
FEC	forward error correction
FFT	fast Fourier transform
FIFO	first in - first out (memory)

FKTG	Fernseh- und Kinotechnische Gesellschaft (German Society for Television, Film and Electronic Media)
FM	frequency modulation
FP	fixed part
FSK	frequency-shift keying
FSS	fixed-satellite service
FT	fixed terminal
FTZ	Fernmeldetechnisches Zentralamt (former German central office of telecommunications)
GF	Galois field
GIF	graphics interchange format
GOP	group of pictures
GSM	Global System for Mobile Communications
GT	grouping terminal
HAVI	Home Audio Video Interoperability
HD-MAC	high-definition MAC
HDTV	high-definition television
HE	head end of broadband cable plants
HEX	hexadecimal
HP	high-pass filter
HTML	Hypertext mark-up language
HTTP	hypertext transfer protocol
HTTPS	hypertext transfer protocol over secure socket layer
IB	in-band
IBO	input back-off
IC	integrated circuit
I²C	inter IC – inter integrated circuit (bus)
ID	identification, identifier
IDCT	inverse discrete cosine transform
IDFT	inverse discrete Fourier transform
IDTV	integrated digital television
IEC	International Electrotechnical Commission
IEEE	Institute of Electrical and Electronics Engineers
IF	intermediate frequency
IMUX	input multiplex
INA	interactive network adapter
IP	Internet protocol
I/Q	in-phase/quadrature phase
IRT	Institut für Rundfunktechnik (research centre of public broadcasting organisations [in Austria, Germany and Switzerland])
ISDN	integrated services digital network
ISI	intersymbol interference
ISO	International Standardization Organization
ITU	International Telecommunication Union = UIT

ITU-R	International Telecommunication Union - Radiocommunication Sector
ITU-T	International Telecommunication Union - Telecommunication Standardisation Sector
IWU	interworking unit
J2SE	Java 2 Standard Edition
J2ME	Java 2 Micro Edition
JPEG	Joint Photographic Experts Group
JSR	Java specification request
JTWI	Java technology for the wireless
LDTV	low-definition television
LED	light-emitting diode
LF	loop filter (of PLL)
LMDS	local multipoint distribution system
LNA	low-noise amplifier
LNB	low-noise block
LP	low-pass (filter or signal)
LQFP	low profile quad flat pack
LSB	least significant bit
MAC	Multiplexed Analogue Component
MAC	medium access control
MAS	medium access scheme
MCP	Multimedia Car Platform
MD5	message digest (version 5)
MDCT	modified discrete cosine transform
MF	multi-frequency
MHEG	Multimedia and Hypermedia information coding Experts Group
MHP	Multimedia Home Platform
MIDP	Mobile Information Device Profile
MMDS	Microwave Multipoint Distribution System
MP@ML	main profile at main level
MPEG	Moving Pictures Experts Group
MR	multiresolution
MR-QAM	multiresolution QAM
MS	mobile station
MSB	most significant bit
MUSE	Multiple Subsampling Encoding (Japanese analogue high-definition television system)
MUX	multiplex, multiplexer
NAB	National Association of Broadcasters
NCC	network control centre
NCR	network clock reference
NHK	Nippon Hoso Kyokai (Japanese broadcasting corporation)
NIM	network interface module

NIT	network information table
NIU	network interface unit
NRZ	non-return to zero
NTSC	National Television Systems Committee
OBO	output back-off
OEM	original equipment manufacturer
OFDM	orthogonal frequency division multiplex
OMUX	output multiplexer
OOB	out-of-band
OSGi	Open Service Gateway initiative
OSI	open systems interconnection
PACF	periodically recurring autocorrelation function
PAL	Phase Alternating Line
PALplus	enhanced PAL system
PAT	program association table
PCB	printed circuit board
PCI	peripheral component interconnect
PCM	pulse-code modulation
PCR	program clock reference
PD	phase discriminator
PDA	personal digital assistant
PDH	plesiochronous digital hierarchy
PDM	pulse-density modulation
PES	packetised elementary stream
PGA	programmable gain amplifier
PH	PES header
PID	packet identification
PLL	phase-locked loop
PMT	program map table
PNG	portable networks graphics
PP	portable part
PPP	point-to-point protocol
PRBS	pseudo-random binary sequence
PS	program stream
PSD	power spectral density
PSI	program-specific information
PSK	phase-shift keying
PSTN	Public Switched Telephone Network
PTS	presentation time stamps
PVRG	Portable Video Research Group
QAM	quadrature amplitude modulation
QEF	quasi error-free
QPSK	quadrature phase-shift keying
RAP	radio in the local loop access profile

RCST	return channel satellite terminal
RCTT	return channel terrestrial terminal
RF	radio frequency
RMI	remote method invocation
ROM	read-only memory
RS	Reed-Solomon (code)
RSA	Rivest / Shamir / Adelmann (asymmetrical encryption scheme)
RSSI	received signal strength indicator
RST	running-status table

SAR	successive approximation register
SAT FW	satellite forward transmission
SAT RT	satellite return transmission
SAV	start of active video
SAW	surface acoustic wave
SCR	system clock reference
SDT	service description table
SDTV	standard-definition television
SECAM	Séquentiel couleur à mémoire (French-developed analogue colour TV system)
SER	symbol-error rate
SHA-1	secure hash algorithm
S&H	sample-and-hold (circuit)
SI	service information
SIF	source input format
SIT	satellite interactive terminal
(S)MATV	(satellite) master antenna television
SMPTE	Society of Motion Picture and Television Engineers
SMS	subscriber management system
S/N	signal-to-noise ratio (in dB)
SNMP	simple network management protocol
SNR	signal-to-noise ratio
SSL	secure socket layer
STB	set-top box
STC	system time clock
STERNE	Système de télévision en radiodiffusion numérique

TA	trunk amplifier in broadband cable plants
TA_A	trunk amplifier of network level A in broadband cable plants
TA_B	trunk amplifier of network level B in broadband cable plants
TCP	transport control protocol
TDMA	time division multiple access
TDT	time and data table
TF	transmission frame
TH	transport-stream header
TLS	transport layer security
TM	Technical Module (of the DVB Project)
TP	transition point

TPS	transmission parameter signalling
TRAC	telecommunications return access concentrator
TS	transport stream
T-STD	transport stream system target decoder
TV	television
TWTA	travelling wave tube amplifier
UDP	user datagram protocol
UER	Union Européenne de Radiodiffusion = EBU
UHF	ultra-high frequencies (470 - 862 MHz, television)
UIT	Union Internationale des Télécommunications = ITU
UNO-CDR	universal networked object-common data representation
UNO-RPC	universal networked object-remote procedure call
USB	universal serial bus
VBS	video blanking synchronisation
VCO	voltage-controlled oscillator
VHF	very high frequency (47 - 300 MHz, television)
VHS	video home system (standard for video recorders)
VLC	variable-length coder
VLD	variable-length decoder
VSB	vestigial-sideband amplitude modulation
WGDTB	Working Group on Digital Television Broadcasting
WRS	wireless relay station
XHTML	extensible hypertext mark-up language
XML	extensible mark-up language
XOR	exclusive-or (function) = EX-OR
ZDF	Zweites Deutsches Fernsehen (one of the German public broadcasters)
ΔM	delta modulator
$\Sigma\Delta M$	sigma-delta modulator

18 Index

ΔM signal 34
16-QAM 214
32-QAM 214
64-QAM 209
128-QAM 214
256-QAM 214
2k OFDM 246
8k OFDM 245
8-packet structure 201
abstract windowing toolkit (AWT) 345
access mode 317
active line 22
additive white Gaussian noise
  (AWGN) 202, 238
Advanced Television Systems Committee (ATSC) 5
aliasing 21
allocation process (time slots) 323
amplitude modulation 156
– noise 374
– resolution 21
– shift keying 155
analogue-to-digital converters (ADCs)
  21
antenna 259
– diversity 269
– pre-amplifier 264
aperture jitter 22
application information table (AIT)
  342
– manager 343
– programming interface (API) 331
asynchronous serial interface (ASI)
  367
– data streaming 287
ATM cell 315, 326
audio CD 29
– coding 45
– -level diagrams 29

auditory sensation area 39
authentication 348
automaton 132
auxiliary data 46

bandwidth 148
baseband signals 22
basis functions 62
BIOP-message 293
bipolar transmission 141
bit allocation 51
– error 112
– -error rate 114
– interleaver 246
– stream formatting 50
block cipher 183
– code 113
– interleaver 142
– -matching 77
bouquet association table (BAT) 100
B-pictures 78
buffer 75
– control 80
burst 329
– error 142
– structure 329

C/N ratio 374
cable modem termination system 318
carrier frequency 156
– phase 199
– recovery 199
– signal 142
car receiver 258
CATV 207
certificate 349
channel bandwidth 203
– coding 111
– estimation 253

circular polarisation 190
clamping circuits 24
class loader 348
code rate 114, 130
coded orthogonal frequency division
    multiplex (COFDM) 244
Comité Européen de Normalisation
    Electrotechnique (CENELEC) 17
common interface (CI) 185
– scrambling system 181
concatenated error protection 193
concatenation 140
conditional access (CA) 181
– access table (CAT) 100
conformance 366
constellation 374
– diagram 154, 197
constraint length 130, 139
continual pilots 252
control application 339
– word 183
convolutional code 113, 129
convolutional code rate 139
– interleaver 142
Costas loop 159, 199
covering radius 206
critical subsampling 45, 49
cyclic redundancy-check code 370

data carousel 289
– container 10, 194
– piping286
– rate 22
– rate: - cable return channel 314, 319
        - terrestrial return channel 330
– streaming 287
decoding time stamp (DTS) 95
delta modulator (ΔM) 32
diameter of the reception antenna 193
dibit 160, 167
differential coding 73
Digital Audio Broadcast (DAB) Java
    359
– Audio Visual Council (DAVIC) 17
– Storage Media – Command and
    Control (DSM-CC) 289, 340
digital modulation 155
– satellite radio (DSR) 29
– serial component (DSC) 28

– -to-analogue converters (DACs) 21
discrete cosine transform (DCT) 60
– Fourier transform 175
dither 34
doppler shift 280
downlink 187
DownloadCancelMessage (DCM) 291
DownloadDataBlock (DDB) 291
DownloadDataMessage (DDM) 291
DownloadInfoIndication (DII) 289
DownloadServerInitiate (DSI) 290
dual-channel audio 37
– -channel sound 30
DVB cable standard (DVB-C) 207
DVB-HTML 339, 351, 352
DVB-Java 339
DVB Measurement Group (DVB-MG) 363
DVB Physical Interface Group (DVB-PI)
    366
DVB Project 4
DVB satellite standard (DVB-S) 187, 206
DVB terrestrial standard (DVB-T) 237
dynamic element matching 34

echo 173
– profile 239
efficiency 128
electronic programme guide (EPG) 285,
    333
empty-channel noise level 30
end of active video (EAV) 26
energy dispersal 191
– dispersal remover 195
enhanced television profile 338
entitlement control message (ECM) 184
– management message (EMM) 184
equaliser 173
equivalent isotropically radiated power
    (EIRP) 190
error correction 111
– rate 374
– vector 122
Euclidean distance 154
European Telecommunications Stan-
    dards Institute (ETSI) 17, 207
event information table (EIT) 101
extensible hypertext mark-up language
    (XHTML) 351
eye diagram 149, 374

fading 173
filter bank 45
flash conversion 24
footroom 29
format 4:2:0 28
format 4:2:2 23
format 4:4:4 22
forward error correction (FEC) 111
frequency modulation 189
– range 188
– shift keying 155

Galois field 115, 127
gap-filling 250
Gaussian channel 202
– impulse 40
generator polynomial 127, 130, 196
geostationary orbit 187
glitches 25
globally executable MHP (GEM) 358
Grand Alliance 5
Gray coding 197
– mapping 247
group of pictures 79
guard interval 178, 245

hard decision 136
HD-MAC 4
head-end 207
headroom 30
hierarchical modulation 270
high profile 84
Home Audio Video Interoperability
  (HAVi) 346
Huffman coding 52, 61, 65
hydrometeorological zone 205

IF processing 261
I/Q amplitude ratio 377
– pre-filter 377
image quality 364
impulse response 151
in-band 313
inner interleaver 246
input back-off (IBO) 189
– multiplex (IMUX) filter 188
Institute of Electrical and Electronics
  Engineering (IEEE) 300
integrated receiver decoder 3, 55

– digital television receiver (IDTV) 257
interaction channel 341, 354
interactive network adapter 306, 313
– services 303
– television profile 338
interactivity 303
interference 374
interleaver 196
interleaving 141, 194
intermediate frequency (IF) 191, 205
internet access profile 351
Internet Protocol version 4 (IPv4) 297
– Protocol version 6 (IPv6) 297
intermodulation 189, 197
interval usage 319
intersymbol interference (ISI) 239, 374
I-pictures 78
IP/MAC Notification Table (INT) 298
irrelevance reduction 36, 60
IRD 213
ISOB-T 9
ISO/OSI layer model 93
ITU 207
ITU –T H.261 60

Java media framework (JMF) 344
jitter 371
Joint Photographic Experts Group
  (JPEG) 59

level 88
– range 21, 24
life cycle 342
linear polarisation 190
linearity 374
Logical Link Control/SubNetwork
  Attachment Point 295
loudness perception 39
low-noise block (LNB) 191

macroblock 75
main level 72
– profile 72
masking 40
– effects 40
– threshold 40
matched filter 151, 154
maximum-length sequences (m-se-
  quences) 254

mean power 164
measurement guidelines 363
medium access control (MAC) proto-
    col 306
    – cable 316, 319
    – terrestrial 330
medium access scheme 329
MHP experts group 357
microwave multichannel/multipoint
    distribution system (MMDS) 206,
    235
middleware 331, 332
minislot 316, 319
mobile information device profile
    (MIDP) 360
    – MHP 356
    – station (GSM) 312
modem:
    – cable 313
    – DOCSIS 318
    – telephone 309
modified discrete cosine transforma-
    tion (MDCT) 53
moduleID 292
moduleSize 292
monitoring 367
motion estimation 74
    – vectors 74
Moving Pictures Experts Group
    (MPEG) 7, 37
MPEG-1 60, 71
MPEG layer 1 49
MPEG-2 60
MPEG-2 systems 91
MPEG-4 72
multicrypt 185
Multimedia and Hypermedia
    Information Coding Experts Group
    (MHEG) 335
    – Car Platform (MCP) 357
    – Home Platform (MHP) 3, 11, 331, 350
multiple subsampling encoding
    (MUSE) 3
multiplex clock 24
multiresolution QAM 273

narrow-band noise 41
navigator 339, 343
network information table (NIT) 99

– interface module (NIM) 267
– interface unit 306, 313
noise bandwidth 381
nonlinear quantisation (companding) 30
Nyquist criterion 148, 197
    – filter 197
    – frequency 149
    – limit 19

object carousel 289
open card framework 355
Open Service Gateway Initiative (OSGi)
    359
Organisation Unique Identifier (OUI)
    300
output back-off (OBO) 189
    – multiplex filter (OMUX) 189
out-of-band 313
overdriving protection 24
oversampling 21, 31, 32

packet identification (PID) 96
packetised elementary stream (PES) 91
parabolic reflector 191
pay TV 181
pay-per-channel 181
    – -per view 181
PCM signal 38
peak-to-average power ratio 377
permission request file 348
personal video recorder (PVR) 3
phase noise 374, 378
    – shift keying 155
plug-in 337, 355
polarisation decoupling 190
polyphase filter bank 45
portable part (DECT) 310
post-filter 24
    – -masking 43
power density at OFDM 378
P-pictures 78
prediction 75
pre-filter 22
pre-masking 43, 51
presentation time stamp (PTS) 97
profile 88
program association table (PAT) 98
    – clock reference (PCR) 95
    – map table (PMT) 99

– stream 91
– -specific information (PSI) 99
protocol tester 363
– testing 366
pseudo-random generator 195
– -random sequence 193
phychoacoustic coding 49
– model 45
pulse shape 148
puncturing 138

quadrature amplitude modulation
    (QAM) 155, 163, 168, 209
– angle 271
– phase shift keying (QPSK) 160, 161,
    168, 189
quantisation 21, 60, 63
– error 21
– noise 22, 45
– steps 21

R-2R resistance network 25
Rayleigh channel 241
redundancy reduction 38, 65
Reed-Solomon code 114
reflection 212
residual bit-error rate 128
return channel 341
return channel:
– cable (RCC) 313
– DOCSIS 318
– LMDS 320
– satellite (RCS) 321
– SMATV 323
– telephone 308
– terrestrial (RCT) 325
return loss 212
Ricean channel 238
roll-off factor 149, 197
rooftop antenna 238
RS code vector 121
running status table (RST) 101

S/N ratio 374
sample-and-hold function of the DAC
    23
sampling 21
– clock 21
– frequency 26

sandbox 347
satellite channel 187
– master antenna television (SMATV) 213
– transmission 187
saturation point 189
scaling factor 50
scattered pilots 252
scrambling 181
section 103
service continuity 205
– description table (SDT) 101
– information (SI) 93, 99
shared medium:
– cable 314
– satellite 322
– terrestrial 329
sigma-delta modulator ($\Sigma\Delta M$) 32
signal-to-noise power 160
– -to-noise ratio 22
– -to-interference ratio 197
silicon tuner 266
simulcrypt 186
simultaneous masking 43
single-frequency network (SFN) 244
SNR scalability 86
soft decision 136, 200
software platform 331
sound-pressure level 39
source coding of audio signals 44
– input format 81
spatial scalability 86
spectral noise shaping 33
spectrum-shaping 47
square-root raised-cosine filter 197
start of active video (SAV) 26
state diagram 132
stationary sound 42
stereo 37
– sound 28
stream cipher 183
subband 46
successive-approximation register 31
superframe 251
surround-sound 37, 55
symbol error 112
– -error rate 114
– interleaver 246
– rate 203
sync-byte detector 199

synchronised data streaming 288
synchronous data streaming 287
– serial interface (SSI) 366
syndrome 127
system dynamics 21
– time clock (STC) 95
SystemSoftwareUpdate (SSU) 299

table structure 103
take-over errors 25
terrestrial transmission 237
test sequence 365
– suites 357
threshold of audibility 39
– of pain 39
time and date table (TDT) 101
– code 30
time-division multiplexing 193
time-division multiple access 316
– multi frequency 322
time frequency slot 327
tonality of a signal 50
transcontrol 186
transition point 209
transmission frame 239
– -parameter signalling pilot (TPS pilot) 253
transponder 188
– bandwidth 203

transport packet 97
– stream 89
– stream system target decoder
(T-STD) 372
transport_error_indicator 370
transport-error indicator bit 194
travelling wave tube amplifier (TWTA)
188
trellis diagram 132
trunk amplifier 208
tuner 260

unipolar transmission 153
Update Notification Table (UNT) 300
uplink 188

vector diagram 374
vestigial-sideband (VSB) modulation 5
video artefacts 364
– drip 344
– timing reference signals 26
– -level diagram 24
Viterbi decoding 133
voltage-controlled oscillator (VCO) 199

white Guassian noise 42

xlet 342